Scientific and Technical Libraries

Volume 1
Functions and Management

LIBRARY AND INFORMATION SCIENCE

CONSULTING EDITORS: *Harold Borko and Elaine Svenonius*
GRADUATE SCHOOL OF LIBRARY AND INFORMATION SCIENCE
UNIVERSITY OF CALIFORNIA, LOS ANGELES

Scientific and Technical Libraries

Volume 1
Functions and Management

Nancy Jones Pruett
Technical Library
Sandia National Laboratories
Albuquerque, New Mexico

With contributions by

Stephen J. Rollins
Zimmerman Library
University of New Mexico
Albuquerque, New Mexico

Sharon Kurtz
Sandia National Laboratories
Albuquerque, New Mexico

Charlotte R. M. Derksen
Branner Earth Sciences Library
Stanford University
Stanford, California

Lou B. Parris
Exxon Production Research Company
Houston, Texas

1986

ACADEMIC PRESS, INC.
Harcourt Brace Jovanovich, Publishers
Orlando   San Diego   New York   Austin
Boston   London   Sydney   Tokyo   Toronto

ACADEMIC PRESS, INC.
Orlando, Florida 32887

United Kingdom Edition published by
ACADEMIC PRESS INC. (LONDON) LTD.
24–28 Oval Road, London NW1 7DX

Library of Congress Cataloging in Publication Data

Pruett, Nancy Jones.
  Scientific and technical libraries.

  (Library and information science)
  Includes bibliographies and index.
  Contents: v. 1. Functions and management — v. 2.
Special formats and subject areas.
    1. Scientific libraries.   2. Technical libraries.
3. Research libraries.   I. Rollins, Stephen J.
II. Title.   III. Series.
Z675.T3P69 1986   026.5   86-10829
ISBN 0—12—566041—3 (v. 1: alk. paper)

PRINTED IN THE UNITED STATES OF AMERICA

86 87 88 89   9 8 7 6 5 4 3 2 1

Contents

Part II   Major Functions

Preface

This work is a guide to the functions and management of scientific and technical (sci/tech) research libraries, primarily in the United States. No other book has focused specifically on all aspects of sci/tech libraries in over fourteen years.

Sci/tech libraries appear in a variety of settings. They occur in the corporate setting, in academia, in the government, and in professional associations. The setting of the library influences many aspects of the library, particularly which of the primary functions are emphasized (e.g., collection development versus information retrieval), which special materials are handled, and what management policies and procedures exist. But no matter what the setting, sci/tech libraries have a common goal: to get the users of the library the information they want. The effectiveness of this information retrieval function depends on four other primary functions: collection development, collection control, document delivery, and current awareness. And in every setting the library's effectiveness is also dependent on secondary functions such as management, automation, and space planning. Thus, this work should be of use in any sci/tech library setting.

Its purpose is to provide an overview of the functions and management of sci/tech libraries and a starting place for information about any particular function. Libraries are complex environments, and the literature is voluminous. This is an attempt to select from the literature and summarize the information most pertinent to sci/tech libraries.

The primary audience for the work is a professional librarian who is knowledgeable in one area of sci/tech libraries but needs to know about the others. Perhaps it is a reference librarian who finds him- or herself in charge of collection development or space planning, a cataloger who is charged with taking over document delivery, or someone in a one-person library who needs to know something about everything. In his or her area of expertise, the work will almost certainly be too basic. But, in the areas where he or she lacks experience, it should be helpful.

ix

An additional audience is the manager of a larger sci/tech library who does not need to *do* everything but wants to know it is being done as well as possible. Sometimes the manager is not a professional librarian, but the best of those will want to know as much as possible about what the library should be doing and may find this work of use as a starting place.

It should also be useful to a librarian who has moved into sci/tech libraries from another environment and needs to know what is different about the sci/tech environment. It may also be useful to faculty teaching special libraries courses or sci/tech library courses in library schools and to library school students. Sci/tech libraries are a continuing market for library school graduates, and good librarians with background or experience in the sci/tech field are in demand even at times that other librarians are not.

Medicine is the only field of science and technology which is omitted. It is well-defined and well-served by its own body of literature. Another related area omitted from consideration is that of information services which are not libraries, such as computer networks, management information systems, word processing centers, records management centers, information analysis centers, and data centers. As we move farther into automation these services are converging, but at this point the library is still distinct. Interestingly, the service personnel in these other centers are finding themselves in the traditional position of the librarian: serving as the interface between the system and the user. They could learn much from the long user-oriented service tradition of libraries, but they are not part of the intended audience.

The work is divided into two volumes, Volume 1 on Functions and Management and Volume 2 on Special Formats and Subject Areas. The arrangement of the volumes is as follows: In Volume 1, Part I, we discuss what sci/tech libraries do, provide descriptions of three sci/tech libraries in different settings, and discuss the characteristics both of the sci/tech literature and of scientists and engineers as information users.

In Volume 1, Part II, we discuss the details of the five primary functions: Information Retrieval, Current Awareness, Collection Development, Collection Control, and Document Delivery. In each case we discuss the function first, then the management aspects.

In Volume 1, Part III, we discuss the secondary functions, those that every library must perform in order to do the primary ones. These include management, space planning, automation, and equipment selection and maintenance.

Volume 2, Part I, includes separate chapters on the various special formats of importance in sci/tech libraries: conference literature, dissertations, government documents, in-house information, journals, maps, microforms, numeric data, patents, software, standards and specifications, technical reports, and translations.

As sci/tech libraries differ from general libraries (those which encompass a broad selection of subject areas), so do the individual sciences differ from each other. In Volume 2, Part II, we have chapters focused on particular subject areas. Chapters on the basic sciences include biology, chemistry, mathematics, and physics. The applied sciences included are engineering, geoscience, and pharmaceuticals.

Any work with a scope this broad is necessarily a compromise between "everything you wanted to know" and simply an outline of the important areas. I have tried to lean toward "everything you wanted to know" when the literature seemed thin (e.g., with the technical reports chapter) and lean toward more of an outline when there seemed to be adequate sources published (e.g., with the management chapter).

Acknowledgments

Many people helped with this book—some directly, some indirectly. Ed Evans, the original series editor, convinced me I should write a book. The initial outline received the benefit of feedback from John Miniter, Alan Benenfeld, Gloria Zamora, George Dalphin, and Calla Ann Pepmueller. Calla Ann was also supportive of me as my department manager, and George (my immediate supervisor) saw me through the whole book with sympathy and interest and found a few tidbits of information for me when I was in dire need.

As the series editor, Hal Borko slogged through all the first drafts and made very helpful suggestions.

Perhaps the greatest help came from the authors of chapters, and I cannot thank them enough: Charlotte Derksen, Suzanne Fedunok, Maurita Holland, Michael Homan, Sharon Kurtz, George McGregor, Lou Parris, Marion Peters, Stephen Rollins, and Camille Wanat. Many of these people also encouraged me, read various chapters, sent me articles of interest, and generally supported the idea of the book. George, Camille, Marion, and Charlotte were particularly influential.

The staff at the U.S. Geological Survey Library were very hospitable during my visit and read over several drafts of the description of the library. Barbara Chappell, Henry Spall, and George Goodwin were particularly helpful, but everyone I spoke with or asked for help was uncommonly gracious.

Several other colleagues read and commented on particular chapters: Dorothy McGarry, Kathy Gursky, Sally Landenberger, Jackie Stack, Larry Cruse, Chris Morgan, and Patricia Newman. Lee Garner read most of the manuscript for the Sandia Book Publication Review Committee and made a number of helpful comments. Other colleagues at Sandia National Laboratories were free with their suggestions of what should be in the book, and the paraprofessional staff was magnificent at getting me copies of needed articles and books. I would especially like to thank Pat Chisholm.

Sharon Gorman, Connie Souza, Janet Padilla, Peggy Poulsen, Vic Dickerson, and Gloria Canon, who cheerfully bore the brunt of my requests for materials and photocopies, and Paula Webb, who helped with the typing and was always available to commiserate. Joy Bemesderfer's willingness to do the index was a blessing, as was her knowledge, thoroughness, and enthusiasm. Bruce Fetzer was a great help with the illustrations. Danielle Brown, Doug Robertson, and Dennis Rowley all became part of Sandia Technical Library management when the book was well toward completion, and their continuing support (and amazement that I had taken on such a large and thankless task on my own time) has been much appreciated.

Whenever I have read other people's acknowledgments, I have always been somewhat embarrassed when people thank their spouses and children for putting up with them, but I find now that I am writing one, I understand. You cannot finish a project this big without rearranging your priorities. I find I cannot finish without saying a public thank you to my friends for not giving me too hard a time when I "stayed home to work on the book," to my parents for seeing to it that I had a good education, to Lew for all the support and encouragement, to Sally for caring, and most especially to Corky for the help, confidence, support, and good food. And for putting up with me.

Contents of Volume 2

Part II   Special Subject Areas

Part I

Scientific and Technical Libraries, Literature, and Users

An information retrieval system will tend not to be used whenever it is more painful and troublesome for a customer to have information than for him not to have it.

Calvin N. Mooers
[Mooers' Law. (1960). *American Documentation* July, p. 2.]

Chapter 1

Introduction

There is no convenient list of sci/tech libraries, so it is difficult to say exactly how many there are and just what they are like. The 1983 edition of the *World Guide to Special Libraries* listed over 15,000 sci/tech libraries out of about 40,000 libraries worldwide. The 1984 *American Library Directory* listed 1986 university and college libraries, 1742 departmental libraries, and 4467 special libraries ("Number of Libraries in the U.S. and Canada," 1984). The 1985 edition of the *Directory of Special Libraries and Information Centers* had 17,500 entries for special libraries, but it is not simple to determine how many of these are sci/tech libraries. If it is the same proportion as in the *World Guide,* then we can estimate that there are approximately 6000 sci/tech libraries in the United States.

These sci/tech libraries occur in a number of settings: (1) in universities with science and engineering programs beyond the undergraduate level; (2) in corporations which have a scientific or technical product (e.g., Exxon, E.J. Lilly, Dow, Lockheed, AT&T, Xerox); (3) in the national laboratories (e.g., Sandia National Laboratories, Argonne National Laboratory); (4) in government agencies with a scientific mission (e.g., the National Bureau of Standards, the U.S. Geological Survey, the National Oceanic and Atmospheric Administration); (5) in the scientific professional societies (e.g., the Joint Engineering Societies, the American Institute of Physics); and (6) in major public libraries (such as New York Public Library).

All sci/tech research libraries have a great deal in common:

1. They are dealing primarily with the scientific or technical literature (in contrast to census data, legal information, Russian literature, or 20th century poetry). Almost any sci/tech library will also include other kinds of information, so the distinction is one of degree rather than an absolute. The importance of primary information, especially journals and technical

3

reports, the importance of currency, and the existence of special formats are just a few of the characteristics of sci/tech literature which will be further discussed in Chapter 3 and Volume 2, Part I.

2. Their primary clientele is scientists and/or engineers. Scientists and engineers have particular information needs. What is known about their needs and information seeking behavior is discussed in Chapter 3.

3. As a group, sci/tech libraries are more dependent on automated services than are general libraries (Chen, 1979). The online bibliographic databases were developed for the sciences and technology first, and continue to be of prime importance.

4. They exist to provide particular functions for their organizations. The primary functions are information retrieval, current awareness, collection development, collection control, and document delivery.

Information retrieval is the activity of retrieving information, both from the organization and from the world's scientific literature. The reference or public services section of the library is characterized by ready reference (or information desk) activities, and by online and manual literature searching.

Current awareness includes all those things that the library can do to keep its clientele aware of new literature of interest. It includes selective dissemination of information (SDI), lists of new acquisitions, displays of new books and periodicals, etc.

Collection development includes selection, acquisitions, weeding, preservation, and other aspects of the content of the collection. A well-selected collection is the most efficient source of information.

Collection control includes circulation, shelving, cataloging, indexing, and other aspects of physical and bibliographic control of the collection. Without good control, the collection is inaccessible.

Document delivery includes delivering the physical document to the user, either from the library collection or from the world outside. It includes traditional Interlibrary Loan operations as well as online ordering, photocopying services, and communication channels such as mail, telefacsimile, etc.

These primary functions will be discussed in more detail in Part II, Chapters 4-8.

How well the primary functions are performed is often dependent on secondary functions such as management, space planning, automation, and the selection and maintenance of equipment. These secondary functions are discussed in Part III.

Although almost every sci/tech library will perform these basic functions at some level, in each setting one or more functions may be emphasized over the others. For instance, in the academic setting, collection development and collection control may claim more resources than information retrieval and document delivery, relying on the user to help himself. In contrast, in the corporate setting the collection is often small, with document delivery and information retrieval performed at a compensatingly higher level. Most sci/tech libraries fall somewhere in between these two extremes. In all libraries the functions are closely interrelated. The detailed descriptions of three earth science libraries provided in Chapter 2 should help demonstrate the differences caused by a different setting.

OTHER DESCRIPTIONS OF FUNCTIONS
OF SCI/TECH LIBRARIES

In *Special Libraries: A Guide for Management,* only three functions of special libraries are delineated: acquiring materials for the library, organizing materials in the library, and disseminating information and materials from the library (Ahrensfeld *et al.* 1981, pp. 8–11). (The authors include current awareness and document delivery in "disseminating information.") Each of these functions has three levels: minimum, intermediate, and maximum. For example, "as a minimum function, the special library . . . makes a card catalog which identifies all major publications in the library by author and/or title. . . . As a maximum level of function . . . the library . . . prints its own catalog cards or obtains them through cooperative cataloging" (Ahrensfeld *et al.,* 1981, p. 9). Although these levels may be helpful as a description of what special libraries are doing, almost everything to do with computers falls into the "maximum" level of function. This is unrealistic given the proliferation of microcomputers in the public. Other studies of functions of company libraries include those by Jackson (1978; Jackson and Jackson, 1980), Bedsole (1963), Strauss *et al.* (1972), and S. Larson (1983).

In 1981, Martha J. Bailey filled a gap in the professional literature by studying functions in special libraries and trying to describe functions unique to special/company libraries. She included statistics from libraries in the following sci/tech industries: aircraft and missiles, chemicals, office machines and computers, petroleum, and pharmaceuticals.

The functions reported by 75% or more in each sci/tech industry were:

Acquire all "library" material.
Provide reference service to company employees.
Order reprints and preprints that are requested.
Provide interlibrary loan service.
Provide manual searches of published literature.
Provide online computer searches of published literature.

Other functions which were listed by 75% or more of the libraries in one or more of the sci/tech industries were:

Compile bibliographies on request.
Prepare a brochure describing library/information services.
Maintain files of government security classified documents.
Abstract or index corporation confidential reports.
Catalog or index all "library" material.
Provide photocopies of material in the collection.
Provide and distribute regularly a new acquisitions list.
Arrange for outside translation of articles.
Provide selective dissemination of information service (SDI).
Prepare and distribute regularly a new acquisitions list.
Route new issues of journals on regular basis.
Maintain files of slides, tapes, and other visuals.

In addition to these functions that most sci/tech libraries said they performed, there were also a plethora of functions that only a few of them performed:

Order personal books and subscriptions for employees at a discount.
Provide manual index to research notebooks.
Handle education and training materials needed by employees.
Maintain files of photos and clippings needed by company.
Operate the photocopy service for the department or installation.
Handle distribution of company reports and publications within company or outside company.
Maintain corporation correspondence files.
Be responsible for records management for company or installation.
Be responsible for company archives.
Translate articles on request.
Distribute tables of contents from current journals.
Verify bibliographic information in company publications.

ORGANIZATIONAL PATTERNS

Two issues frequently center on the organizational patterns of sci/tech libraries. The first is the placement of the library in the organization. This is of most interest in the corporate setting, and although there is continued interest in the question, no meaningful pattern has emerged about where libraries are placed (Ferguson and Mobley, 1984, pp. 87–96).

The second issue is whether to centralize or decentralize services. The centralization/decentralization debate has been of most interest in the academic setting, where the question of whether or not to have a separate

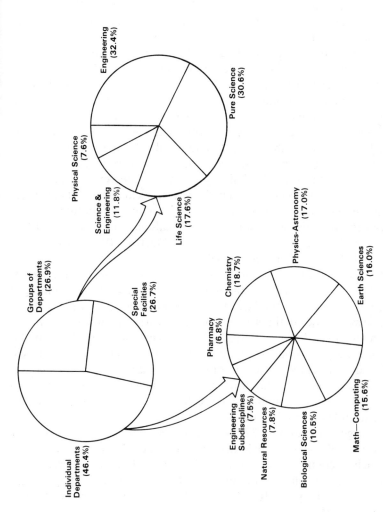

Fig. 1.1. Organization of 633 academic sci/tech libraries in larger institutions. [Based on data from Stankus and Schlessinger (1979, 1980).]

sci/tech library or a subject branch often arises. However, increased automation has made it appear as an issue in the corporate environment as well. Now that a reference librarian can, from one terminal, search the library's catalog online, catalog materials, and also search thousands of commercial databases, all with nothing but a terminal and a telephone, the librarian should be located close to the users.

Users tend to prefer decentralization (because it usually means the library is closer to them and smaller, thus more accessible), and library administrators tend to prefer centralization (since it reduces duplication of staffing and materials and thus costs). A large number of factors influence what choice is made, including politics, perceived accessibility, and the availability of space and funds. Normally the decentralized system is perceived as more accessible, but if a centralized collection can have longer hours, more materials, better surroundings, or if there is a considerable amount of crossover (interdisciplinary research), the centralized system might be accepted by users. No scientific way of deciding has yet appeared, although the question has been discussed widely in the literature, particularly in regard to the academic setting (Waldhart and Zweifel, 1973; Cooper, 1968b; Wells, 1961; Shera, 1961; Marron, 1963; Legg, 1965; Walsh, 1969; Bruno, 1971; Mount, 1984).

Although there is little research to support a decision to centralize or decentralize, there are some data on what kinds of patterns had emerged in academic libraries by the late 1970s. Stankus and Schlessinger (1979, 1980) reviewed entries in the *American Library Directory* and in the *Subject Directory of Special Libraries and Information Centers* to develop statistics showing the proportion of sci/tech libraries which were branches. They found that sci/tech branch libraries existed at 185 large academic institutions (633 individual libraries) and at 37 small colleges (46 individual libraries). The average number of sci/tech branch libraries was 3.4 in the larger institutions and 1.3 in the smaller.

In the larger institutions, 27% of the branch libraries served several departments, 46% served single departments, and 27% served special facilities (such as contract research centers). Figure 1.1 shows the breakdown of these categories. The most likely branches are chemistry, physics–astronomy, and geology–earth science (in that order) (Stankus and Schlessinger, 1979, 1980).

OBJECTIVES AND EVALUATION OF SCI/TECH LIBRARIES

All libraries serve as an interface between the consumer of information and the world's recorded knowledge. Given this role, reasonable library goals are to maximize accessibility of materials to users and to maximize exposure of users to materials (Chen, 1979; Lancaster, 1977; Hamburg *et al.*, 1974). The library as a whole and each of its functions can be evaluated against these goals.

"Accessibility is a multidimensional concept encompassing physical access to the source, the interface to the source, and the ability to physically retrieve potentially relevant information" (Culnam, 1985, p. 302). Thus it can be increased by many means. One of the primary ways is to locate the library as centrally as possible. Figure 1.2 shows the great difference in number of visits to the library depending on transit time (King *et al.*, 1984). Since perceived accessibility is related to familiarity, the chances are that those who are closer will perceive the library as more accessible.

Within the library, a good space layout, obvious arrangements of materials, good access points in the catalog, circulation policies which get materials back in the collection so they are available on the shelves, publicity about services, quick turnaround times, functioning equipment (e.g., photocopiers), and friendly and competent staff can all influence accessibility.

However, the movement toward the paperless society means we cannot

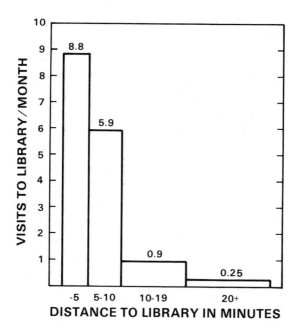

Fig. 1.2. Number of visits to the library per month as a function of distance to the library in minutes. [From King *et al.* (1984), p. 28.] Note that a distance from the library of 10 minutes seems to be a dividing factor.

concentrate only on the physical environment. The user who can search the library catalogs from his or her office, send requests for material via electronic mail, and have the information delivered to the office (let alone downloaded to the terminal) may well perceive the library as accessible even if he or she is physically quite removed from it.

This goal of maximizing accessibility can help us make decisions in all the functional areas of the library. And statistics about the number of items provided per person or per library staff member (a focus on output measures) can show clear progress toward maximizing exposure to materials (Manthey and Brown, 1985; Kantor, 1981a,b). Keeping this goal firmly in mind can prevent short-sighted decisions (such as cancelling a widely read update bulletin in a time of personnel shortages, or accepting a remote location for a new library building). But it does not allow us to answer the real question of evaluation: what is the value of a sci/tech library to its organization?

Although the economic value of information is not well understood, there are several cost–benefit studies which help point toward answers to this question.

King Research has published two technical reports which bear on the issue of the value of information (King et al. 1982, 1984), and a book entitled *The Use and Value of Special Libraries* was in press when this chapter was written.

Figure 1.3 is from the first report and provides an overview of the factors related to increasing the value of information. The library, by serving as the facilitator of access to both primary and secondary information, can increase the use of information and thus its value.

Both King Research reports were done on contract to the Technical Information Center of the Department of Energy. The first was an attempt to establish the value of the Energy Data Base; the second studied the effect of intermediary organizations (e.g., libraries) as well. Both studies focused on scientists and engineers, how much they read, and the value of that reading. Many of the scientists and engineers indicated that a recent reading of a technical report or article led to a savings of time or equipment or both. There was some savings reported for 25% of all article readings and 75% of all technical report readings. By averaging the savings over all the readings, they found an average of $590 per article reading and $1280 per report reading. The second study found $385 for reading a journal article and $706 for reading a technical report.

Libraries could do a cursory test of cost–benefit by using library statistics such as number of photocopies distributed (from the library and from interlibrary loan) and report circulation. By multiplying these figures by the values above and comparing the total value to the library's budget, a case

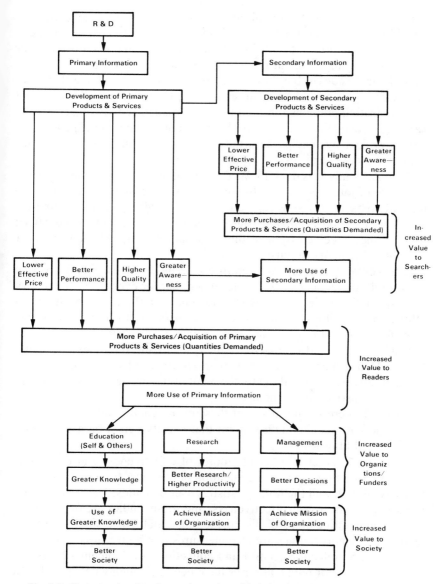

Fig. 1.3. Factors related to increasing value of information. [From King *et al.* (1982), Fig. 1.1, p. 2.]

can usually be made for the library being a bargain to the organization. One library found a ratio of 16 : 1 (value of services to cost of services), and this calculation does not include answers to reference questions or the reading of anything but reports and journal articles. Caution in using these figures is advisable, however, because they are so large that management may dismiss them as unreasonable. However, the importance of not duplicating previous research is widely recognized, and management can easily recognize that the avoidance of duplicating one expensive research project by reading an article can average out to a high average value for reading articles.

A similar study was done by Weil (1980) in order to determine whether the value of journals was worth what they were costing, at a time when journal costs were rising very quickly. Although the participating researchers were only able to assign a dollar value to 2% of the beneficial impacts they received from reading, the dollar values yielded an 11 : 1 ratio of net benefits to outside costs.

A survey of the benefits of literature searches at the Boeing Aerospace Group Library (Kramer, 1971) determined that for every hour the library staff spent literature searching, the technical staff saved 8.86 hours of effort, and that it would have taken 130 engineers to find answers that were found by only five library staff members.

Magson (1973) did cost–activity analyses of the functions at the ICI Central Management Services and compared the results to similar analyses for alternatives. He found the total benefit of having the Information Unit was 115% versus not having it.

Although there are not many of these cost–benefit studies, what there are support Rothstein's (1985) point that society gets a bargain from librarians and libraries.

There are a larger number of studies of effectiveness and cost-effectiveness, two other aspects of evaluation of libraries. Effectiveness is the measure of how well the library satisfies user demands. User surveys are the usual way of testing effectiveness, and these studies are of particular use for pinpointing areas of dissatisfaction. Catalog use studies and collection availability studies are also measures of effectiveness. (These are covered with collection control and collection development.) Measures of effectiveness for public libraries have been thoroughly developed, and they cover (among other things) the extent to which the library's community is aware of the library services available, and use of services (McClure and Reifsnyder, 1984). These are both areas which sci/tech libraries may wish to measure.

Cost-effectiveness studies normally compare one way of doing something to another way to determine which is the least costly. For example, when online searching was first introduced there were a number of studies comparing the costs of manual and online searching. Now, a more common study would be to determine which of two online services offering the same database is the least expensive, or which method of producing a new acquisitions list is the least expensive.

The importance of evaluation was pointed out by Matarazzo (1981), who studied corporate library closures and found that the lack of evaluation of library services was found to be a factor in the closure of the library.

Evaluation of particular functions is covered with the chapters on the functions, and evaluation as a check on how the organization is meeting its objectives is covered in the chapter on management. For further information on the evaluation of libraries, consult Lancaster (1977), McClure and Reifsnyder (1984), Kantor (1984b), or Ferguson and Mobley (1984). For further information on the economic value of information, Lamberton's (1984) review in *ARIST* is a good starting place.

THE FUTURE ROLE OF SCI/TECH LIBRARIES

This is an exciting time to be a librarian. New technology is changing the face of the publishing world and the methods of doing business. The revolution in the way society produces, disseminates, and consumes information is forcing a reexamination of the role of the library and whether our solutions really respond to information access needs. Access to information resources is reaching the individual independent of institutional, organizational, or professional affiliations. Databases and files are becoming equivalent to books and journals. As Matheson (1984, p. 210) says, "The critical issue is how we will control the management and distribution of information within institutional networks. Connecting our online bibliographic databases and circulation systems will not suffice for long."

What is the library's role in the revolution? Matheson (1984, p. 213) thinks we must move toward providing information which is problem-specific.

We all understand that the time has passed for building definitive collections of books and materials. But we seem to be engaged in compiling definitive collections of bibliographic data instead. We must question whether an unqualified list of bibliographic citations is responsive in an educational or problem-solving environment like a university. As time goes on, and people gain more familiarity with databases and other information sources, the realities of time and economics will exert great pressures on libraries to provide data that are problem specific.

In addition to the increased need for selected information, there is a need for a systematic approach to the management of information. There is a burgeoning awareness of the importance of information and that the librarian's skills in organization and designing systems to meet user's needs are valuable. A number of writers are urging libraries to take on the role of information resources management, a role which combines library and information science, computer science, records management, and electronic publishing. Many special libraries find themselves organizing records management, software libraries, and education and training centers. Some libraries are experimenting with information analysis, a "value-added service," in response to the need for condensed and evaluated information rather than just a flood of articles and reports. There will continue to be a need for the traditional role of the library—that of an interface between the user and the world's recorded knowledge—in fact, there may be more need than ever. But the shape of that role is changing daily.

Although the remainder of this book deals with the more traditional library functions, the reader should always keep in mind that the environment is changing. As long as we focus on our user's needs for information, and keep accessibility firmly in mind, we should be able to meet the challenges of this age.

Chapter 2

Three Libraries in Different Settings

Although sci/tech libraries have much in common, they also vary depending on their setting and the objectives of the overall organization to which they belong. In this chapter we provide detailed descriptions of three libraries. The Branner Earth Sciences Library at Stanford University, Stanford, California, described by Charlotte R. M. Derksen, serves as an example of a branch library in an academic setting. The Information Center at Exxon Production Research, Houston, Texas, described by Lou B. Parris, is an example of a library in the corporate setting. The U.S. Geological Survey Library, Reston, Virginia, is an example of a sci/tech library in the government setting. Each of these is recognized as an excellent library, and since all three are in the earth science subject area, the differences because of setting and differing objectives should be more obvious. One would predict an emphasis on collection development and cataloging to national standards in the academic and government settings and an emphasis on information retrieval and current awareness in the corporate setting. We would also expect more activity in bibliographic instruction in the academic setting and a longer history of automation in the corporate setting. Indeed, these patterns hold true.

These detailed descriptions should also serve as an introduction for those who are less familiar with the day-to-day activities of sci/tech libraries. Several other descriptions of libraries are worth reading to get a sense of the variety of sci/tech libraries and the types of objectives, services, and functions which are performed. For instance, Kennedy (1978) described the Bell Laboratories Library Network; Mount (1982) described the Engineering Societies Library in great detail; and the John Crerar Library was described on the occasion of its move to the University of Chicago (Holton, 1985; Brandchoff, 1985).

BRANNER EARTH SCIENCES LIBRARY, STANFORD UNIVERSITY

Charlotte R. M. Derksen
Branner Earth Sciences Library
Stanford University
Stanford, California 94305

Background

Stanford University, founded in 1885, is a private university serving both graduate and undergraduate students. The university has emphasized both applied and theoretical sciences since its inception. The primary goal of the libraries at Stanford is to serve the information needs of the present and future faculty, students, and staff of the university. The earth sciences library, one of seven branch libraries in the science department, serves primarily the four departments of the School of Earth Sciences: Applied Earth Sciences, Geology, Geophysics, and Petroleum Engineering. There are 42 tenure-track faculty in the school, and an additional 40 emeritus, adjunct, and part-time faculty, as well as research associates, assistants, and visiting scholars from around the world. The earth sciences students using the library include 270 graduate students (doctoral, masters, and engineer degree candidates) and 180 undergraduate majors, plus students taking classes within the school to satisfy general degree requirements. Other members of the university community also use the library frequently, especially those from the Civil Engineering Department. Access to the library collections and services is unrestricted for all members of the university community.

The library has open stacks; therefore anyone, Stanford affiliate or not, can walk in and use the collections. However, the general public is not entitled to either reference service or borrowing privileges. Since the library is a depository for materials from both the U.S. Geological Survey and the California Division of Mines and Geology, the public is free to use any of the depository materials held in the library.

The collection consists of materials in the fields of economic geology, geochemistry, geology, geomathematics, geophysics, geostatistics, geothermal engineering, hydrogeology, invertebrate paleontology, petroleum engineering, petroleum geology, and remote sensing. The largest portion of the collection is in the form of maps, bound volumes, and journals. Microforms, floppy disks, manuscripts, seismic logs, and data tapes are also collected. So far, aerial photographs have not been collected; however, this may change as the needs of the faculty change.

History

The first faculty member hired by the once-fledgling Stanford University was geologist John Caspar Branner. Branner brought his personal library with him in two box cars and subsequently sold the collection to the university. The original collection was augmented by materials donated by other early faculty, such as the James Perrin Smith library on cephalopods and the Montessus de Ballore collection on seismology. The present collection includes an amalgamation of the collections of several former faculty and small departmental libraries. The library has had a very stable collection development history, since it has been directed by only four librarians, including the present one, throughout most of its history.

Organization

The staff consists of a librarian, who is also the bibliographer, 3.5 library specialists, and 100 hours per week of student help. The organizational structure is shown in Fig. 2.1. The library specialists include a full-time

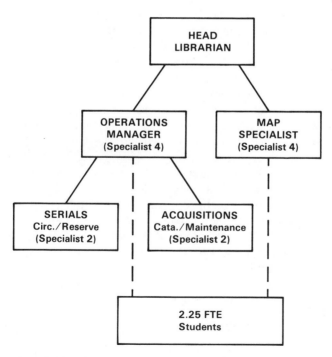

Fig. 2.1. Organization chart, Branner Earth Sciences Library, Stanford University, October 1985.

serials and circulation/reserves specialist and a half-time acquisitions/cat-alog maintenance specialist, both classified as level two in a four-step ranking scheme. They report to an Operations Manager, a level-four specialist, who is responsible for the day-to-day management of the library. The fourth member of the staff is the map specialist, also a level four. Both of the specialist fours report to the librarian. The librarian-bibliographer reports to the Chief and Curator of the Science Department, who in turn reports to the Director of Research Services. The Director of the Stanford University Libraries, of which the Science Department is a part, reports to the Provost of the university.

Although there is no formal reporting structure between the librarian and the Dean or the faculty of the School of Earth Sciences, the librarian works closely with them. Copies of the branch library's annual report and library memoranda of major significance are routed to the Dean for his information.

A faculty–student library advisory committee meets annually to advise the librarian regarding changes in policy. For example, when journals had a 1-month circulation period with indefinite renewal privileges, there were many complaints that journals were not on the shelves when needed. Several alternative arrangements were explored with the committee, including making the journals noncirculating, as is the policy in most of the other Science Department branch libraries. The option selected was a 3-day circulation period for journals, with no renewal, except for paleontology journals. If a faculty member or student needs a journal for a longer period in order to examine fossils under a microscope and compare his or her specimens with plates in the journal, he can request permission for a longer loan, with concurrence from a faculty member. This new policy has worked very well.

The chairperson of the committee is appointed by the Dean, and one additional member is appointed by each of the four departments. One graduate student also sits on the committee. All members serve from the time of appointment until they ask to be relieved; thus there is a certain amount of continuity from year to year. In addition to the annual meeting, the librarian often consults with the chair or other members on matters arising during the interim.

In addition to informally reporting to the Dean and consulting with the library faculty advisory committee, the librarian joins the faculty from each department at least once every 2 years during one of the regular departmental meetings to talk over their library concerns. By far the most contact with the researchers using the library comes, however, as a result of daily contacts in the hallways and requests for information.

Major Functions

Collection Development

The collection is composed of three main parts: 90,000 maps, 75,000 bound volumes, including books, journals, and monographic series, and 1000 technical reports. There are about 3000 journal and series titles, of which about 1500 are current subscriptions. Collection development and management absorb the major part of the time of all of the staff, with the exception of the operations manager.

The collection development policy is based on the teaching and research interests of the faculty. In order to find the research interests, several steps were followed:

1. Brochures published by the various departments were studied.
2. Forms were sent to the faculty members asking them to delineate their interests.
3. Recent publications by members of the faculty were skimmed.
4. Titles of currently held grants were noted.

This information was indexed according to the *Library of Congress Subject Headings* and then compiled on the campus mainframe in a file entitled Faculty Interest Profile (FIP).

In order to obtain the teaching interests the course catalog for each department was studied. This information was also added to the FIP file, coded to indicate teaching, rather than research, interest. After all information was assembled, each faculty member's profile was discussed with him or her for accuracy. The librarian continues to scan new publications and check for new course offerings and so update the profiles on a regular basis. A formal collection development policy based on the information in the FIP file has been written. Thus decisions on whether or not to acquire material are based on current or potential interests of the faculty and their students.

In order to acquire the materials needed, approval plans are maintained with three vendors: Yankee Book Peddler (for University Press publications), Blackwell North America, and Harrassowitz (for both English-language and German-language western European publications). These plans cover mainly materials published by mainstream publishers, which is not where the majority of earth sciences materials are published. However, with approval plans in place, staff time is not taken up with ordering these easily acquired materials, but can be expended on materials more difficult to acquire.

The acquisition of petroleum engineering and geothermal engineering materials is different from that of the other earth sciences materials. There are publishers (e.g., Pennwell) which cater to the energy industries. Approval-plan vendors rarely handle these publishers (except possibly as part of their form approval plans) because the books are too expensive, they must be prepaid, many are nonreturnable, and the academic demand for them is not high enough to justify the effort. However, getting on the mailing list for the publishers' blurbs and watching the book reviews in the major energy industry journals keeps one abreast of the majority of these needed titles.

Petroleum engineering technical reports are acquired as they are requested. An attempt is being made to build a strong collection of geothermal engineering technical reports. The library is on the mailing list of several agencies that produce reports, and thus receives anything (on certain topics delineated by the library) published by those agencies. There are also exchange agreements with several foreign agencies producing geothermal reports. The faculty select topics of the reports to be sent that they or their students will be needing for their research.

For other earth sciences materials, selection and acquisition are very time-consuming. Since the faculty's areas of research and those of their students are widely scattered about the globe, it is necessary to acquire materials in many languages, from all areas of the world. In fact, an effort is made to obtain the research publications from the national geological surveys of every country in the fields of economic geology, geology, geophysics, micropaleontology, and petroleum geology. An attempt is also made to secure the research-level publications from all state geological surveys in these five fields, as well as in hydrogeology, remote sensing, and geochemistry. It is possible to set up exchange or depository agreements or subscriptions with some surveys. For the majority, however, it is a case of being alert to a new publication or a new issue and writing to them either to order it or to remind them that we should have received it due to our subscription or agreement.

Societies are another major source of earth sciences publications. Acquiring these materials is similar in difficulty and in procedure to getting those from geological surveys.

In an effort to find out about the publications from these diverse sources, all issues of journals and series that are received in the library are scanned by the bibliographer as each is received. This process takes an average of 2 hours per day. For many titles, the advertisements and the articles, as well as the book reviews, are scanned for selection purposes.

It is most efficient for the bibliographer to search the various files in the Research Libraries Information Network (RLIN) databases, while scanning the journals, publishers' blurbs, etc. One can determine at once whether a book has already been received or is on order, either for Branner or another location at Stanford, and whether, if the title is of marginal interest, that title is held at or is on order for a nearby library. Occasionally the full cataloging record, if available, yields some information that will influence the decision about whether to buy the item. If an older edition of the same title shows up on line, one can note the call number and check that the older edition has indeed been useful, before deciding to buy the new edition. The review may be for a non-English title, but an online search may reveal a preferable English translation.

A search of RLIN is particularly valuable for geologic series. For example, the title of the "book" may turn out to be, in reality, the title of a volume of a series; the search for the series reveals which library has the series, and a decision can be made as to how valuable having the particular title immediately at hand is to the earth sciences clientele. Furthermore, searching the series in the books file may yield information about other volumes of the series that might be potential candidates for acquisition.

One of the biggest projects undertaken in the library every year is the bindery project. This is done during the summer, when the majority of the users are usually out in the field. As a by-product of collecting the materials to send to the bindery, the current holding status of each series title pulled is checked, and those that have not been received recently are marked; during the winter the serials specialist goes through the file, searching the RLIN files for each title that has been marked as having lapsed or as having missing issues. Claims or orders are placed for all items for which proof of publication can be found.

Each summer one-eighth of the bound volume collection is examined for weeding, inventory, and conservation purposes; during the evaluation of this collection segment, all series in this section are evaluated and lapsed titles or missing issues are noted; these are included in the specialist's winter project.

Approximately once every 3 years, the department secretary makes an effort (as time permits) to go through the file of publications lists from geological surveys and societies. She then sends out letters to all from whom a recent publications list has not been received. The letter (in one of the following languages: English, German, Spanish, French, or Portuguese) asks that the library be placed on their mailing list for future mailings, and that they send one copy of any of their recent publications that they might care to send.

Collection Control

The bound volumes of the collection, including books, journals, theses, and government documents, are fully cataloged. Unfortunately, a large number of the monographic series are available in the catalog only under the name of the series; especially for older materials, individual analytics are not available.

Until 1970, these materials were classified according to the Dewey schedules; since that time, the Library of Congress scheme has been used. The titles that have been acquired since 1973 or that have had maintenance done on their cataloging records are in machine-readable form and can be found in the RLIN databases and in the Stanford online catalog, called SOCRATES. A project has been planned and funded to load records for all of our pre-1973 titles from the manual catalog into RLIN and SOC-RATES. Cataloging is done in the main library by the Technical Services staff, who use RLIN to capture cataloging data. The microfiche, datatapes, and floppy disks have also been cataloged and thus are available in SOC-RATES. So far, the datatapes and floppy disks are shelved behind the circulation desk for security reasons.

Unlike the cataloged materials, which are received, targeted for the security system, stamped for ownership, and assigned call numbers in the main technical services unit, both the maps and the technical reports are completely processed in the branch. The possible alternatives for building an online index to the largest segment of the collection, the maps, is being investigated.

The technical reports collection was new to the library in 1982. There is access to author, title, report number, and agency name online (in MARC format), in a separate file on SOCRATES. The inputting for this file is done by the student assistants who are staffing the circulation desk on Saturdays (the slowest day of the week). The reports are shelved by report number.

Information Retrieval

Reference service is available from 9 in the morning to 9 in the evening, Monday through Thursday, and until closing at 6 p.m. on Friday. Although the library is open 7 hours on Saturday and 9 hours on Sunday, no reference service is provided. During summers and intersessions, one staff member at a time covers both the circulation and the reference desks. All members of the regular staff spend some time (varying between 4 and 26 hours per week) at the reference desk. The librarian and both of the specialist fours have strong subject backgrounds.

In addition to helping faculty and students with their SOCRATES questions and searching the RLIN databases to retrieve citations for patrons, the reference staff frequently perform quick reference searches on the BRS (Bibliographic Retrieval Service, Inc.), DIALOG, CAS/STN (Scientific & Technical Information Network), and QUESTEL systems. Faculty or students frequently request full literature searches to be carried out on one or more databases on any or a combination of the commercial systems. These are generally scheduled for late in the day (with the exception of QUESTEL) in order to take advantage of quicker response time. The costs to the library for computer time, prints, and telecommunications for these requested literature searches are charged to the patrons. Literature searches on SDC and NEXIS are available but are seldom requested. Frequently, literature searches are also performed on the RLIN databases, particularly on the books and the maps files. There is no charge to the patron for the RLIN searches.

Bibliographic instruction and orientation sessions, carefully tailored to subject and research interests, are offered each fall to all incoming graduate students in the School of Earth Sciences. These sessions include online demonstrations of SOCRATES, RLIN, and one or more commercial databases, as well as an introduction to reference tools and abstracting services valuable for the group. Groups usually consist of four to six students with similar subject interests.

Several course-specific bibliographic instruction sessions are requested by the faculty throughout the year. Examples include petroleum engineering, marine geology, geochemistry, and hydrogeology. Each session is fitted to the specific research and/or class needs of the graduate and upper-level undergraduate students in the class and includes appropriate online demonstrations. Since students in one of the introductory classes must write short research papers citing maps and other materials unfamiliar to freshmen, a short bibliographic session is given to all 130 of the students in this class, during their laboratory sessions.

Classes in which faculty and graduate students are taught how to search *Chemical Abstracts* online on the STN system are also given several times a quarter. The trained users then sign up and come in to the library to do their own literature searches on the library's equipment, using the library's account.

Document Delivery

The central interlibrary loan department in the main library serves the whole library system. This unit acquires the more traditional library materials for the earth sciences patrons from the Research Libraries Group libraries, from the University of California, Berkeley, and from other libraries with whom the university has broad cooperative agreements.

In addition, the science department has set up a cooperative loan agreement with several science or technical special libraries in the neighborhood. The Operations Manager handles all of the loans between our branch and these special neighbors. For materials that are not available through either of these channels, that can not be acquired through purchase, or that would take too long to borrow or to buy, deposit accounts have been set up with several information brokers or document delivery sources. The acquisitions specialist handles these transactions.

Current Awareness

Unfortunately, due to recent cuts in the budget, a monthly acquisitions list is no longer produced. However, all materials received in the library, other than topographic maps, are put in a display area for 1 week, so that the researchers can scan these shelves once a week and be fairly sure that they are seeing all new materials. Each new piece has a sign-up sheet attached so that any who want to may check out the item as soon as it is released for circulation. In addition, there is a bulletin board for displaying both book jackets of new books (with the call number on the jacket) and announcements about all new publications written by faculty of the School of Earth Sciences. Finally, display cases in the entrance of the building alert patrons to special honors by researchers, highlight the works of one faculty member or department, or exhibit rare materials.

Selective dissemination of information or current awareness searches are available on DIALOG or other systems, but so far few faculty are taking advantage of this opportunity. However, there is an informal SDI system in place. The librarian and the senior specialists work so often with the same patrons answering reference questions and performing literature searches, and the librarian and the map specialist spend so much time selecting materials for purchase based on faculty interests, that they know what will interest their users. Therefore, as a new title comes in, or as a journal article is received that is relevant to a search just performed or to a question just answered, the library staff notify the patron of the possibly interesting material.

Issues and Trends

Space

Space planning is, and probably will continue to be, a big issue. It affects weeding priorities, acceptance of gifts, collection development priorities, installation of new equipment in the library, and so on. It is especially important in the earth sciences libraries, where materials do not obsolesce as quickly as many other science or engineering materials do. The paleontology materials, containing detailed illustrations, and the items containing maps are not good candidates for microfilming, so it is necessary to keep hard copies in the library. Patrons often need to look at the complete run of a data journal. The reference staff seeking to find answers must often look at the actual pieces of monographic series, because series are not analyzed in the card catalog. The lack of truly comprehensive indexes in the earth sciences, the length of the average geological journal article (perhaps 50 pages for the older titles), and the long half-life of geological literature all contribute to making off-site storage of materials impractical. One solution to the space problem is to convert study areas to shelving. Another somewhat more expensive but perhaps better long-term solution is to convert some of the shelving space into compact storage. A combination of these two alternatives, melded with a rigorous weeding program and selection of low-use materials for off-site storage, is presently being considered for Branner library. Figure 2.2 is a floor plan of the library.

New Formats and Technologies

The librarian is talking with the school about access to their computer network so that reference questions, circulation problems (overdues, recalls, etc.), requests for new materials, and other standard library transactions can be handled by electronic mail, and announcements can be placed on electronic bulletin boards. Acquisition of information stored on disks and data on tapes has already begun, albeit slowly, but it is expected that the pace and kinds of these items will grow greatly. The remote-sensing laboratory, the geophysics department, and the petroleum exploration group, in particular, are already working with large data files stored on tapes. Some of these might be moved into the library. Petroleum engineering literature is beginning to come in the form of floppy disks or of serial issues or books with disks in the pockets.

Access to nonbibliographic databases, such as the EROS files or the U.S. Geological Service (USGS) map files, will probably be available in the near future. Since so much of the geological literature is produced, however, by small geological societies or by geological surveys in Third World countries, books, journals, and maps will be used, purchased, and cataloged for the foreseeable future.

Already, faculty and students are able to search SOCRATES, the online catalog, from their offices or dorm rooms, if they have registered for a (free) account. End users are doing their own literature searching on some of the more friendly systems. However, students and faculty will still need to come into the library to access the largest and most valuable parts of the collections: the maps and the unanalyzed monographic series, neither of which is presently available in the online catalog or adequately covered by the online indexing services.

Staffing Trends

More and more of the routine clerical work is now handled by the student assistants, leaving the staff to handle more complex problems. Students

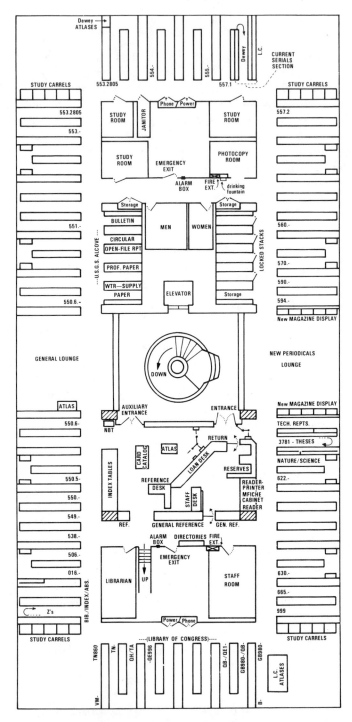

Fig. 2.2. Floor plan, Branner Earth Sciences Library, October 1985.

are now responsible for all of the circulation functions: overdues, recalls, holds, filing circulation cards, and notifications. They also type orders, file cards in the card catalog, type new book cards, place items on reserve, check in documents, input technical reports into the file (from the item directly, not from a worksheet), and maintain statistics. As more of these functions are handled automatically, the students will take over other work, leaving the staff freer to concentrate on collection development, the difficult aspects of bibliographic control, and information retrieval. Students no longer consider the library an easy place to work, where they can study on the job. A lot of training and retraining is required, because students do not normally stay very long and are gone during intersessions.

More and more staff time is taken up with literature searches, training patrons to do their own online literature searching, and providing bibliographic access to materials such as technical reports. Therefore, even though students are taking on more "staff" work, the staff members still find themselves facing a growing backlog of work.

Public Relations

Maintaining good relations with the faculty, staff, and students whom we serve is extremely important. To that end, we often attend school functions such as colloquia, graduations, and so forth. In addition, it is helpful for them if we provide support during special functions such as Alumni Day, job fairs, and Associates Day. Even though persons not affiliated with Stanford are not entitled to reference service, it is wise to be helpful to those who might become donors to the school or to the university.

The role of the Branner Earth Sciences Library is to serve the earth sciences information needs of the present and future university community. This includes developing the collection, maintaining it, retrieving from it when necessary, keeping abreast of issues and trends in the earth sciences and the library sciences, and utilizing the new technologies to the best advantage.

INFORMATION CENTER, EXXON PRODUCTION RESEARCH COMPANY

Lou B. Parris
Exxon Production Research Company
Houston, Texas 77001

Background

Exxon Production Research Company (EPR), located in Houston, Texas, is Exxon's major research organization for petroleum exploration and production. The principal roles of EPR are (1) to provide Exxon affiliates worldwide with the technology needed to carry out exploration for and production of oil, gas, and other hydrocarbons, (2) to assist its affiliates by applying the latest technology to solve specific problems, and (3) to provide technical training, both for EPR staff and for the staff of its affiliates.

The EPR Information Center provides information services to support EPR's 600 scientists and engineers, and to assist other Exxon affiliate companies that need information about petroleum exploration and production. With other major information facilities at Exxon Research and Engineering Company, Exxon Corporation, Imperial Oil, and Esso Resources Canada, the EPR Information Center shares responsibility for providing timely information services and products to Exxon Corporation's locations worldwide.

In the course of providing services and products, the Information Center performs all the five major functions of scientific and technical libraries (collection development, collection control, information retrieval, document delivery, and current awareness) and also has responsibility for records management. Computer-based systems support every function.

Organization

Exxon Production Research Company

Exxon Production Research Company (EPR) is a wholly owned subsidiary of Exxon Corporation, with a President as its chief executive officer. Functional Vice Presidents direct the major research areas—exploration and production—and lead 7 technical divisions, each headed by a Manager. Support services are organized under a General Manager—Administration, who reports to the President. These include Employee Relations, General Services, Controllers, and the Information Center's division, Technical Information. Rounding out the organization are divisions providing computer and communications services, legal services, strategic planning, and training.

Technical Information

Technical Information is comprised of two sections: the Information Center, and Publications and Graphic Services. Within Publications and Graphic Services are the Illustration, Publications, Word Processing, Reprographics, Photography, and Editing groups. The organization of Technical Information is shown in Fig. 2.3.

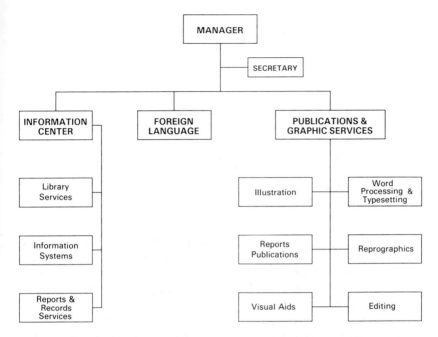

Fig. 2.3. Organization chart, Technical Information, Exxon Production Research, August 1985.

Assigned to the same division as Publications and Graphic Services, the Information Center is close organizationally to activities associated with the production of research reports—the permanent documentation of EPR research activity. After a report is written and produced, the Information Center completes the continuum by providing distribution, indexing, storage, retrieval, and archival retention.

Information Center

The Information Center is organized into three groups (Library Services, Records/Reports Services, and Information Systems) and headed by a Supervisor. Figure 2.4 shows the organization of the Information Center. Library Services staff (a coordinator, three research librarians, two reference librarians, and four library assistants) are responsible for collection development and control, document delivery, information retrieval, and current awareness for published materials.

Reports/Records Services staff (a coordinator, one information spe-

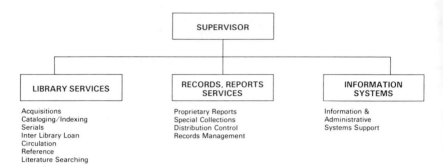

Fig. 2.4. Organization chart, Information Center, Exxon Production Research, August 1985.

cialist, and three library assistants) are responsible for the company's record management program, distribution control, proprietary reports, laboratory notebook control, map collection, and two special collections—the Environmental Information Center (EIC) and the Offshore Information Center (OIC).

In the Information Systems group, two research librarians utilize their computer-programming and systems expertise to support the Information Center's computer-based resources. In addition to responsibility for the Information Center's systems, this group provides systems support to a number of other activities within General Administration.

Facilities

The Information Center is housed in a 10,000-square-foot area located on the central corridor near the building's main lobby. Office spaces for the staff are either permanent or open plan with Datapoint work stations in most offices. Approximately two-thirds of the space is allocated to published material and one-third to proprietary material. The published material is divided into reference materials, books and series, journals, and technical reports. Both the published technical reports and the company reports are shelved in mobile, high-density shelving to conserve space. Figure 2.5 shows the floor plan for the Information Center.

Major Functions

Collection Development

The collection reflects the company's focus on petroleum exploration and production. The largest part is devoted to geology and geophysics; other areas of emphasis are petroleum engineering, ocean sciences, phys-

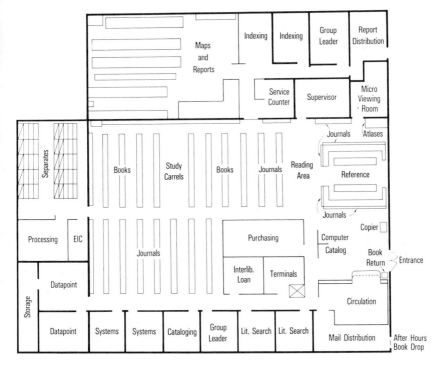

Fig. 2.5. Floor plan, Exxon Production Research Information Center, August 1985.

ics, mathematics, chemistry, mining, minerals, and metallurgy. It is divided into two parts: published materials, and proprietary or company materials.

As of December 1985, published documents totaled 48,500 items, including books, items in series, and reports. Currently, the Information Center maintains 722 journal subscriptions. The company proprietary collection includes 40,500 documents (company reports, maps, and two special collections relating to offshore and environmental information). Most of the collection is in printed form, although microform versions are used when practicable.

Collection development objectives differ between the published and proprietary segments of the collection. For published material, the objective is to maintain a core collection of materials needed to support the research effort, supplemented by a high-volume interlibrary loan service.

Collection development for published material is facilitated by written acquisitions guidelines, which describe in some detail the subject areas to be covered and establish retention periods. The need for guidelines became evident when EPR underwent a period of rapid growth during the late 1970s, and the increased size and diversity of the research staff dictated a more structured approach to collection development.

Since there was a clear need for assistance from the research staff, an Information Center Advisory Group consisting of one or more representatives from each division was appointed. Some divisions elected to have only one representative; others chose to have a representative from each of the division's sections. The resulting group of 24 assisted in developing the guidelines, and individual members continue to provide selection advice on an ad hoc basis. The group also assists with an annual review of the journal subscription list to assure that renewals continue to reflect the current research interest.

The collection development guidelines generally stress recent materials, but geologic documents pose a special retention problem because they have a very long useful life and typically contain significant nontextural information (plates, maps, seismic sections) not easily reproduced by photocopy or printed from microfilm. Using high-density mobile shelving allows retention of paper copy even though the materials are not frequently used.

Materials are chosen by the Library Services Coordinator with assistance from the research librarians. Selections are made from publisher announcements, published reviews, conference publication announcements, and announcements in secondary services. Recommendations by the EPR research staff are, of course, encouraged.

Purchasing practices vary, depending on the type of material. Journals, with a few exceptions, are handled through a subscription agent. The Information Center maintains a number of standing orders for either regular annual publications or series publications and utilizes the services of a jobber for about 10% of other acquisitions. However, since requirements are quite specific, neither standing orders with jobbers nor approval programs with publishers are helpful. In fact, an analysis of recent purchasing practices showed that in a given time period, we purchased approximately 3000 items from 565 vendors, averaging only six items per vendor.

Proprietary materials are acquired through the company's report production and distribution procedures. Circulation and file copies of all EPR reports are deposited in the Information Center. Since each Exxon affiliate is responsible for the permanent custody of its own reports, no emphasis is placed on collecting any but those produced by EPR. Close coordination with the Publications and Graphic Services Section assures the integrity of the research reports collection.

Maps are ordered either at the request of the research staff or selected from publisher information. Materials for the Offshore Information Center (OIC) are received as a result of joint efforts between EPR and educational or research institutions. Several interested EPR research groups select materials for the Environmental Information Center (EIC) collection. Participants acquire the documents and prepare subject indexing utilizing a controlled subject vocabulary. The Records/Reports group creates an index and provides storage, retrieval, and alerting services.

Collection Control

For collection control, conventional cataloguing and classification systems are combined with computer-based indexing systems. In the published collection, books and series items are catalogued using a modified version of Library of Congress classification and the Information Center's own controlled vocabulary. This vocabulary, originally Library of Congress subject headings, has migrated toward a more technical indexing vocabulary based on the University of Tulsa's *Petroleum Abstracts* thesauri, which are more suited to computer retrieval.

The published technical reports, known here as "Separates," are not classified. They are arranged in an accession number order file and indexed with the same subject vocabulary as the books/series.

Proprietary materials are classified by an EPR report number and indexed with a controlled vocabulary quite similar to that used for the published material. The maps, EIC, and OIC collections are filed by accession number and each has its own vocabulary, although a degree of uniformity is maintained throughout all vocabularies because they all utilize the University of Tulsa's *Petroleum Abstracts* thesauri as the basic language control.

All cataloging and indexing are original and do not conform to the MARC format. Data capture from other bibliographic systems does not meet our needs at present, although we do utilize some OCLC and CIP (Cataloging in Publication) information. Of course, we will continue to follow network developments and participate in mutually beneficial activities.

The EPR Information Center was an early advocate of using computers to sort and match documents. Beginning in the 1960s with uniterm indexes—laboriously produced and difficult to use—we have evolved through a series of improvements to today's minicomputer-based, integrated system. The system utilizes a minicomputer system configured as a local area network and ELMS (Exxon Library Management System) software. Two public terminals provide user query access, and 13 additional terminals support staff activities. The system, originally designed for public libraries, was modified for Exxon and is installed at several other Exxon locations.

This system is completely integrated; that is, data captured in the acquisitions module is utilized throughout the other functions of cataloguing, retrieval, and circulation. A journal control module with purchasing, check-in, claiming, routing, and binding features will complete the system.

All published documents circulate except the current issues of journals and reference material. Proprietary materials circulate as well, but require approvals appropriate to their level of sensitivity. ELMS provides information about the current status of any document, prepares overdue notices, and generates circulation statistics.

Information Retrieval

The EPR Information Center places a high priority on providing information retrieval services for the research staff from both the published and proprietary segments of the collection.

Reference. A collection of reference materials is readily available to the research staff. In addition, reference questions are answered by one of the reference librarians, or referred to the research librarians if necessary. Typically, requests are made in person or by telephone. While many of the requests can be answered using documents in our collection, those that cannot are handled using expertise within the research center, collections, and expertise of other Exxon information centers, or publicly available resources.

Literature Searching. Two research librarians, devoting the major part of their effort to literature searching, complete over 2000 searches each year. Requests are initiated in person or by telephone from the laboratory staff and by telephone, letter, or cable from affiliates. The process begins with a search interview using a structured form (completed by the research librarian) to obtain an understanding of the search requirements. During this interview, the searcher and the requester agree as to the level of complexity, depth, time and language coverage needed. Requesters are encouraged to describe the project or application for the search and to elaborate on technical terms and concepts. The various database possibilities are discussed; however, users are not expected to be familiar with all pertinent databases, and the research librarians use their discretion in selecting the most suitable.

For the most part, requesters are not present when the searches are done. However, in cases where the requester cannot give the searcher a clear idea of his expectations, he participates in the session to provide immediate feedback. Typically, within 2 days the requester can expect to receive his completed search with information about the search parameters and logic used and with instructions for ordering documents. Urgent searches, of course, take precedence and can be done more quickly. No written evaluation is solicited, but requesters are encouraged to provide feedback and especially to return for another search if the first is not on target.

The cost of providing the searching service is shared by the research staff and the Information Center. Online and print charges are billed to the requester, while the Information Center absorbs the costs of equipment, searcher time, training, and demonstrations.

To reduce the effort needed to distribute search charges when the invoice is received, all searches are logged into a computer file designed by the Information Center systems staff. This system generates a report to accompany the vendor invoice showing how charges are to be allocated, and it can also be used to identify previous searches or to review the literature-searching activity of an individual or a group.

Literature searches are not limited to published database services; requesters are offered the option to search the Information Center's published and proprietary collections, as well. These are done on the ELMS system, which allows Boolean operations and has left and right truncation and some proximity features. Among the searchable parameters are personal and corporate authors, title words, keywords, dates, document types, call numbers, and language. The research staff are encouraged to utilize the public terminals for the more straightforward document retrievals and to refer long or complex searches to the research librarians.

Except for searches of the ELMS database, the research staff are not encouraged to do their own searching, and up to this time little interest in this capability has been expressed. We continue, of course, to monitor progress in system interfaces and expect end-user searching to have a place at EPR when it can be done in a cost-effective manner.

Document Delivery

A very active document delivery service complements the information-retrieval services. As noted above, each literature search is accompanied by instructions for requesting documents, while document requests not resulting from searches are submitted on standard forms. The procedure is the same for handling either. The collection is checked for availability, and the document, if available, is copied or loaned as appropriate.

Since the collection development guidelines emphasize a core collection, approximately 30% of the documents requested must be obtained from other sources. Typically, if the document request results from a search and the database containing the document offers document delivery, the document is requested from that source. The University of Tulsa *(Petroleum Abstracts)* and the American Geological Institute (GeoRef) are examples of database document delivery services we frequently use. Principal local area sources are Rice University (R.I.C.E.) and the University of Houston. Other useful resources are document delivery services such as IOD (Information on Demand), and private, academic, and government libraries. Use of other corporate libraries is minimal since we do not, in most cases, make our collection reciprocally available. Other Exxon corporate libraries are occasionally used, but for the most part, their collections do not support EPR's interests. The OCLC interlibrary loan (ILL) subsystem is used to identify lending libraries, and sometimes to transmit requests, but we more often use our own system. This system, also designed by the Information Center systems staff, prints the request on the standard ILL form or on one of several special forms, if the lending library requires it. After the order is placed, the system allows status update and query. When a document is borrowed and forwarded to the requester, the system maintains a calendar and issues a notification as the end of the borrowing period approaches.

In addition to utilizing interlibrary loan services, we also purchase documents for either the EPR staff or for affiliates without libraries. Since the Information Center has the established mechanisms and the greatest expertise in document purchasing, it processes all EPR publications requests. Document purchases are expedited by placing telephone orders confirmed by written orders. Costs for purchases and interlibrary loans are charged back to the requester, utilizing the same programs as are used for literature-search charge distribution. Since timeliness is a high priority, we utilize high-speed mail, delivery services, and facsimile transmission to expedite the actual delivery of documents or photocopies.

Current Awareness

EPR maintains a large number of current journal subscriptions for its research staff, which the Information Center handles in addition to its own journal subscriptions. To provide additional access to current literature the Information Center offers a variety of other services:

A monthly bulletin, *InfoUpdate,* announces new company reports and published acquisitions. New Information Center services are also featured.

Petroleum Abstracts, the University of Tulsa's weekly alerting service specifically developed for the exploration/production segment of the industry, closely follows EPR's areas of interest. Weekly issues containing approximately 200 abstracts of literature and patents related to petroleum exploration and production (E&P) are sent to requesting individuals or sections. Subscriber participation on the service's Advisory Committee assures that coverage reflects current industry interest.

Literature update profiles developed on database vendor systems supplement *Petroleum Abstracts* in related areas of science, technology, and business.

Individual scientists and engineers request subscriptions to such current-awareness publications as *Current Contents, CA Selects,* and the American Petroleum Institute abstract bulletins.

In addition to the major functions common to all scientific and technical libraries, the EPR Information Center is responsible for several other activities.

Records Management

The Records/Reports Services Group administers the company's Records Management Program. This program provides for the systematic flow of those records requiring removal from local office file locations to remote storage when these records no longer require frequent or immediate access, and for the destruction of company records. In connection with this responsibility, we have developed a records-management manual which includes retention guides and procedures for handling records. Using this manual, the Records/Reports Group works with the division Records Coordinators to implement procedures and coordinate records transfer to and from storage, as well as handling the orderly destruction of records that are no longer required.

Distribution Control

The Records/Reports Services Group also provides distribution control services for EPR. An Exxon-developed software package handles the creation and generation of mailing lists and labels. The primary application for this software is the distribution of EPR reports, both within the laboratory and to Exxon affiliates. Utilizing this system, the Records/Reports staff generates customized distribution lists for all new reports, handles the actual distribution of the reports, and maintains a permanent record of that distribution. The service is available, as well, for any group at EPR needing to routinely distribute documents.

Systems Support

Automated systems are an integral part of all the main operations in the Information Center. In addition, we make our programs, hardware, and computer-programming services available to other EPR groups.

The proprietary bibliographic database programs formerly used to produce the Information Center's indexes are also available to other EPR groups and local Exxon affiliates. The Information Center provides both systems support and database development assistance to any research group or affiliate using this system to produce its own document database.

The ELMS system is also available in a microcomputer version. While the Information Center has no direct responsibility for supporting this version, we will provide guidance to any Exxon group interested in considering its use.

The Information Center's minicomputer system and Information Systems staff provide support service to several other General Administration groups. For example, a job-control system developed for Graphic Services allows step-by-step tracking of a job and its costs. A database of papers and presentations by EPR authors provides reports for Strategic Planning, as well as creating an additional resource for the Information Center. A number of inventory and tracking systems have been developed to support activities in General Services and Employee Relations.

Other Activities

The preceding section describes the Information Center's major functions. In this section, we discuss several key areas of activity that combine with the major functions to assure the efficient and effective operation of the Information Center. These are public relations, staff development, stewardship, and cooperative activities. The section concludes with comments on our service philosophy.

Public Relations

For many years, the Information Center did not greatly concern itself with public relations, since in a smaller organization and a more static environment, the library's services were well known through word-of-mouth. With the growth and diversification of EPR's research interests and the rapidly changing arena of available information resources, we took another look, and concluded that there was a need for a more active information program about our products and services. While the approach is still low-key, there are now several mechanisms used to announce new services and promote old ones.

To acquaint the new employee, the Information Center participates in the new employee orientation program, during which new employees tour the Information Center, meet key staff, and view demonstrations of services. A brochure describing the collection and services reinforces the orientation. To keep the research staff up-to-date, a monthly acquisitions announcement, *InfoUpdate,* includes feature articles about new databases or other services.

For more general updates, we make presentations at the company-wide seminars. In addition, talks or demonstrations tailored to specific interests are available to any company group. On occasion, company publications carry feature stories highlighting Information Center activities.

Because the Information Center is considered one of the company's resources for research support, we are sometimes asked to provide tours and demonstrations for affiliate visitors or U.S. and foreign government representatives. This exposure has the additional benefit of updating our own technical staff as they host visitors.

Staff Development

The entire Exxon organization has a strong commitment to developing its human resources. The Information Center staff participates in courses offered by EPR to develop technical knowledge of EPR's research activity and enhance supervisory, communication, and interpersonal skills. Other courses available include such self-paced courses as speed reading, microcomputer competence, and time management. Exxon's Educational Refund Program is an incentive to staff to pursue career development coursework offered outside EPR.

To specifically address growth in library/information skills, we have developed a reference course designed primarily to provide our library assistants and reference librarians with both skills in handling reference questions and knowledge of our reference resources. Intracorporation task-force assignments provide the research librarians and coordinators with opportunities to develop special expertise. Staff members also participate in database and software training offered by vendors and attend association conferences and symposia. Participation in professional associations adds another dimension to staff development, and members of the Information Center staff regularly hold leadership positions in library/information associations at the local and national levels. They also serve in advisory capacities to such information services as the American Geological Institute's GeoRef and the University of Tulsa's *Petroleum Abstracts*.

Stewardship

An annual objectives program and the capital and expense budgets provide the stewardship controls for the Information Center. Each year we develop a set of specific objectives, the completion of which enables us to provide better, more cost-effective service. Typically, developing and promoting services, acquiring and enhancing systems, improving facilities and equipment, and reducing costs are considered appropriate objectives. The entire staff participates in the development of these, and in addition sets personal objectives in line with the organizational plans. Monthly reviews and semiannual and annual reports document progress. The annual report is reviewed by the senior company management and offers an opportunity to discuss the Information Center's contribution to the research effort.

The annual budget is, of course, reviewed throughout the year and provides a mechanism for containing costs and monitoring activity levels. Particular emphasis is given to achieving vendor cost savings through corporate contracts.

Cooperative Activities

Exxon's widespread operations dictate many information units, varying greatly in both size and purpose. There is no central control of all information functions, but an appreciation of the need to maximize the use of information resources encourages cooperation.

Within Exxon, ITUG (Information Technology Users Group), formed in the early 1970s, acts as a forum for technology utilization. The mission of ITUG is to identify suitable technologies and to maximize their application in Exxon information centers. Through ITUG, systems and equipment are identified and the cost of acquisition can be shared.

When ITUG considers the need for a new or improved system, it typically appoints a task force of several information professionals drawn from various information centers. This task force develops specifications, investigates available systems, analyzes costs versus benefits, and makes recommendations. If the ITUG managers approve the project, the task force maintains close contact with either the outside contractor or with Exxon's Computer and Communication Division as the system is developed and installed.

While participation in any given project is optional, EPR's active involvement has been a key factor in our technology utilization. Through ITUG the computer-based systems previously and currently in use were developed, and ITUG continues to play a valuable role in determining the future direction of information technology within Exxon information centers.

Since each affiliate's interests are quite different, shared serials lists and catalogs have not been in demand. It has proven more effective to use a person-to-person network to tap the resources of our affiliates and they, ours. To facilitate this relationship, and to share information and address mutual problems, the Exxon librarians in a particular geographic area occasionally meet as a local group. ITUG task-force projects also bring staff members from several centers together, thus facilitating the personal network.

Service Philosophy

A description of the EPR Information Center would not be complete without mention of the philosophy that guides its operations. We are proud of our reputation as a "can do" organization, and work to assure that that attitude is part of every staff member's approach to the job. All aspects of work—meeting deadlines, locating obscure references or documents, identifying unique information resources, solving system problems, or reducing costs—are accepted as challenges to be carried to a successful conclusion.

Challenges

As a part of the petroleum industry, the EPR Information Center shares in the challenges that test the entire industry—expensive exploration, uncertain prices and markets, and dwindling reserves. These overall industry conditions give heightened importance to the more specific challenges encountered in the delivery of information services.

Currently, a major challenge facing the Information Center is the efficient utilization of space. The collection has outgrown the available space, and while we have several mechanisms for deleting obsolete documents, the nature of the technical material imposes certain limits, especially in the areas of geology and paleontology. High-density shelving and remote storage provide partial solutions, and technologies such as digital disk storage will merit consideration.

Another challenge is the geographic separation of a number of our research groups. Because the laboratory site can no longer accommodate the entire staff, several groups are located in buildings some distance away. The Information Center is making a concerted effort to deliver services to these groups in ways that minimize their distance. For example, to make catalogs and indexes accessible to these remote sites, we plan to deploy at least one public access terminal in each location. Interfaces that allow personal computers to access the ELMS systems are also being considered.

A third challenge is the identification and delivery of new information services or enhancement of present services to respond to changing needs. Two examples illustrate our efforts to enhance our existing services.

In a recent experiment, we extended the literature search a step further to document analysis. After the documents were identified in the search, Information Center staff scanned them for specific data content to be entered into a database matrix.

In another experiment, a computer-produced current-awareness search was enhanced by scanning some additional documents manually, selecting relevant documents from both manual and computer searches, and providing photocopies organized in a binder and cross-referenced by subject. Our costs to provide this service compared favorably with an existing commercial service and provided the additional benefit of involving the Information Center more closely in the research effort.

The fourth challenge is one voiced by a chorus of information professionals—managing the wealth of ever-changing technologies. As part of a research organization in which technology is the raison d'être, we are especially committed to investigate and innovate in the technological arena.

However, our approach has been evolutionary rather than revolutionary and always involves careful consideration of benefits versus costs. The Exxon ITUG organization previously discussed has been an important factor in our use of new information technologies.

These four major challenges (and of course there are many others of importance) and the stimulating environment of petroleum research provide the Information Center with an exciting outlook and reinforce our commitment to providing a quality information service for EPR staff and Exxon affiliates.

THE U.S. GEOLOGICAL SURVEY LIBRARY

Nancy Jones Pruett
Technical Library
Sandia National Laboratories
Albuquerque, New Mexico 87185

The U.S. Geological Survey (USGS) Library, established in 1882, is the premier geological collection in the United States. The main library is in Reston, Virginia; three branch libraries are located in Denver, Colorado; Menlo Park, California; and Flagstaff, Arizona. Together they contain more than 1.2 million volumes and 105 full- and part-time staff members.

The Library's primary responsibility is to support the research of the U.S. Geological Survey, carried out by its 13,000 staff members. The Survey, a part of the Department of Interior, was established by an Act of Congress in 1879 and charged with responsibility for "classification of the public lands, and examination of the geological structure, mineral resources, and products of the national domain." This involves "collecting, analyzing and publishing detailed information about the Nation's energy, mineral, land, and water resources" (Rabbitt, 1979). Further information about the Survey can be found in two publications: *The U.S. Geological Survey* (Library of Congress Environmental Policy Division, 1975) and the *United States Geological Survey Yearbook* (U.S. Geological Survey, 1984b).

Although its primary mission is to be responsive to the research requirements of U.S. Geological Survey scientists and researchers, the Library system traditionally has made its resources available to various components of the U.S. Department of the Interior, other Government agencies, State geological surveys, universities and colleges, research organizations and the general public by means of interlibrary lending, reference service, and onsite use of its collections. (Chappell and Goodwin, 1984, p. 17).

Between 1980 and 1984, 41% of the library's service was given to outside users. The Denver branch, because of its close location to numerous mineral and oil industry companies and consultants, has a particularly high outside use.

Although the collection is devoted to all aspects of the geosciences, the major subjects of interest are geology, paleontology, petrology, mineralogy, geochemistry, geophysics, ground and surface water, cartography and mineral resources. The current topics of special interest reflect the Geological Survey's interest in the environment, Earth satellites and remote sensing, geothermal energy, marine geology, land use, planetary geology, and the wise use and conservation of resources. (U.S. Geological Survey, 1984a, p. 3)

The USGS library is a federal library. Employees are civil servants, and the budgeting and hiring processes and general direction of the agency are influenced by the same factors that influence all federal agencies. In the 1980s an issue which had an impact on federal libraries was the Office of Management and Budget (OMB) Circular A76. It forced federal agencies to consider which of their activities (including libraries) could be contracted to the private sector.

Within the Geological Survey, the library is part of the Office of Scientific Publications in the Geologic Division. Although the library serves all divisions of the Survey, 80% of the use is from the Geologic Division. Statistics on use are kept so that the other divisions can be assessed in proportion to their use.

Six sections report to the Chief Librarian (Fig. 2.6): the three branch libraries, and the three sections within the Reston library: Reference and Circulation, Catalog and Classification, and Exchange and Order. There is also a translations service and an Assistant Chief Librarian reporting to the Chief Librarian. A floor plan of the Reston facility is included as Fig. 2.7.

Menlo Park has a staff of 14 full- and part-time staff, divided among five sections: Reference, Catalog, Photo Services, Technical Services, and Support Services.

Denver has a staff of 20 full- and part-time staff organized in four sections: Photo Services, Field Records, Technical Services, and Reference and Circulation.

Flagstaff has a staff of three, a librarian and two technicians.

The major functions will be discussed primarily from the point of view of the headquarters library.

Collection Development

The main library and the three branches together contain over 1.2 million bound and unbound monographs, serials, and government publications; 355,000 pamphlets and reprints; 230,000 maps; 80,000 field-record note-

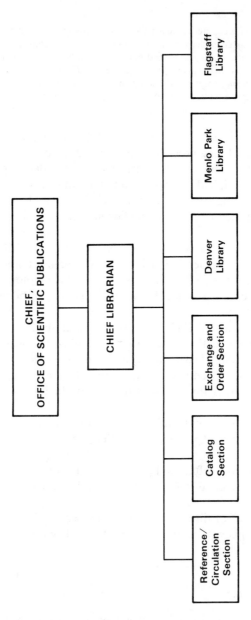

Fig. 2.6. Organization chart, U.S. Geological Survey Library, February 1985.

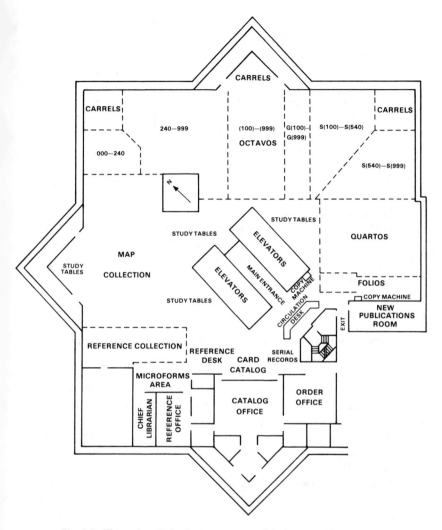

Fig. 2.7. Floor plan, U.S. Geological Survey Library, February 1985.

books and manuscripts; 300,000 album prints, transparencies, lantern slides, and negatives; and 304,000 microforms consisting primarily of doctoral dissertations and technical reports.

The collection philosophy has been to be comprehensive: that is, the library has attempted to obtain at least one copy of every useful publication within the subject areas collected. In Reston, publishers' flyers and new issues of all journals are reviewed daily by the librarians in the Reference

Section; materials of earth science interest which are not in the collection are ordered. Recommendations from Survey staff are also encouraged. In addition to these normal channels for collection-building, the USGS Library relies heavily on a program of exchanges with other geological surveys and organizations around the world. Over 75% of the serials collection is received on exchange rather than purchased. The Library has about 2000 exchanges with foreign organizations, and about 670 with U.S. organizations. These exchanges bring in about 7500 of the 10,000 serial titles. The other 2500 are purchased. The exchange program is particularly successful because the USGS publications that the library has to offer are of high scientific quality and highly valued by other organizations.

There are some other avenues of systematic collection development, too. The library has subscriptions with the National Technical Information Service (NTIS) for receiving microfiche of geological reports and with University Microfilms for all earth science dissertations.

Preservation is a concern in the library because of the age of the collection, as well as the special materials and rare books.

Collection Control

Collection control is a challenge in a library which has material in almost every language of the world and in nearly every format.

The library has its own classification system (U.S. Geological Survey Library, 1972), which was devised in 1900, even before the Library of Congress classification scheme was available. It uses a scheme of geographic divisions, which can be added in parentheses after any subject number. The library joined OCLC in late July, 1975. Since 1976, the library has cataloged all titles on OCLC, with Reston being the central cataloging operation. Although standards are followed for descriptive cataloging, some changes are made in LC subject headings, in order to meet the specific needs of the library clientele. For example, the Library of Congress would use the subject heading Geology—Ohio; the USGS library reverses these headings to Ohio—Geology. (If it did not, much of the catalog would be under "geology," and it would be more difficult to answer the common question of what exists on a particular area.)

The OCLC archival tapes provided the initial 85,000 records for loading into the Integrated Library System (ILS) installed in 1984/1985. Retrospective conversion projects are in process to convert another 100,000 records representing all the USGS publications and all the serials into machine-readable form. The Library has a turnkey contract through the Federal Library and Information Center Committee with Avatar (now part of OCLC) for installation, maintenance, and development of ILS.

There is a backlog of uncataloged book and journal material of approximately 3000 titles. Most microform materials are not cataloged. Pamphlets are not classed and are kept alphabetically in vertical files. Probably no more than 10% of the map collection is cataloged. Technical reports are classed and cataloged along with book materials. Translations of articles and portions of books are not cataloged, but are treated like pamphlets in vertical files. Whole-book translations are classified and cataloged.

Libraries and individuals outside the Survey have two sources for access to the bibliographic records of the USGS library collection. The *Catalog of the United States Geological Survey Library* was published from 1964 to 1976 by G. K. Hall & Co. Since the USGS library is the source of most of the materials indexed for GeoRef, the online database prepared by the American Geological Institute, the GeoRef database can be used as a catalog of material in the Survey Library (without location information, of course). The dial-up access of the ILS system will provide a third source.

Circulation is currently being automated. The Reston Library circulates about 30,000 items per year. Some over-the-counter circulation is done by means of American Library Association (ALA) Interlibrary Loan forms. All materials, including journals, circulate for loan to Survey staff after they are released from the New Book room; they are recalled when needed by someone else.

Information Retrieval

Service to outside users has been steadily growing. Figure 2.8 shows the increases in questions answered (both total and to outsiders) and interlibrary loan requests from 1970 to 1984. The increase has been handled

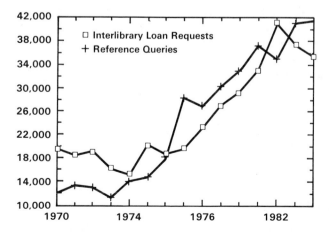

Fig. 2.8. Trends in service, 1970–1984, U.S. Geological Survey Library.

so far, despite staff reductions and without instituting user fees. Handling the future growth is one of the serious challenges the USGS library system faces.

In 1984, the American Geological Institute (AGI) began a new document delivery service, with one of the primary sources as the USGS library collection. There is some hope that the use of AGI's service will take some of the burden of actually filling requests from the public services staff in Reston and the branch libraries, but it does not solve the long-term problem of increasing use of the collection.

The reference librarians at each branch spend some time each day on the reference desk, and later follow up any questions they received which were not completely answered while they were on the desk. Statistics are kept according to the type of user, since funding is prorated to the other divisions. Online searching on DIALOG, SDC, and the DOE RECON system is available to Survey personnel, but not to outsiders.

Current Awareness

The prime means of current awareness in the Reston library is a New Book room, where every new item is placed for at least a week. Certain journals on display shelves remain there for a month. Users who wish to have materials routed to them after they are released from the New Book room can sign their names on a slip in the book. Menlo Park does produce a new acquisitions list, but none of the other libraries do. SDI profiles are not offered.

Document Delivery

The Interlibrary Loan unit is a part of the Reference and Circulation Section. Both borrowing and lending are handled primarily through ALA Interlibrary Loan forms, federal library interlibrary loan (ILL) forms, and the OCLC ILL subsystem. Also, some items are acquired through document delivery services such as the British Lending Library and Chemical Abstracts. Over 5000 items are borrowed through Interlibrary loan each year. This is low compared to the 30,000 loaned, but it still shows that no collection, even the best, is ever complete.

Within the Survey, regular mail is the channel of delivery currently being employed. Most onsite users come to the library to get their own materials.

Issues and Trends

The USGS faces a number of serious problems in the next 5 years. All of the issues must be dealt with in the climate in which all federal libraries are currently operating: the specter of increasing cuts in budgets and staff-ing, pressure to contract out to the private sector, potential downgrading of librarians, and possible modifications of federal pension and retirement benefits.

Whether the library can continue to obtain the necessary specialized personnel (e.g., Slavic and other foreign-language catalogers, map catal-ogers, geologic information specialists) in this climate is a critical concern, as is the overall question of what resources will be available, both for collection-building and for service. A 20% cut in the budget for collection development was received in FY85, the first such cut since the mid-1970s.

Space is another serious long-term problem. When the Survey library moved into the National Center in 1974, the space seemed generous, but even then some of the collection had to be stored in the basement. Re-cently, the basement was claimed for office space during a reorganization of the Survey, and the collection was then stored in a nearby warehouse about 20 minutes away by car. Deliveries are made to and from the ware-house twice a week. Almost one-third of the Reston collection is now in that warehouse, including a large number of maps.

Service to the public is also an abiding question, particularly since over 40% of the use of the library is by non-Survey personnel. Although in-stituting of charges for borrowing and interlibrary loan has been discussed, the Survey's position so far has been that it is part of their mission as a public agency to disseminate information.

Perhaps the most interesting of the challenges facing the USGS library is how its Integrated Library System will interface with other information systems within the Survey, such as those in the National Cartographic Information Center, the Public Inquiries Offices (PIOs), the Water Re-sources Division's Distributed Information System (WRD DIS), and the newly developing Earth Science Information Network. It is too early to predict what form such interaction will take, but the PIOs and the WRD DIS network already are accessing the library system's ILS capability.

Chapter 3

Scientific and Technical Literature and Its Use

It is not the purpose of this chapter to review the characteristics of the sci/tech literature [which has been done thoroughly by Subramanyam (1981) and others]. Forms of literature which are particularly important in sci/tech libraries (e.g., technical reports, conference literature, patents, standards, etc.) are covered in detail in Volume 2, Part I. This chapter will focus on the user of sci/tech literature and summarize patterns of sci/tech literature use which have implications for library and information use.

The answers to numerous questions of sci/tech library management lie with the users: Why are libraries used relatively infrequently? Why do users complain that the information they can get from the library is not current enough? Why are the most powerful faculty (the ones who would be most useful to approach for library support) often not library users? Why do libraries seem to be used more by pure scientists than by applied? What different forms of material and approaches to service might be designed to better serve the needs?

The use of libraries is a complex sociological and psychological phenomenon and not fully understood in spite of a large number of use studies and a voluminous literature. However, some patterns of use have been clearly established and can be used to improve sci/tech library services.

The relevant research falls into three broad areas: (1) communication behavior of scientists, (2) information-seeking behavior of scientists, and (3) the behavior of engineers in contrast to scientists. There are also variations in information use by subdiscipline (e.g., how chemists differ from physicists), but these are covered in the special subject chapters gathered together in Volume 2, Part II.

COMMUNICATION BEHAVIOR OF SCIENTISTS

The communication behavior of scientists and how scientific progress is made has been widely studied (for example, see Allen, 1977; de Solla Price, 1963; Garvey, 1979; Hagstrom, 1965; Nelson and Pollock, 1970). Most researchers divide scientific communication into formal and informal, as does Garvey, whose 1979 work *Communication: The Essence of Science* is an excellent introduction to scientists' communication behavior.

In order to understand the scientist's view of communication, let us look at Fig. 3.1, a time scale for a typical publication process. This scale was derived from a study of psychology, and the exact dates will shift somewhat depending on which scientific specialty we study. (For example, Table 3.1 shows that the elapsed time between article submission and publication varies from 8.3 to 17.2 months, depending on the field of science. But it serves to illustrate the timing of the different forms of communication.)

Note that it is about 28 months from the initiation of a work to its publication in a journal. The communication during that 2½ years is in the informal domain. Once the work is published in a journal, it enters the formal domain. From then on it is public, unambiguously retrievable, and

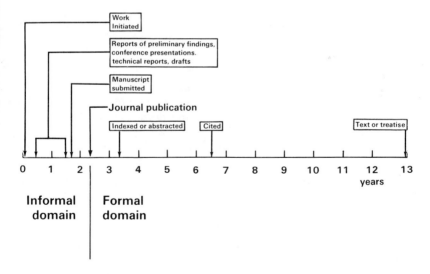

Fig. 3.1. Time scale for the typical publication process. [Based on data from Garvey, (1979). Reprinted with permission from "Communication: The Essence of Science", Copyright 1979, Pergamon Press Ltd.]

TABLE 3.1.   Elapsed Time between Article Manuscript Submission and Publication by
Field of Science—1977 and 1984[a]

	Elapsed time between submission and publication (months)	
Field of science	1977	1984
Physical sciences	8.0	8.9
Mathematics	20.5	17.0
Computer science	10.6	12.1
Environmental science	14.4	17.2
Engineering	9.0	9.9
Life sciences	12.1	12.7
Psychology	12.1	15.1
Social sciences	10.3	13.1
Other sciences	5.8	8.3
Total	10.3	12.9

[a]From King et al. (1984), p. 13.

carries with it the stamp of approval of the scientist's peers. Scientists behave differently and apply different sets of standards in each domain (Garvey, 1979, p. 22). Let's look at the two domains and the function of the scientific journal in more detail.

THE INFORMAL DOMAIN

There is a much larger volume of communication before publication than there is after publication. Prepublication communication includes personal correspondence, conversations and presentations at professional meetings, drafts that are circulated for comments, etc. The communication is an exchange, and at this phase it is under the control of the scientist, who may get feedback on the same work many times. The work may also appear differently in various communications when it is still informal. Communication in the informal domain thus is redundant and unstable. It is also under the control of the scientist, who can choose how much and just what he or she will share. What he or she receives back can also be filtered and evaluated for personal relevance. The informal domain includes the communication which actually enters into the creation of scientific information, and is thus of the most interest to the scientist.

The informal domain is characterized by invisible colleges, which were first described by de Solla Price (1963) and later studied in detail by Crane (1972). An invisible college consists of a group of people with similar interests who use a number of informal communication channels, including exchange of preprints, reprints and manuscripts; telephone calls; personal visits; conversation at conferences (both national and international); and hosting visiting scholars and guest speakers. Members of invisible colleges tend to cite each other's papers, and the colleges can often be delineated by citation studies. Often, invisible colleges are held together by highly influential scientists who have accumulated a large group of former research students (Crane, 1972; D. Wood, 1971).

The importance of personal contact in communication of scientific information cannot be overemphasized. For example, one paleontologist listed "keeping up with colleagues" as the most important of approaches to keeping current (Bambach, 1981). And a Department of Defense study of the origins of information and ideas that were important in the development of 20 successful weapons systems (Project Hindsight) showed that personal contact was the means through which information was introduced into the using organization in 70% of the cases (Sherwin and Inemson, 1966).

The types of information included in the informal domain are conference presentations, preprints, technical reports, drafts, theses and dissertations, in-house publications (reports, bulletins, or memos), and conference proceedings. Essentially, any publication which has not been reviewed is informal. Conference proceedings are an interesting case of an informal publication appearing formal, and they are troublesome for that reason.

THE FORMAL DOMAIN

The types of publication which are considered in the formal domain are the journal, the treatise, and the text. Note from Fig. 3.1 that it takes an average of 13 years for work to appear in a text or treatise and that the scientific journal marks the publication boundary between the informal and formal domain. Publication is the prime product of science. Thus, the entering of a work into the formal domain is essentially the completion of the research.

The characteristics of the formal domain are that it is public, orderly, and stable. Journals are public in two senses: anyone can submit articles, and also anyone has access to the published product, either in libraries or by subscribing. They are orderly, because the selection of articles is based on their scientific merit and relevance. They are stable, because once the articles are published in journal form, the same research is not normally published in any other form. Thus, it can be unambiguously cited and retrieved (Garvey, 1979, p. 69).

The prime purpose of publishing in a scientific journal is *not* communication, it is the attainment of professional goals. The scientist publishes in the journal to gain and maintain visibility among his or her peers. The journal also is the socially accepted medium for establishing priority of discovery. It is who publishes first which decides who gets credit in the annals of science for a discovery. The second use for the scientific journal is for the establishment of a public body of knowledge.

Another part of the formal domain of particular interest to information specialists is postpublication processing. This includes indexing, abstracting, and citation analysis. Postpublication processing, in contrast with the informal domain and the publication and review process, has been primarily in the hands of librarians and information specialists. In the last 10 years we have made great strides in automating indexes and cutting down the time it takes to index an article. There is no question that these improvements were necessary, but it is clear from Garvey's work that even if articles were indexed the moment they were published, the index would be too slow for the scientist.

INFORMATION-SEEKING BEHAVIOR OF SCIENTISTS

Estimates of the amount of time scientists and engineers spend searching for information range from 20 to 25% of their working time (Lufkin and Miller, 1966; Amsden, 1968; Hazell and Potter, 1968; Wood, 1971).

The sources for information are many. They include colleagues (both internal and external); libraries (both personal and organizational); and journals, reports, handbooks, manufacturer's catalogs, standards, and other printed materials. For any given information need, the choice of which source to employ is complex and depends on such factors as job function, size of organization, qualifications, place of employment, and academic discipline (Wood, 1971, p. 12). It also depends *greatly* on the convenience of the source. In fact, convenience is much more important than the amount or quality of information expected (Allen, 1966, 1968a,b, 1977; Utterbach, 1969; Rosenberg, 1967). For example, Rosenberg (1967) asked 96 research and nonresearch personnel to rank information-gathering methods according to ease of use and amount of information expected, and to indicate which method they would use to solve three hypothetical problems. In order of preference, the methods were:

1. Search personal library.
2. Search material in building where you work.
3. Visit a knowledgeable person nearby.

4. Telephone a knowledgeable person.
5. Use a library that is not within your organization.
6. Consult a reference librarian.
7. Write a letter.
8. Visit a knowledgeable person 20 miles away or more.

There was no relationship between preference ranking and the amount of information anticipated. For example, use of one's personal library was ranked seventh when the methods were arranged in order of anticipated value. Allen (1966, 1968a,b) found similar results when he studied the problem-solving methods of engineers in industrial environments by having each engineer record his use of a number of information channels during a period of about 15 weeks. When these were ranked on the basis of perceived accessibility, perceived ease of use, perceived technical quality, and amount of use, the investigators found that accessibility determined both frequency and priority of use and that the perceived quality of the channel had no bearing on how often the source was used.

Because information-seeking behavior is so complex, detailed studies have tended to produce more useful information than broad surveys. Garvey's 1979 work summarized a number of such detailed studies. He considered 10 kinds of information needed, 11 stages of scientific work, and seven sources of information. By combining the various categories, he could determine, not only that "journals" and "local colleagues and students" were the most often successfully used sources, but also that different formats were used to satisfy different needs. Journals were most useful in providing information needed to place the work in context with similar work already completed and to integrate findings into current scientific knowledge. Local colleagues and students were least often used for those reasons, and most often used to provide information needed to select a design for data collection, to design equipment or apparatus, or to choose a data-gathering technique. Books were effective in providing general information needed to formulate a scientific solution and also specific information needed to choose a data-analysis technique. Reprints and meeting presentations were most effective in providing information needed to relate a scientist's work to ongoing work in his area. Technical reports seemed to be used most often to design equipment and to select data-gathering techniques (Garvey, 1979, p. 265).

Garvey also found variation in information needs depending on the stage of research. He noted that most scientists were engaged simultaneously on several "different" research projects at once and were generally in different stages in each. The only information need that was constant through all stages was the need "to place the work in context."

Other factors that Garvey noted include variations in use between:

1. Physical and social scientists. Meeting presentations and technical reports were more useful to the physical scientists. Books and local colleagues and students were more useful to the social scientists.

2. Those working in a new subject area and those working in the same area. Journals seemed especially useful in satisfying information needs of scientists working in a new area.

3. Less experienced and more experienced scientists. The least experienced scientists had greater information needs than the most experienced scientists. They also found local colleagues and students and journals especially useful.

4. Basic scientists and applied scientists. Applied scientists use all media less than the basic scientists, except that they seem to find technical reports more useful (Garvey, 1979, p. 273).

Another point about scientists' information-seeking behavior was brought out by Stoan (1984), who admonished librarians not to think that library search strategy and research are the same thing. To a scientist,

> The genuine core of a research project . . . consists of essentially uninterpreted data, many or all of which may be gathered outside of the library altogether, as in a laboratory or archive, or from a questionnaire or case study. For the scholar, library use comes into play for the gathering of some primary data in some disciplines and for the gathering of secondary literature, that is, the books, articles, and research reports in which are reported the results of research. This secondary literature of the scholar is what librarians call primary literature. Since scholars must master this primary literature in their disciplines, it follows that library use is one aspect of research. (Stoan, 1984, p. 100).

Stoan points out that scholars have a preference for following footnotes and bibliographies in the subject literature rather than using indexes and abstracts, and he postulates that "the primary literature indexes itself, and does so with greater comprehensiveness, better analytics, and greater precision than does the secondary literature" (Stoan, 1984, p. 103).

COMMUNICATION AND INFORMATION-SEEKING BEHAVIOR OF ENGINEERS

Garvey's research showing a difference in information use between applied scientists and engineers is supported by numerous other use studies. Engineers use informal information more than scientists, they use the formal literature (and libraries) less, and they rely more on oral, internal sources (Allen, 1977; Slater and Fisher, 1969; Gralewska-Vickery and

Roscoe, 1975; Bickert *et al.* 1967; Wood and Hamilton, 1967). For example, Rosenbloom and Wolek (1967) found engineers used internal information sources 68% of the time, while scientists use them only 21%, and that engineers used oral sources more (69% of the time, and scientists only 41%). They also used formal literature less (9% versus 46%).

Engineers are also more likely to use certain forms of literature, including handbooks, standards and specifications, technical reports, and trade literature. For example, Davis (1965) studied 1800 engineers and found that 83% used handbooks, about 62% used standards and specifications, 58% used research reports, and 85% used manufacturers' catalogs, referred to them at least once a week, and kept a personal collection of catalogs. In contrast, their use of abstracts and indexes is much lower (for example, Davis found approximately 33%).

Engineers seem to want digested information, and if they are given too much unevaluated information they will select based on authors they know (Gralewska-Vickery, 1978, p. 277).

Engineers overall read less, and they use literature and libraries less. Even information services which are directly oriented to the engineer are little used except by academics. This may be because the literature of technology is not cumulative, as science is. There are fewer citations to previous papers or patents, and these are more often to the author's own work (Allen, 1977, p. 39).

Allen, in *Managing the Flow of Technology* (1977), insists that much of the confusion in information use studies comes as a result of not differentiating between scientists and engineers. Engineers differ from scientists in family background and in personality characteristics such as goals and values. The educational process and their socialization processes are also different. For example, Krulee and Nadler (1960) studied science and engineering undergraduates and found that science students placed a higher value on independence and on learning for its own sake than did the engineering students, who were more concerned with success and professional preparation. The engineering students also expected their families to be more important than their careers as major sources of satisfaction, whereas the science students expected the reverse.

The education process itself provides a long socialization for the Ph.D. scientist, in contrast with the shorter educational process of the engineer, who most often has a baccalaureate degree. The different patterns remain after graduation. For example, Ritti (1971) found differences in work goals. The scientists valued publication of results and professional autonomy highly, while the engineers valued them least. And their career advancement was seen as coming from performance in different arenas. For the engineer, advancement is tied to activities within the company, and the scientist advances based on his or her reputation established *outside* the organization.

In addition to differences between engineers and scientists as people, there is a difference in the nature of science and technology.

In science, as we have seen, the literature is the repository for all scientific knowledge, and it is permanently recorded. To a technologist, the document is not so important. The creation stands for itself.

> The names Wilbur and Orville Wright are not remembered because they published papers . . . the technologist's principal legacy to posterity is encoded in physical, not verbal, structure. Consequently, the technologist publishes less and devotes less time to reading than do scientists. (Allen, 1977, p. 40)

The communication structure of technology is also different. Rather than invisible colleges, technologists rely on communication with co-workers in their own organization to keep them informed. Partly this may be because the engineer's organization has a need to protect proprietary information.

> The technologist is not quite free to communicate the results of his R & D effort . . . even if the results of the technologist's work are amenable to presentation in a paper, the information generated may be of a proprietary nature and may have to be kept confidential to protect the interests of the corporation. In the case of sponsored research, confidentiality may be a contractual requirement stipulated by the sponsoring agency. The engineer is more likely to write a technical report to be submitted to the sponsoring agency, or to apply for a patent to protect his invention from unauthorized exploitation. (Subramanyam, 1981, p. 13)

In Allen's studies of engineering information use, he found that the source for the piece of information which resulted in a better solution to a design problem was most often an internal consultant rather than literature or an external source. Those internal sources often turned out to be "gatekeepers," the exceptions to the rule, people who were high communicators, read the literature, and had extensive outside contacts. These people translate the "outside" information into terms that are applicable "inside."

Allen also pointed out that because of the importance of personal communication, turnover plays an essential part in furthering technology. When an engineer changes jobs, he carries practical information with him. In fact, the most successful technology transfer has occurred when someone involved in the scientific or technological development left the research organization and went into business.

> Technology . . . builds on its own prior developments and advances quite independently of any link with the current scientific frontier, and often without any necessity for an understanding of the basic science underlying it. According to Price, the results of

work at the research front are passed along to technologists only after having been packed down into textbook form. The process, of course, involves two rather long delay periods. The first one, during which original research is published in primary journals to await extraction into textbooks, may require several years; the second one, the period between education and the utilization of knowledge, is variable and one of the major causes of the technical obsolescence problem now being faced. (Allen, 1966, p. 1057)

IMPLICATIONS FOR INFORMATION SERVICES

What does our knowledge about scientists and engineers imply for information services?

First, the formal literature is enormously important to scientists for reasons of prestige and priority. Formal publication is an essential part of the progress of science. One implication of this is that libraries must continue to buy journals. But there is also an implication for the future of publishing which has not been clearly recognized by the information industry. Attempts to speed communication or decrease costs of publication will not be successful if they undermine prestige or the review process. For example, many people hear scientists' complaints about publication delay times and propose automated solutions: for example, unreviewed electronic journals which are available instantly. They do not understand that scientists view this as "garbage in = garbage out" and that the review process and the formality of the publication process are essential.

Second, the informal domain is also very important in scientific communication. One reason the most elite faculty members are often not heavy library users and that no one seems to find the library current enough is that they are focussing on the informal domain. Even an index produced at the moment a journal came out would be too slow. Libraries cannot help much in the informal communication process (and may not even be welcome, since one of the characteristics is that communication is under the scientist's control). However, literature searches can provide names of people working on similar topics (who can then be contacted by the scientist). And libraries can do their best to acquire conference literature and to be sure scientists are aware of conferences in their areas of interest.

The importance of openness to the scientist's entire communication process helps us understand the threat posed by increased control of scientific communication for national security purposes (Relyea, 1986).

What are the implications from our knowledge of engineer's behavior? For the short term, libraries which serve technologists should seek out the gatekeepers. Mote (1971) designed a methodology to do just that. It included tracking circulation, current-awareness publications, inquiries, and interlibrary loans, and noting individuals who were heavy users.

Once we have identified the gatekeepers, we should make sure they know about available information services such as computerized literature searching, SDI, displays of new books, etc. And we should talk to them as sources of information about what their colleagues are doing. In the long term, we should be working toward digested, evaluated information. Our aim should be some kind of information system which retrieves the way an internal consultant does, applying broad knowledge to a specific problem.

All studies of how scientific research is done emphasize that it is a personal and creative process and requires browsing and serendipity. Thus any library policies such as storage facilities, compact shelving, and large microform collections may seem to get in the way of scientific research and should be considered carefully.

Although our knowledge of our users is incomplete, the more those providing information services understand, the more likely they are to provide services that meet the user's needs.

Part II

Major Functions

Chapter 4

Information Retrieval

The information-retrieval function includes activities directly related to getting the user the information he or she wants.

Users may ask the library any kind of question at all. Most libraries take it as a challenge to attempt to answer all questions and will at least try to put the person in contact with someone who might know the answer, even if the answer is not published. However, libraries are most successful in answering the kinds of questions that the tools they have are designed to answer.

These tools provide access to the world's published literature, the prime resource the library has to answer questions. If the collection development function has been accomplished well, a large proportion of questions will be answerable from resources at hand.

But much of the world's literature will not be in the library: what the library must have is the means to find out what exists and to acquire it if needed. Searching the secondary sources (indexes and abstracts) should lead to literature available outside the library.

Also, personal contacts with other librarians and subject specialists are invaluable in answering questions which are outside the scope of the collection (Stursa, 1985). Membership and active participation in professional organizations for a librarian is a necessity, not a luxury.

Thus, indexes and abstracts are a very important part of the library's collection. As more and more secondary services are available online, even small libraries can have access to indexes to the world's literature with only a terminal and a telephone. The management of online services has become a large part of the information-retrieval function and is considered separately in this chapter.

The information-retrieval function overlaps with all the other functions, of course. Current awareness can be thought of as anticipatory information retrieval. Rather than locating existing information in response to a request, we anticipate the request and provide a means of notifying the requestor of new items he or she has not asked for but which might be of interest. Information retrieval is dependent on the collection development function to provide the collection to answer most of the questions. Information retrieval is dependent on the collection control function to provide the cataloging and indexing which allow access to the collection. The document delivery function delivers the actual item, either from the library's collection or from an outside source.

There have been several attempts to divide up the kinds of questions that users may ask, usually for some particular purpose. Voigt's (1961) division into the everyday approach, the comprehensive approach, and the current approach was particularly useful in considering sources, since certain sources are best for keeping current, others for doing exhaustive searches, and others (e.g., handbooks) for everyday information. But as more and more sources are appearing online, the division of sources is becoming blurred.

Another common division is between directional questions, ready-reference questions, and more complicated questions. Based on studies by Wilkinson and Miller (1978) and Lynch (1978), 30–35% of questions at an information desk are directional. Another 47–52% are "one-step" questions, what we will call ready-reference type questions. Only 22% require either more than one step or more than one question to define (Thomas *et al.*, 1981, p. 99). The libraries which these studies focused on were general libraries, rather than sci/tech. They do not actually perform literature searches for their users; instead, they lead the users to the sources and let them do their own searches. Nevertheless, , it seems reasonable to estimate that even in sci/tech libraries, 20–35% of the inquiries will be more complicated than directional or ready reference.

For our purposes, it is useful to ignore the directional questions and consider the two other areas (ready reference and literature searching) in more depth. Ready reference includes determining whether the library has a book, report, journal, etc. and providing addresses, phone numbers, biographical information, definitions of terms, and specific handbook-type information on request. Not all these questions are simple to answer, of course: they are just simple in format.

Literature searching is using indexes, abstracts, and bibliographies to get a list of citations to articles, books, reports, etc. on a particular subject.

Usually the information-retrieval staff provide the actual information requested in response to a ready-reference question. In the case of literature searching, what the library usually provides the user is a list of citations rather than the actual "information." Providing the actual information has been a document delivery function rather than information

TABLE 4.1. The Eight Steps of Information Retrieval

1. Understanding what the user wants (via the reference interview)
2. Choosing the important elements of the question and their relationship to each other (search strategy development)
3. Choosing the source(s). This requires knowledge of possible sources and their accessibility
4. Putting the search in the appropriate terms for the source (vocabulary development)
5. Executing the search
6. Evaluating the results and doing another iteration if necessary
7. Delivering the information
8. Evaluating the results with the user

retrieval. But with the growth of full text and numeric databases, these lines may blur.

There is also a step beyond literature searching: information analysis. An information analyst might choose appropriate items, gather the copies, and provide them to the user. Or he or she may read and digest the items and write a summary. Information analysis normally requires subject experts. The Congressional Research Service is an excellent example of an information analysis organization, and there are many separate information analysis centers in the sci/tech world, usually dealing with numerical data. Although some libraries have experimented with information analysis, few libraries have the resources to function as information analysis centers. However, as the need grows, more libraries may begin this as an added-value service.

The eight steps of information retrieval are listed in Table 4.1. The first step, understanding what the user wants, is critical to all aspects of information retrieval. This step is usually encompassed in the reference interview (sometimes called the reference transaction). After we discuss the reference interview, we will discuss the remaining steps as they apply to ready reference and literature searching separately.

THE REFERENCE INTERVIEW

Understanding what the user wants is not a simple matter. In addition to the intellectual analysis of the question, all the aspects of interpersonal communication, both verbal and nonverbal, come into play. Particularly in a sci/tech environment, the background of both user and information-retrieval person can be critical.

The reference interview or reference transaction is the common term for the critical interaction between the user and the information-retrieval person. It is during the reference interview that the staff member determines what the user wants and tries to translate that into terms which the library system can answer.

Any practicing reference librarian can site many cases where what the user asked for was not at all what was wanted. Often the user begins with a general question when seeking a specific answer. For example, he may say he needs a book on fluid dynamics, when he actually needs a mathematical description of the bathtub vortex.

Why is it that the question is so often not what the person really wants? There can be many reasons.

1. Sometimes the person asks the question he thinks the staff member can answer rather than the one he needs the answer to. Thomas *et al.* (1981) say that even directional questions may be "testing the water" to see if there is a receptive climate for a more difficult question.

2. The person may feel he or she should do his own "research." Students have been taught in our school libraries that they should know how to use the library. Even Nobel Prize-winning scientists hear the echo of "have you checked the card catalog?" every time they enter our doors. In fact, Mellon (1986) found that 75–80% of students in library research courses described their initial response to library research in terms of fear. This early training gives people the idea that they should be able to find things themselves. Even though the complexity of modern libraries (especially sci/tech libraries) makes that an unrealistic expectation, they approach us with guilt and inadequacy.

3. He may be embarrassed to reveal his ignorance of the specific subject, but feels it is OK to ask for directional-type assistance, or help in finding a particular reference that might help him. The user tries to convey the question but also to preserve his or her self-image.

4. He or she may think you do not have time to answer the real question.

5. He may not really know what he wants.

6. Especially with technical or scientific questions, he may think that the information-retrieval staff person may not understand the real question, or may not be competent to help (Mount, 1966).

For all these reasons, the user may not state what he or she really wants. And even if he does, the information-retrieval person may misunderstand or not understand the context. There are three areas of skills which will improve the staff member's chances of understanding: (1) nonverbal communication; (2) verbal interpersonal skills; and (3) background in the area of the question. In addition, the similarity between reference service and counseling suggests that empathy, attentive behavior, and content listening—techniques developed in counseling work—should also be used in reference work (Peck, 1975). In fact, the University of Pittsburgh teaches a course designed to make the information-retrieval person an "information counselor."

Nonverbal Communication

What the information-retrieval staff member does nonverbally can encourage or discourage the user. Experts in nonverbal communication tell us that only one-third of a message is verbal; over half is projected in the body or face. However, there is not a large body of literature on nonverbal behavior specifically in the library setting. Jennerich and Jennerich (1976), Lopez Munoz (1977), Boucher (1976), Jahoda and Braunagel (1980), and Thomas et al. (1981) all provide some discussion of the influence of nonverbal behavior on the reference interview, and the article by Kazlauskas (1976) gives some specific guidelines.

Kazlauskas found that the following behaviors were positive: use of the eyebrow flash to indicate immediate acknowledgement of a patron; immediate eye contact with the patron when he or she moves into the business transaction space and then a follow-up with a positive verbal contact; using evaluative gestures, such as nodding, indicating that the request is being understood; and a cheerful disposition as found in facial expressions, such as slight smiling.

He found that the following nonverbal behaviors were negative: lack of an immediate nonverbal acknowledgement of a patron waiting to ask a question; no perceptible change in any body movement upon the movement of a patron into the business transaction space; a staff member in a sitting position with hand held on the brow covering the eye vision and engrossed in reading, filing, or some other activity; tapping the finger(s) on the counter when a request is made and twitching the mouth upon movement to fulfill the request; and pacing behind the counter when the patron is using an item at the counter (Kazlauskas, 1976, p. 133).

Sometimes the reference transaction will take place on the phone. All the same factors still exist: nonverbal behavior considerations become tone of voice rather than posture.

Verbal Interpersonal Skills

In the area of verbal communication, what the staff member says to reassure and draw out the user can be critical to the amount and quality of information received about a search. One needs a combination of comments indicating a willingness and ability to help, such as "we have lots of information on that" or "I'm sure we can find some information to help you," with some questions to get at a better definition of the information need, or to get at the why of the question.

The best results come when the information-retrieval staff member has enough background to narrow down the subject and give the user some subgroups to reject or accept (see below). However, even without background there are several phrases that work well to draw out the user:

I'm sure we can find that. Is there some particular aspect of Halley's comet that you're interested in?

I think there's a lot of information on artificial intelligence. What is it you want to know about it?

How did you hear about this product?

How did you come to be interested in lightning?

Is this related to some project you've been assigned? (or. . .some broader research you're working on?). It might help me find the right kind of information if I knew more about the whole project.

I don't have much background in that subject. What other terms are possible synonyms?

Could the authors of articles about it be calling it anything else?

What general field is this in? or What kind of people would be writing about that? (e.g., chemists, electrical engineers, etc.)

Note that all of these questions are "open" questions. Open questions generally elicit longer responses that reveal more information than do closed questions (see G. King, 1972). Auster and Lawton (1984) found that the types of pauses also had an influence on the success of the transaction. They should be neither too long (e.g., more than 10 seconds) or too short (e.g., the user can't get a word in edgewise).

Skills Based on Background

The best question-negotiation procedure requires some background in the subject in order to postulate possible subsets or contexts. In this way, the information-retrieval staff member demonstrates knowledge and interest and can get the user talking.

The following dialog provides an example of using background to ask questions which will help to elicit what the user wants.

USER: I need some information on fiber optics.

LIBRARIAN: (Nodding) We have lots of information on fiber optics. Can you give me some more background on what you want? What are you interested in using fiber optics for?

USER: Well, I'm using it for CCTV.

LIBRARIAN: Do you want to know about other people using it for closed circuit tv? Possible suppliers? The theory behind it?
USER: No, what I'm really looking for is lifetime data.
LIBRARIAN: You mean data on how often and when fiber optics fails in a CCTV application?
USER: Yes, or any other data communication type link.

In Summary

Part of the process of eliciting the real information need is dependent on background and understanding the context of the question. This is what Thomas *et al.* (1981) call the intellectual component of the reference transaction. But the verbal and nonverbal interpersonal skills to make the person feel comfortable exposing the real question are essential. The critical skills in understanding what the user wants are:

1. The ability to give appropriate reassurance, both nonverbal (immediate acknowledgement of the person's approach, nodding, etc.) and verbal ("We have lots of material on fiber optics").
2. Enough background to ask intelligent questions to narrow the field and to get at the "why." Or superior interpersonal skills to compensate for the lack of knowledge.

We have discussed the skills necessary to understand what the user really wants in some detail, because the understanding is absolutely critical to the delivery of effective information services. But it is only the first step. Once the understanding exists, the query must be translated into terms the "system" can answer. This is discussed in more detail in the sections on ready reference and literature searching.

READY-REFERENCE ACTIVITIES

Ready-reference information includes (a) addresses and phone numbers for companies, book publishers, government agencies, scientists, etc.; (b) biographical information; (c) definitions of terms; (d) specific information that is usually in handbooks (e.g., the melting point of copper); (e) internal information (e.g., information about who does what in the organization); and (f) library services, procedures, policies, etc. Ready-reference staff also commonly determine whether the library has a copy of a particular book, periodical article, map, or report. This includes checking or helping the user check the library catalogs. It also includes deciphering citations and abbreviations of periodical titles in order to be able to check the library catalogs. Often the question "do we have this?" requires verification of the citation in order to check it against library catalogs. If the item isn't in the catalogs, verification is necessary in order to borrow or buy it. Verification means finding the citation in a standard source such as abstracts and indexes, the shared bibliographic networks, the *National Union Catalog*, or *Books in Print*. Verification gives some reassurance that the author, title, date, etc. are correct.

Providing ready-reference information includes all eight steps of information retrieval shown in Table 4.1 (p. 65). However, in ready-reference questions, steps are frequently combined or shortcut. For example, let's say the user wants the altitude of Albuquerque. We just go to the *World Almanac*, look in the index (which we remember is in the front, not the back) under "altitude, U.S. cities", turn to the right page, and voilà! There is the answer: 4945 feet.

In fact, all the steps have been done. We chose the important elements of the question (altitude, Albuquerque) and their relationships to each other (and). We chose the source (the *World Almanac*). We put the search in the appropriate terms for the source (altitude, U.S. cities). We executed the search (we looked on the page, found Albuquerque and the column for altitude). Since we found the answer, we do not need another iteration, so we can deliver the information and check with the user to be sure it is what he or she wants.

Hardly anyone consciously goes through all the steps for a question like this. The experienced reference librarian is able to make lots of shortcuts (which is the reason reference librarianship is an art, not a science), but understanding the steps involved is useful when a search is unsuccessful for some reason. It is particularly necessary in online literature searching, as we will see later.

There is a thin line between ready-reference type information and a more in-depth request. Sometimes when a person asks "do you have a book by Anderson on circuit analysis?" what he or she really wants is to find a description of a particular circuit, which may or may not be in that particular book. The information-retrieval person who only answers the asked question has not necessarily dealt with the person's information need. So a question to see if the user got what he or she wanted is critical.

Usually a formal follow-up for quick reference questions is unnecessary but information-desk personnel should be cautioned to always ask something like: "Is this what you wanted? Is there anything else I/we can do for you? Does this answer your need?"

As an example, what if the person who asked for the altitude of Albuquerque in our example were really planning to set up some kind of experiment at the corner of Eubank and Candelaria and needed the altitude of that point to within about 10 feet. The altitude of Albuquerque actually varies from about 4700 to 6200 feet because the city slopes toward the Sandia Mountains. In this case, our answer from the *World Almanac* is not accurate enough. The best source would be a topographic map.

If we had had a lot of background with this sort of question, we might have asked what part of Albuquerque and how close do you need the answer. But if we did not ask earlier, at least at the evaluation stage, we must ask, "is that good enough?" or "is that what you wanted?"

Sources

Much training of librarians and paraprofessionals concentrates on sources, and knowledge of sources is certainly a critical point in the steps in information retrieval. Many of the sources for quick reference in a sci/tech library are the same as those in other libraries. A scientist or engineer is just as likely to ask for the address of Academic Press or what the inflation rate has been since 1968 as any other kind of patron. Thus, the standard lists of good ready-reference tools (e.g., Cheney and Williams, 1980) are useful for sci/tech libraries, too. In addition, special tools will have to be added depending on the library's setting. Every sci/tech library should have tools in the following categories:

Definition. Dictionaries are necessary ready-reference tools. They fall into the following categories: regular English-language dictionaries; overall scientific or technical dictionaries (e.g., *Van Nostrand's Scientific Encyclopedia*), overall scientific and technical encyclopedias (e.g., the *McGraw-Hill Encyclopedia of Science and Technology*), dictionaries and brief encyclopedias for particular subject areas, and foreign-language dictionaries, both for science and technology as a whole and for particular subject areas. Tools that define acronyms and initialisms (e.g., the *Acronyms, Initialisms and Abbreviations Dictionary*) also fall into this category and are critical for sci/tech libraries.

Directory information. Directories which provide addresses and phone numbers in these categories are essential: (1) associations (e.g., the *Encyclopedia of Associations*); (2) research laboratories (e.g., *Research Centers Directory* and *Industrial Research Laboratories*); (3) companies (e.g., *U.S. Industry Directory*); (4) manufacturers and products (e.g., the *Thomas Register*); (5) government (e.g., the *Washington Information Directory*); (6) general (e.g., phone books for major cities or Phonefiche); (7) authors (e.g., *Who Is Publishing in Science*); (8) biographical information (e.g., *American Men and Women in Science*); and (9) other libraries (e.g., the *American Library Directory* and the *Directory of Special Libraries and Information Centers*). Directories have been appearing online with increasing frequency. For example, the *Encyclopedia of Associations,*

American Men and Women of Science, and the *Electronic Yellow Pages* are available on DIALOG.

Statistical information. Sci/tech libraries should also have publications which can answer requests for statistical information (e.g., *Statistical Abstract of the United States* and sources for the particular subject area of interest).

Factual information. Ready reference also needs sources of factual information. This category includes general facts such as those collected in sources like the *World Almanac*, but also the very important category of handbooks for the various subjects. Handbooks like the *CRC Handbook of Chemistry and Physics* are vital tools in sci/tech libraries.

The library's catalogs. The whole library collection can serve as a ready-reference source, and the catalog is the index to it.

Internal information. A ready-reference service will often need to refer people to sources within the organization. At the very least, the sources include the organization's phone book. Any other directories of services and sources of information within the organization should also be gathered.

Verification. For deciphering citations to periodicals, any quick reference desk should have *Ulrich's International Periodicals Directory*, the *Standard Periodicals Directory*, *Chemical Abstracts Service Source Index* (CASSI), and the *Union List of Serials*. Periodicals lists for the major abstracting and indexing services in the subject area should also be available. To actually verify periodical citations (in contrast to deciphering them), the best sources are the abstracting and indexing services, either in hard copy or online.

For verifying books, the most-used tool is *Books in Print* (which is available online as well as in hard copy). The bibliographic networks (OCLC, RLIN, WLN, and UTLAS) discussed in Chapter 7 are also useful (Farmer, 1982), as are the *National Union Catalog* and the REMARC and LC-MARC files on DIALOG.

For verifying reports, the best sources are the reports databases online (and their hard-copy equivalents): NTIS (Government Reports Announcements and Index), EDB (Energy Reports and Announcements), NASA RECON [Scientific and Technical Aerospace Reports (STAR)], and DROLS (Technical Abstract Bulletin). These are covered in detail in Volume 2, Chapter 12. Verification is also discussed in Chapter 8.

Steinke (1985) surveyed a sample of sci/tech librarians and found that the five reference books that they had found most useful over the years were the *CRC Handbook of Chemistry and Physics*, Kirk and Othmer's *Encyclopedia of Chemical Technology*, the *McGraw-Hill Encyclopedia of Science and Technology*, CASSI, and the *Merck Index*.

Many of these various kinds of sources (directories, statistics, etc.) are available online. Also, the bibliographic utilities (OCLC, RLIN, UTLAS, WLN) are increasingly being used for ready reference in categories other than verification. A list of examples of various kinds of uses compiled by the RASD MARS Direct Patron Access Committee (1984) included searches for addresses of corporate authors, the meanings of acronyms, publisher's output (truncated search on ISBN), availability of translations, where journals are indexed, and founding dates of organizations. These are in addition to the use of the files as verification tools, to find out who has a copy, and to make up bibliographies on topics or of an author's works.

This list of the types of tools that are needed in a sci/tech library for ready reference is only meant to be suggestive. Detailed lists of reference tools abound and can be used to expand this cursory list for a particular sci/tech library. These are listed in Table 4.2.

TABLE 4.2.   Sources Listing Reference Tools for Science and Technology

Chen, Ching-Chih. (1977). "Scientific and Technical Information Sources." MIT Press, Cambridge, Mass.

Doyle, James M., and Grimes, George H. (1972). "Reference Sources: A Systematic Approach." Scarecrow, Metuchen, N.J.

Grogan, Denis J. (1976). "Science and Technology: An Introduction to the Literature," 3d ed. Clive Bingley, London.

Houghton, Bernard. (1972). "Technical Information Sources: A Guide to Patent Specifications, Standards and Technical Report Literature," 2d ed. Shoe String, Hamden, Conn.

Malinowsky, H. Robert, and Richardson, Jeanne M. (1980). "Science and Engineering Literature: A Guide to Reference Sources," 3d ed. Libraries Unlimited, Littleton, Colorado.

Mildren, K.W., and Meadows, N.G. (1976). "Use of Engineering Literature." Butterworths, London.

Mount, Ellis. (1976). "Guide to Basic Information Sources in Engineering." Norton, New York.

North Atlantic Treaty Organization. Advisory Group for Aerospace Research and Development. (1974). How to obtain information in different fields of science and technology: A user's guide. Neuilly-sur-Seine, France: AGARD LS-69; AD 780061.

Primack, Alice Lefler. (1984). "Finding Answers in Science and Technology." Van Nostrand Reinhold, New York.

Sheehy, Eugene P. (1976). "Guide to Reference Books," 9th ed. American Library Association, Chicago.

Walford, Albert J. (1980). "Science and Technology," vol. 1, "Guide to Reference Material," 4th ed. Library Association, London.

Weiser, Sylvia G. (1972). "Guide to the Literature of Engineering, Mathematics and the Physical Sciences," 3d ed. Johns Hopkins University Applied Physics Laboratory, Silver Springs, Maryland (technical memorandum TG 230-B3).

Many published updates of reference books in the library world ignore science reference books, but the notable exception is *American Reference Books Annual* (ARBA), published annually by Libraries Unlimited.

In addition to these sources, which cover all of science and technology, there are also guides to the literature of individual sciences, which are useful as lists of sources for reference material. These are included in the subject chapters in Volume 2, Part II.

Management Aspects of Ready-Reference Service

The management aspects of a ready-reference service include organization, staffing, budgeting, record-keeping, evaluation, training, and the use of online databases.

Organization

Where should this service be physically? How many people should staff it? What level of staff? Should it be integrated with a "reference desk" and also handle in-depth requests? If not, how will referrals be handled? What information will be given over the phone? Does a user standing at the desk take precedence over a phone call? If so, what do you do about the ringing phone?

None of these questions have absolute answers: the answers depend on the organizational setting.

One organizational arrangement is to combine the ready-reference and literature-searching aspects of information retrieval. Reference desks are staffed by professionals, who refer ready-reference questions to paraprofessionals, giving them directions about where to find the answer if necessary. Because so many "easy" questions are not what they seem, this arrangement has the advantage of putting the professional in the question-negotiation stage right away, providing less chance that the real question will go unanswered. Another arrangement, more common in the corporate setting, is to have an information desk staffed by high-level, experienced paraprofessionals who refer questions they cannot answer and those which require literature searches (usually about 20–35% of the questions) to the technical-information specialists. This has the advantage of freeing the professional staff from dealing with directional and ready-reference questions. However, there may be a tendency for the paraprofessionals to give out less than the best information in order not to appear incompetent by asking for help. Deciding which arrangement to use will depend on many factors, including the size of staff and the amount of turnover among the paraprofessionals.

Larason and Robinson (1984) discussed the physical arrangement of the reference-desk area to make it more approachable and less of a barrier to the user. In their case study, one information desk, close to the catalogs but not flanked by bookcases or files, was used much less than an information desk farther away. Thus, convenience is not the only factor; people must be able to identify the desk. For instance, a counter is perceived more readily as a service point than a desk.

Staffing

There are few figures which one can use to calculate how many questions a staff member should be able to handle in an hour or a year, what is the proper proportion of staff members to users in the community served, how long one person should staff the desk, or whether or not to put paraprofessionals or professionals at an information desk. When Emmick and Davis (1984) surveyed a large number of academic libraries, they found that the average reference query rate was 5–10 per hour (p. 80), and Larson (1983) found that in 1981 three reference librarians in the U.S. Senate Library responded to more than 12,000 requests for assistance and an additional 1000 requests for searches (Larson, 1983, p. 478). These studies are an indication of volume that can be handled—but the variability in local situations makes all these factors hard to generalize.

Another consideration in staffing is to provide some extra activity for the information desk personnel to do in slack times—something that will not interfere with service. Checking bibliographies and working on collection development are usually compatible as long as the staff members do not become so involved that they forget to look up and serve the user.

Budgeting

Budgeting for the information-desk function should cover these categories: collection, online time for verification, cost of training staff, staff salaries and fringe benefits, telephone costs (including long-distance calls to get answers to questions), and equipment (at least a desk and chair and handy shelves for reference tools, but probably also a search terminal and typewriter).

Record-Keeping

Almost every library keeps some record of the number of questions asked. Everyone realizes that this record does not really tell us what we want to know. Who are the users? Are they calling or coming in? What is the workload on the information-desk staff? Are there patterns of daily or seasonal peak times for incoming questions? Are we giving out correct answers in the shortest possible time?

Indeed, a simple count of transactions answers none of these questions.

It also can be misleading because one question can take 2 hours to complete and another can take 2 minutes. And the same question could take one staff member half a day and another staff member (who had had a similar question before, perhaps) just a few minutes.

So why count transactions? Well, collecting in the detail that is necessary to answer the questions we really want to answer is pretty time-consuming. Sampling helps, but many of the factors are too variable to be sampled successfully. If you recorded everything you want to know about a question, it could take longer to count the question than to answer it. Since libraries are almost always understaffed, they have usually chosen to answer more questions and keep less detail, rather than cut production perhaps in half.

The simple count provides two advantages. There is at least some statistic representing the information desk use, and if it is kept accurately (which, unfortunately, it usually is not), it can show growth or shrinkage of the use of the service over time and peak periods seasonally. Also, it provides the staff people a psychological reward for finishing a difficult question: they get to make that little mark on the paper indicating they did something.

Library Statistics: A Handbook of Concepts, Definitions and Terminology (1966, p. 109, published by ALA) recommends:

> The reporting of the number of reference questions handled is recommended as a recurring statistic, but only as an indication of the size of the operation. It tells little about the scope of the service rendered and should be supplemented by more detailed studies at intervals, possibly on a sampling basis. These would be directed toward a level of detail which would make qualitative analysis of the service closer to realization.

A number of forms in use for recording information service activities were included in the article by Ciucki (1977).

Evaluation

The factors for evaluating ready reference activity should include the satisfaction of the user, the correctness of the answer, and the speed of the answer. Only the last of these is easy to measure.

User Satisfaction. The users are notoriously easily satisfied, especially with a free service. In library surveys, user satisfaction is usually 80–90% or above (Myers, 1983, p. 7). In fact, Kantor (1976) showed that users are satisfied in proportion to the amount of time that someone spends with them, whether they got the right answer or not.

Correctness of the Information. A number of studies of reference service in both public and academic libraries have proven that there is a weighted average of only 56.4% of questions answered correctly (Crowley, 1968, 1985; Crowley and Childers, 1971; Myers and Jirjees, 1983; Gers and Seward, 1985; Hernon and McClure, 1986). These studies used unobtrusive measures and were based on a standard list of questions asked (sometimes in person, sometimes by phone) as if they were legitimate questions. The researcher had the correct answer and compared the library's answer with the correct answer. The studies which had good methodology have found a consistent ratio of correct answers around 50% (Crowley, 1985).

Although sci/tech libraries have not been specifically tested, any library could test its own ready-reference service by making up a list of questions, and getting someone (or several people) to call and ask the questions. However, there is an ethical issue. If the staff knows that a test is in progress, they may perform better than usual and thus throw off the results, but if they do not know it's happening, they could be very resentful later.

A more simple (but less accurate) way libraries have used to check for accuracy, especially when staff are in training, is to keep records of the questions and the answers, noting all sources checked. These are then examined by someone else.

Bunge has developed a two-part form for gathering data on answering success and the various factors in the reference situation (one part is given to the user, one to the librarian). Data from a field test of the form in 15 academic libraries found that 55.16% of patrons asking reference questions reported obtaining just the information they needed and being satisfied (Bunge, 1985). Since this is in the same range of satisfaction that the unobtrusive reference studies found, this may provide another possible method for self-study, although nothing but the unobtrusive study really gets at the factor of the *correctness* of the answer.

Speed. Turnaround time is easier to measure than either satisfaction or correctness, but it is a function of the question and the information-retrieval person's training, so it is not clear what significance speed should have in the evaluation except that it is probably related to user satisfaction. Nevertheless, a study of turnaround time can be useful in administering the service: we can keep records of the time it takes to answer, and establish an average and a guideline for turnaround time. Staff can be instructed to ask for help on anything that takes longer than the guideline.

Interest in evaluation of reference is high in the mid-1980s. In 1984, Powell reviewed research of reference effectiveness, and the Silver Anniversary issue of *RQ* [Fall 1985, **25**(1), 6] had several articles on evaluation, including one by M. White (1985) with a methodology for evaluating reference interviews and one by Young (1985) on evaluation of the performance of reference desk personnel, recommending a behaviorally anchored rating scale rather than management by objectives. Cronin (1985) also examined performance measures for public services.

Training and Continuing Education

Courses in library school designed for information-desk staff include courses in basic reference and courses in the bibliography of science and technology. In-house courses or on-the-job training focused on sources can also be useful.

Bunge (1982) reviewed methods being used in large and medium-sized public and academic libraries for keeping up-to-date on reference materials and techniques. The common techniques include reviewing professional literature, attending professional meetings, and using reference staff meetings to review questions and discuss alternative sources and approaches, or to review new tools.

Most discussion of training for the information desk focuses on learning and keeping up-to-date with new sources, and, to a lesser extent, on the reference interview. A neglected aspect of the necessary training for information-desk personnel is orientation to the library and to the overall organization. Many questions about who does what, both within the library and in the organization, will come to the library, and the more the library staff know, the better the referrals will be.

Use of Online Databases

In many ways, online sources are simply another source for answering ready-reference questions. However, because of the administrative questions (funding, locating the terminal, training staff, etc.), they are often treated as a separate topic.

With online databases, it is possible to quickly identify a correct citation to a reference for which incomplete information is presented (a partial title, no year or page numbers).

Not all libraries are using online databases for ready reference. Hitchingham et al. (1984) surveyed 180 libraries about their use of online databases for reference use. Of 1290 librarians, only 741 (57%) had used online databases for reference (as opposed to "formalized online searching"). Of the 177 libraries responding to a question about how they began using online databases for ready reference, almost half (49.2%) replied that their use of databases for ready reference evolved informally, although it was now recognized as a regular part of reference service. Only one in five had a fixed sum of money allocated for ready-reference use. One in four had a terminal dedicated to ready-reference use.

Some cases when online searching is an advantage in verification include:

1. When the date is unknown. You can usually search a large number of years online more quickly than in print indexes and abstracts.

2. When the date is known, but is recent enough that the printed index is not yet cumulated and you would have to go through a number of monthly issues.

3. When you know a lot about the citation: online you can limit to specific authors, publication, year of publication and other fields. In print indexes, you can normally only search by one of these at a time.

4. When the item is very recent. The print index may not be available yet.

5. If you have several items to verify. Sometimes for one item the time involved in dialing and logging in may be longer than going to the print index.

Jones (1981) has done a preliminary study showing that 50% of incomplete citations could be verified on OCLC or RLIN with an average time of 5 minutes. Using print indexes, fewer items were found (20% in the main library, 50% in the Science Library), and the average time was 15 minutes.

Most time studies and cost studies of manual versus online are hard to generalize because they are so dependent on the local situation. Whether the staff time saved is equivalent to the cost of the online service depends on the salary of the person searching, the cost of the online file, the cost of the printed tool, and local factors such as the availability of the printed tools and the online terminal. Often the cost of the online search is less (Roose, 1985), but even if the cost of online is greater, there is almost always a time advantage to online services, so it may be worth it because the benefit to the user is so much greater. By increasing the speed of the answer, we increase the effectiveness of the service.

As mentioned above in discussing sources, online sources are readily available in two areas of ready reference: verification and directory information. The availability of these sources online raises a number of management questions: Should these online sources be available to information desk staff? Which ones? Do they replace print or supplement it? Where do we locate the terminal? If at the information desk (as described by Becket and Smith, 1986), what do we do about interruptions? What do you do if you are searching online spending $2 a minute and someone else approaches with a question? If instead we locate the terminal away from the desk, we are introducing a different kind of problem. Either the information-desk person has to refer the person to someone near the terminal (for just the kind of information the information desk is designed to handle), or the information-desk person has to leave, and then who covers the information desk? There are no perfect answers to any of the questions.

Budgeting can also be a problem, especially in an environment where the cost of online searching is generally recharged to the user. Recharging for ready-reference searching should be avoided if at all possible. The costs of recharging will almost always be greater than the cost for doing the search. It would be better to budget for a certain amount each month. Keeping a running total of the balance on log sheets would be an adequate control over the amount spent.

LITERATURE SEARCHING

When the user wants a comprehensive search of everything that has been done on a particular subject, he or she needs a literature search. Whether the library does it for him or whether the user does it himself is a matter of the setting of the library, the level of service offered, the user's preference, and whether the best source is online. Even libraries which do not do manual searches for patrons are doing online searches.

Manual Searches

Because the advantages of online searching in the sci/tech environment are so strong, most manual searches currently are done in indexes which are not available online, such as the back volumes of *Chemical Abstracts* or *Mathematical Reviews*.

Procedures and suggested forms for doing a manual search and keeping records of what is searched are detailed in *Scientific and Technical Libraries* (Strauss *et al.*, 1972, pp. 274–296). The important items to record are the title of the source index or abstract, the terms searched, and the volumes or dates searched for each term.

Citations discovered in the search should be recorded (or photocopied) on separate cards or sheets with the source noted, so if there is any problem later the citation can be rechecked.

Generally it is best to proceed from the most current volumes to the older ones to avoid tracking down superseded information. Also, the searcher may come across reviews or bibliographies which may expedite the search by making searching the earlier volumes unnecessary.

Online Searching

Advantages of Online Searching over Manual Searching

The advantages of online searching over manual include speed, currency, convenience, access, ability to combine concepts and search elements to do both more complex searches and more specific searches, ability to modify the search strategy immediately, and the ease of setting up an SDI profile. Also, most authors have found online searching cost-effective compared to manual searching (Buntrock, 1984; Naber, 1985; Bivans, 1974; East, 1980; Jensen *et al.*, 1980; Magson, 1980). For example, Elman (1975) found the average searching time for manual searching was 22 hours at $250; for online searching, it was 45 minutes and $47.

Information is retrieved rapidly online from large bibliographic files. The online files are almost always more current than the hard-copy indexes. The online printout is much more convenient than copying or photocopying citations out of hard-copy indexes. Also, some hard-copy subject indexes (e.g., *Engineering Index*) refer from the subject index to an abstract number. The searcher must record the abstract number and then look up the full citation in another volume. Searching online provides the full citation directly.

Boolean logic capability of online files means one can effectively search on various elements of a citation, such as an author and a journal name, and on multiple concepts (e.g., radioactive waste disposal and granite and Britain). Some files are available only online, either because the library does not have the hard-copy equivalent or because one does not exist.

Disadvantages of Online Searching

Online searching is not effective for certain types of questions. These include questions on subjects for which databases do not exist, and those on subjects for which the literature of interest was written in the 1940s or 1950s. Also, "what if" questions, controversial questions, pro and con questions, and questions which require degrees of quality or quantity or time qualifications qualify as difficult (Dolan, 1979). Certain searches also produce false drops because one cannot usually specify the relationships between the terms. For instance, a search on methods for quality control of software will also produce articles where software is used for quality-control purposes. Unfortunately for the engineering profession, practical questions also do not work too well. For example, if someone needs to know how to connect their widget to their gadget, a database may not be the best approach.

Moreover, computers are dumb. When you do a manual search, your eye automatically cuts out irrelevant items under the subject heading you are looking at. But the computer sometimes puts in magnificently irrelevant citations. For example, a search on locks (for security purposes) may produce material on the kind of locks used to transport ships between bodies of water of different levels.

The policies of the database producers can also affect online search success. The producers may exclude the information you need by policy, their indexing may obscure what you are looking for, and they may make mistakes in indexing (Pemberton, 1983).

History and Background of Online Searching

The earliest databases were developed by government organizations such as NASA (National Aeronautic and Space Administration), AEC (Atomic Energy Commission), and the National Library of Medicine (NLM). The government contracted out the design and development of the online searching systems, and these government systems formed the basis of many of the commercial systems we see today. For example, the NASA RECON system developed by Lockheed provided the basis for the DOE RECON system and for DIALOG. And System Development Corporation (SDC) developed the ELHILL system for NLM, which formed the basis for ORBIT (Neufeld and Cornog, 1986).

Commercial online databases developed primarily as a by-product of printed indexes. In the late 1960s and early 1970s, the most efficient way to produce a printed index or abstract became to create a computer tape, which was used by the printer to set the type and print the index. Once the tape existed, it was possible to mount it on a computer and search it directly. The timing of this development explains why so few databases go back online earlier than the late 1960s. The earlier records were not in machine-readable form as a result of the production process, and the process of conversion is expensive.

When these tapes first began to be available, many sci/tech libraries purchased the tapes for files they used often from the vendors and searched them in-house. Some still do. But a much larger number search through one of the vendors like DIALOG, BRS, and SDC. These vendors [or "databanks," as Pemberton (1984) prefers] make a large number of databases searchable via a common protocol. They write the programming and provide the computer storage necessary for allowing access to the databases. The libraries which search through the vendors have a wide variety of databases available, and they pay only for what they use. Having to learn a new protocol for each database would decrease the use of many of the lesser-used databases. Nevertheless, many database producers, seeing the shift from print to online and the size of the growing online market, are trying to set up their own search networks in order to not share profits with the vendors and to provide more hand-tooled features (Neufeld and Cornog, 1983). Database producers are also looking at the possibility of marketing whole databases on optical and video media (e.g., CD-ROM). The first of these appeared on the market in late 1985. For libraries, this kind of product might mean a return to an environment which

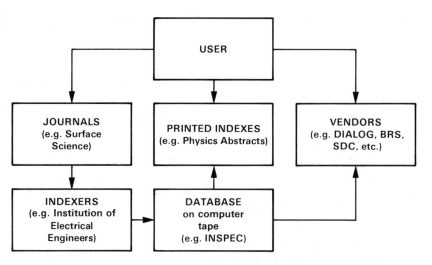

Fig. 4.1. The relationships among database producers, vendors, and users.

allows unlimited searching for a single subscription price. For the database producers, the product means they will have more control over the searchability of their product—they may regain the control which they had to allow the vendors on online services.

If you look at it from the user's point of view, he or she has three basic approaches to finding literature on a subject. As shown in Fig. 4.1, he can look through journals, books, reports, etc. looking for articles of interest. He can approach the printed indexes (either himself or through an intermediary). Or he can search the online files, either himself or (more often) through an intermediary.

The growth of online services has been nearly exponential. In the 1969 issue of *A Guide to the Selection of Computer-based Science and Technology Reference Sources in the U.S.A.* (American Library Association, 1969), only 28 machine-readable databases were listed, and all were sci/tech except *Psychological Abstracts*. The growth in the number of available databases, database producers, and vendors from 1979 to 1985 is represented in Fig. 4.2, based on figures from the *Directory of Online Databases* (Cuadra Associates, Spring 1985). The *Directory* includes numeric databases as well as bibliographic, and their growth may exaggerate the exponential aspect of the growth. The growth in the number of bibliographic databases seems to be slowing down in the mid-1980s.

There are a number of terms for different types of databases in common use. The division into two types, "bibliographic" and "nonbibliographic,"

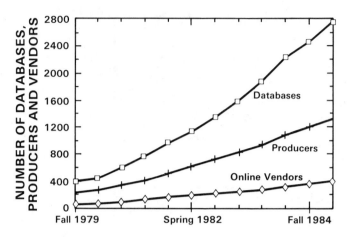

Fig. 4.2. Growth in the number of databases, 1979–1985. [Based on data from the *Directory of Online Databases* (Cuadra Associates, 1985).]

shows the bias of those familiar with bibliographic files who throw everything else into one big "other" category. Another common division is between "bibliographic" and "numeric," with "numeric" being a less bibliomorphic name for the "other" category. Conger (1984) stays with bibliographic, but divides nonbibliographic into "numeric," "text," "directory," and "dictionary." Bibliographic databases have citations to articles, reports, etc. and are thus secondary reference sources; "the researcher must seek out the cited references before retrieving the desired information" (Conger, 1984, p. 94). Numeric databases, in contrast, are primary resources. They include databases like the I. P. Sharp files, which provide answers to statistical questions without recourse to other reference tools. Text databases include databases such as NEXIS (which has full text of the *New York Times* and other newspapers and newsletters), and a growing number of databases which include full text of journal articles (e.g., the *Harvard Business Review* and the American Chemical Society journals). Directory databases are similar to bibliographic databases except that they include people, organizations, companies, or products, rather than citations to books or reports. Dictionary databases are primarily tertiary reference sources. They contain the terminology needed for searching secondary databases. An example is Chemname on DIALOG, which gives synonyms, formulas, registry numbers, etc. for use on CASearch.

In contrast to Conger's five types of databases, Cuadra's *Directory of Online Databases* uses a classification scheme of four types: reference (made up of bibliographic and referral) and source (made up of numeric and textual–numeric). Probably it will be a while before the terminology settles down, because there are so many innovations in the databases themselves.

The Eight Steps of Information Retrieval

As we saw in the section on ready reference, in answering ready-reference questions, the eight steps in information retrieval (Table 4.1 on p. 65) are often short-circuited or unconscious. They are more often explicit in literature searching, and particularly so in online literature searching. In our discussion of these steps below, we will be specifically considering online literature searching and assuming that the searcher is trained. Further information about training searchers is considered later under Management Aspects.

1. Understanding What the User Wants. Understanding what the user wants is critical in literature searching, because of the amount of time and money which can be spent before the iteration back to review results.

Computerized literature searching has somewhat formalized the reference interview; most search services have some kind of search request form. Sample search forms are available (Daniels, 1978; Palmer, 1982). Most request the following information:

A statement of the question.
Keywords and synonyms.
A date range.
Whether the search is to be comprehensive or for "just a couple of good articles."
Citations to good articles, if the requestor has them.
Names of authors of relevant articles.
The anticipated number of citations.
Cost limitations or database limitations, if any.

The request sheet can be used as a basis for the online interview. The factors affecting the interview include the location of the requestor, whether the requestor will be at the terminal during the search, the searcher's knowledge of the subject, the subject's specificity, and the searcher's knowledge of the database. For any interview, Somerville (1982) suggests using interpersonal communication and negotiation skills, discussing the subject with the user, determining if a computer search is the appropriate way to answer the question, making sure that the question is understood, determining the comprehensiveness of the search question, identifying limits, and discussing confidentiality. If the user is a new or infrequent user, it is desirable to add a discussion of the benefits/limitations of computer searching, briefly describe or review software features (if not already mentioned earlier in the interview), and describe potential databases. Somerville (1982) and Palmer (1982) both give samples of extended reference interviews for online searching. They both include explaining the choice of databases and the search strategy and evaluating the search results as part of the interview.

2. Choosing the Important Elements of the Questions and Their Relationship to Each Other (Search-Strategy Design). Search-strategy design consists of figuring out the concepts that are being searched and linking them with the logical connectors "and," "or," and "not." Most online databases search by Boolean logic (that is, by combining sets which share or do not share certain properties).

There are four basic search strategies: building blocks, successive fractions, citation pearl growing, and cited reference searching. With building blocks, each concept is formulated separately and the separate blocks are combined at the end. This strategy is logical and easy to follow, but sometimes too inflexible to take into account unexpected developments online. With successive fractions, a large set is made on a general topic, and specific concepts are intersected with it until the set is the right size. This is a particularly good technique when the topic is vague or broad or when a series of useful but not essential restrictions are possible. With citation pearl growing, the searcher starts with the pearl (a known relevant citation), retrieves it, and examines its indexing. Then he or she plugs the good indexing back in to get other (hopefully) similar articles. To use cited reference searching as a strategy, we start with the citation to a good article or book and search SciSearch (Science Citation Index) to see who has cited that article since it was published.

Much has been written about search strategy design, particularly as it relates to online searching. The basic theoretical ideas of search strategy are presented by Buntrock (1979), L. Smith (1976), Taylor (1968), and Bates (1979a,b). Fenichel (1980–1981) reviewed the literature of search strategy. Hawkins (1982) and Dolan (1983) added some other strategies to the basic repertoire, one of which is the failsafe approach (planning a different strategy in case anything goes wrong). In another column, Dolan also offers a flowchart of the search formulation process which is useful, if not all-inclusive (Dolan, 1979).

It is useful to have this theoretical understanding of search strategy (especially in case something goes wrong), but the usual approach is simply to pick out the important concepts in the searcher's request and determine the boolean relationships (and, or, and not) between them. A simple diagramming approach is taught by both Jo Maxon-Dadd (in training for DOE/RECON) and Anne Farren (in workshops on BIOSIS). Figure 4.3 shows

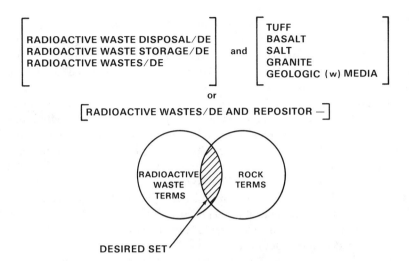

Fig. 4.3. Diagram and representation of Boolean logic of a search on radioactive waste disposal in geologic media.

a diagram of a search on radioactive waste disposal in different types of geologic media. This diagramming method is similar to Boolean diagrams but allows for vocabulary expansion and for fairly complicated relationships.

3. Choosing the Source(s). Choosing the best database is limited by what is available. Some subjects are covered in several databases (with distressingly little overlap); some are not covered in any. One of the main responsibilities of the literature searcher, and one of the main things professionals are trained in, is knowledge of sources. Learning all the sources was never an easy task, but in the last 10 years with the explosion of online sources, it has become even harder. One of the main tasks in the reference librarian/technical information specialist's continuing education is to keep up with the appearance of new databases, to evaluate them, and to keep up with changes in the databases he or she already knows.

Although the online scene is changing constantly, the vendors of publicly available databases of most relevance to a sci/tech library include the following: DIALOG Information Systems, Inc. (3460 Hillview Avenue, Palo Alto, California 94303); SDC Information Services, 2500 Colorado Avenue, Santa Monica, California 90406; BRS, 1200 Route 7, Latham, New York 12110; CAS/STN, Chemical Abstracts Service, 2540 Olentangy River Road, P.O. Box 3012, Columbus, Ohio 43210; Mead Data Central, P.O.

Box 933, Dayton, Ohio 45401; Pergamon Infoline, Inc., 1340 Old Chain Bridge Road, McLean, Virginia 22101; and Telesystemes-Questel, 83085 boulevard Vincent Auriol, 75013 Paris, France. These and a number of others (Wilsonline, CAN/OLE, INKA) were discussed by Saffady (1985a) in *Library Technology Reports*.

The government search systems are also important to many sci/tech libraries. These are normally available only to organizations which have contracts with the agencies involved. They include the Department of Energy's RECON system (contact Office of Scientific and Technical Information, Technical Information Center, P.O. Box 62, Oak Ridge, Tennessee 37830); the NASA RECON system (contact NASA/STIF, P.O. Box 8757, WW1 Airport, Maryland 21240); and DROLS, the Defense RDT&E Online System (contact DTIC, Defense Logistics Agency, Cameron Station, Alexandria, Virginia 22314). These government databases are discussed in more detail in Volume 2, Chapter 12.

Other vendors may be important because they provide a particular database of interest, and they can be located through the directories of online databases discussed below.

Each of these vendors has a large number of databases available. Some of the many databases of interest in a sci/tech environment are CASearch (Chemical Abstracts), INSPEC (Physics Abstracts, Electrical and Electronic Abstracts, Computer and Control Abstracts), Compendex (Engineering Index), and NTIS (U.S. Government Research Abstracts). The others which might be useful are quite numerous. Williams list 2805 databases in the 1985 edition of *Computer-Readable Databases: Science, Technology and Medicine* (Williams, 1985a), and if we do not limit ourselves to sci/tech subjects there are even more. Hawkins reviewed online physical science and mathematics databases for *Online* (e.g., Hawkins, 1985), and more detail about databases for particular subjects is also available in the chapters in Volume 2, Part II.

In choosing which database to search, searchers rely heavily on the materials provided by the major vendors. There are printed materials describing the databases, and there are also online indexes (DIALINDEX on DIALOG, BRS/CROSS and Database Index on SDC) which enable the searcher to search a subject and find the databases which cover it.

For other vendors and databases, the database directories are the best source, and unfortunately there are a growing number of them, many aimed at the end-user market.

The three leading database directories are *The Directory of Online Databases* (Cuadra Associates, 1979–), *Computer-Readable Databases* (Williams, 1985a), and *Database Directory* 1985–1986 (Knowledge Industry Publications). Williams's directory has longer entries about each database, but covers fewer nonbibliographic databases. These and other directories were reviewed in *Online* in July 1983 (pp. 73–75) and May 1985 (pp. 78–79), by Smith (1980), by Tenopir (1983a–e, 1984a–d, 1985a–c), and by Borgman and Case (1984). The online versions were reviewed in "The Search for Online Data. . . ," 1986.

These directories can be helpful in discovering the existence of databases which cover the subject of the question. But knowledge of sources cannot merely be whether the database covers the subject or not. As with ready-reference sources, choice is based on accessibility, quality of information, appropriateness of indexing, etc. A knowledge of the special features of both the system and the database are necessary.

Databases are commonly reviewed in *Online, Database,* and *RQ.* A database description, whether it appears in a directory, in the vendor's information, in reviews in the professional literature, or in the advertisements, should contain the following items of information if it is to be useful in the choosing process: dates of coverage (and frequency of updates), subject scope, size (and number of records added per update or per year), types of material and languages covered, availability, cost, and what aids to searching are available. Depending on the kind of search, the searcher may also need to know policies for author entries, whether journals listed as included are indexed cover-to-cover or selectively, the types and numbers of descriptors assigned, what terms are included in the vendor's "Basic Index," and how foreign titles, abbreviations, acronyms, and journal names are handled. If the database is available from more than one vendor, the searcher will want to know what the differences are in access points and cost.

If the description does not include these details, the searcher can consult the chapters about the database provided by the online vendor, call the producer, or test the file online. Most searchers also keep a file of professional literature where the author has provided detailed comparisons of coverage of various databases for different subject areas, such as the article by Tenopir (1982a) investigating subject coverage of emergency management and the article by Derksen (1984) investigating coverage of geoscience.

If the library has access to several vendors which have mounted the same database, there is an additional element of choice. Costs for connect time, telecommunication charges, and online and offline prints have to be considered, as well as whether special search features on one vendor or another will make a particular search more cost-effective. Detailed average cost comparisons between vendors were published by Saffady (1985a), but averages are of limited use in decisions about a particular database.

As with ready reference, choice of source is only the third step of eight, and if the choice is not fruitful, the searcher may have to return to this step and choose another database.

4. Putting the Search in the Appropriate Terms for the Source (Vocabulary Development). After the choice of database and system is made, the literature searcher looks for synonyms and for appropriate assigned descriptors. Many databases have controlled vocabulary (for example, instead of using "mill refining process," the Energy Data Base uses "pyrochemical processing"). And databases may use categories or numbers to represent subjects or concepts (e.g., CASearch's registry numbers, BIOSIS's concept codes). For example, if someone asks for literature on the interaction of sodium with liquid metals, the searcher may want to list the liquid metals separately, and will certainly want to look up the registry numbers for the search in *Chemical Abstracts*.

The tools to help with vocabulary development may include thesauri, other term lists (like frequency lists), database chapters produced by the vendors, handbooks or other instructional materials produced by the database producer, and articles in professional journals. With some databases (e.g., *Chemical Abstracts*), the hard-copy index is useful at the vocabulary development point. DIALOG publishes a useful list of search tools available to help search the databases it makes accessible (Search aids for use with DIALOG databases, SA-1, 1985).

5. Executing the Search. The efficiency of this step in an online search is dependent on the searcher's experience with the system. Efficiency can be destroyed by problem equipment or phone lines, too. Response time on the various vendors is better at certain times of the day or night.

6. Evaluating the Results and Reiterating if Necessary. At some point the searcher will get a set of citations which meet the criteria of the search. It is common practice to review a few of them for relevance before printing the set offline. Often the requestor is a big help in determining relevance if he or she is present during the search. Also, the searcher should look at the indexing terms of the relevant hits in order to see if there are other good terms that were missed (if so, return to vocabulary development).

The two factors used to evaluate online searches are precision and recall. Precision is a measure of the proportion of items in the set which are relevant to the search as stated. Unfortunately, the determination of relevance is subjective, and almost any measure of precision will be open to criticism. However, clearly a search which recalls items on canals when the search question was about door locks has less precision than one in which all the citations are related to door locks. Recall is a measure of the proportion of relevant citations available which are actually pulled out by the search. Increasing precision normally decreases recall and vice versa.

Often searchers print sets if the set looks mostly good (many of them judge it "good" if three out of five look good to the user). If the set does not look good, they revise the search strategy or the vocabulary, or try another database.

In the ideal, literature searching is an iterative process: we give the user some material, look at it together, revise, and search again. Especially when the interview is not successful (some people simply cannot or will not explain what they want), one of the techniques that usually works is to give the user something and ask if it is what is wanted and if not, why not. Sometimes in explaining what they do not want, people can explain better what they *do* want.

At any step it may be useful to go back and redo a step. You may need to interview again in order to determine what is needed. You may need to redesign the strategy. You may need to examine the vocabulary; you may discover terms you did not enter but which look good. After these modifications, you also need to review the output again and perhaps modify again, all within whatever cost constraints exist.

7. Delivering the Information. The immediate product of a literature search is a bibliography of citations with or without abstracts on the topic of interest (or mostly on the topic of interest). The first decision is whether to provide offline or online prints.

Offline prints are printed the same day by the vendor and are mailed, which requires about a week. Most vendors offer options in printing (e.g., sorting by author, report number, journal title), which should be offered to the user. If the requestor is in a hurry, it may be cost-effective to print at the terminal if you have a 1200-baud terminal and print in short format (Boyce and Gillen, 1981; A. Stewart, 1978). Many libraries have begun downloading the set to disk in their microcomputers. This gives them the capability of giving the requestor the search in machine-readable form or of reformatting the prints to remove obvious irrelevant ones before printing the citations for the requestor. However, there are copyright problems associated with downloading. Downloading is discussed further in the next section, management aspects of literature searching.

8. Evaluating the Results with the User. As with ready-reference questions, there must be an evaluation by the user and instructions about how to get the actual documents. New users or users who were not present

for the search need to have the process explained to them, as well as how
to interpret the printouts. A follow-up is an excellent idea, especially for
prints sent by mail. "Did you get what you wanted? Did we miss anything?"

MANAGEMENT ASPECTS OF LITERATURE SEARCHING

Organization

Factors that influence the choice of organization include the physical
arrangement and space resources available, the uses to be made of the
online equipment, the number of staff using the equipment, the philosophy
of searching, the financial setting (are costs recharged?), the volume of
business, the type of questions, and the services used (e.g., the DROLS
system from the Department of Defense requires a security-classified location).

Lamb (1981) reported that all of the 50 academic libraries she surveyed
had access to DIALOG and almost all to ORBIT, that online searching
was almost always part of public services and often part of reference, and
that almost all charged the patron a fee for the service. However, there
was much more variation in what the search services were called, the
number of searches and number of searchers, the ways of funding, the
groups served, the equipment used, and in organization, promotion, and
procedures.

In all three settings, a coordinator for search services could be assigned
the specific responsibility for monitoring and managing the library's online
services. Among other things, the coordinator could handle contracts and
relations with the vendors, ordering and updating manuals and other search
tools, accounting, and statistics.

Staffing

How does one staff for online searching? There are some figures in the
literature which can be used in the beginning planning of an online service,
but they should be checked as soon as possible against local records, and
the local records should be used to forecast from then on. For example,
the Wanger *et al.* (1976) survey found a median time of 40–50 minutes,
with one-third for presearch activity, one-third actually online, and one-
third for postsearch activity. Werner (1979) found that health sciences
libraries had an average of 35 minutes per search, and that "in hospitals
and professional schools 1 FTE might be expected to handle roughly 1,400–
1,600 searches per year." (These are mostly Medline searches and the

number is database accesses, not requests.) Hawkins (1981) has also contributed information about average search times. He reported that at Bell Laboratories the average online time per search decreased from 1974 to 1979 from 39 to 12.8 minutes, while the number of databases accessed per search increased.

In academic libraries, where librarians formerly did not conduct literature searches for patrons but rather instructed them in how to use indexes themselves, the addition of online services is an additional service. However, the service provided the user is considerably improved, and the status, prestige, and morale of the librarians is also considerably enhanced.

In special libraries, manual searches were commonly done for the users. The rationale was that it was more economic for a librarian to search than a more highly paid scientist or engineer. Partly this is because the librarian is more efficient, but also because he or she is less well-paid. In the special library setting, online searches are clearly an opportunity to increase productivity. Indeed, search requests increase dramatically when online services are introduced. A sustained 20% per year increase in searches is not unusual. Hawkins (1980) found the number of searches performed at Bell Laboratories between 1974 and 1979 increased *2136%* over the 5 years.

The traits which make a good searcher have been widely discussed (Harter, 1983; Van Camp, 1979; Dolan and Kremin, 1979; Jackson, 1982; Hock, 1983). Many of the attributes of a good reference librarian are equally important for an online searcher. In addition, online searchers should not be terrified of spending money, should have typing skills, should react quickly to the unexpected, and should have the ability to "stay loose" under pressure. Bellardo (1985a, p. 241) found that

> . . .differences in searching performance can be attributed, to a small degree only, to general verbal and quantitative aptitude, artistic creativity, and to an inclination toward critical and analytical creative thinking. The findings also raise doubts, however, that high intelligence and other attributes cited by writers in the field are *necessary* for high performance.

King Research conducted a research project for the Department of Education called "New Directions" which developed competencies for information professionals and a planning framework for keeping them up-to-date. The competencies, as reported by Griffiths (1984), comprise the three major components of knowledges, skills, and attitudes, and are grouped by the following dimensions: type of work setting, functions performed (e.g., reference, cataloging, abstracting), professional level (e.g., entry-level, mid-level, senior-level), and prevailing trends affecting the information profession. Subsidiary dimensions include types of users served, tools and techniques used/applied, and types of materials handled. The sample competency list for a senior-level reference librarian (over 10 years of experience) in a special library appears as Table 4.3.

TABLE 4.3.   **Competency for a Senior-Level Reference Librarian**[a]

Knowledges

Knowledge of:

- the overall structure and organization of libraries and information centers
- the various functions performed by libraries and information centers
- the range of services offered by libraries and information centers.
- the expanding roles of the information community in meeting information needs
- alternative methods for organizing information and their implication for retrieval
- reference/referral services
- reference/referral tools
- how to use reference/referral sources and tools
- the basic principles of computer-assisted retrieval
- how to prepare, maintain, and use special-purpose files (e.g., vertical files, picture files, etc.)
- how to use and instruct others in the use of library tools, including catalogs, bibliographies, indexes, directories, encyclopedias, almanacs, etc.
- various techniques to promote the library or information center and its services
- performance evaluation methods and techniques
- personnel management methods and techniques
- the mission, goals, and objectives of the organization served
- the role of the library or information center within the organization
- users' information needs and requirements
- the collection and related collections
- the primary subject fields of the users

Skills

Ability to:

- establish rapport with users
- communicate well by written, verbal, and nonverbal means
- negotiate a reference interview
- analyze information requests
- formulate query, and collect, analyze, and interpret data
- use a variety of vocabulary devices
- select appropriate search strategy based on available resources, time constraints, costs, etc.
- search online databases effectively
- when appropriate, evaluate and advise patrons as to the currency, accuracy, and sufficiency of information retrieved or received so that patrons can evaluate usefulness of that information
- devise and publicize pathfinders, booklists, displays, etc., which will ease access to collections and will motivate use
- market the library and its services
- perceive the needs of the organization and not just the library
- anticipate long-range needs of reference service and accumulate materials accordingly
- analyze community information processes and translate findings into improved information services
- teach staff

TABLE 4.3 (*Continued*)

- provide instruction as necessary to a wide variety of patrons, in a wide variety of topics, both formally and informally
- select methods for providing current awareness services
- develop and present library programs
- develop special tools which provide access to information not readily available, e.g., community resources, special collection
- determine method(s) for locating desired materials that could not be found in available lists of holdings
- make effective, timely, and well-informed decisions
- isolate and define problems and develop the necessary criteria and action for their solution
- apply methods of measurement and evaluation
- manage time effectively
- apply quantitative skills in budgeting, projecting, supplying
- make recommendations on system improvement
- assign job responsibilities
- develop plans
- develop and implement policy and interpret policy both to parent organization and to staff
- assign staff to develop specific objectives and new programs from general objectives
- supervise staff
- evaluate adequacy of personnel utilization based on available positions, tasks to be performed, cost analysis, etc.

Attitudes

Practitioner should demonstrate:

- a positive attitude toward the profession, the organization served, the library, and its users
- a positive attitude toward working under various constraints, e.g., time, cost, staff
- a desire to satisfy the needs of user
- a desire to be cooperative
- a sense of responsibility
- a willingness to help people
- a willingness to keep up-to-date through professional reading, etc.
- a willingness to fail
- a willingness to learn
- tenacity/determination
- inquisitiveness/curiosity
- diplomacy
- patience
- resourcefulness
- flexibility
- confidence

*Reprinted by permission of the American Library Association, the Sample Competency List from Jose-Marie Griffiths, "Our competencies defined: A progress report and sampling," *American Libraries* January 1984, © 1984 ALA.

The Role of Subject Specialists

Although a good searcher can search for information on anything and becomes skillful at getting the user to explain what he or she is looking for, subject background is always a plus. Since satisfaction with a literature search or any information process is so subjective, there is a lot to be said for the easy communication with the user which is facilitated by background in the subject.

The background is useful in understanding the question, since if you know something about a field and how people work, it is easier to understand what is behind a question or what the limits might be. It is also useful in establishing rapport, in choosing the right database(s), in vocabulary development, and in evaluating (it is easier to judge whether a search result is too large or small if you have some background in the subject). Roth (1985), Kreutz (1978), and Girard and Moreau (1981) all support the need for subject expertise, especially in searching the chemical literature.

Although ideally there should be one subject specialist for each community need (Katz, 1982, p. 34), in reality it is rarely possible to have a large enough staff to have good subject background in all areas of interest. Broad backgrounds, willingness to learn, flexibility, and the ability to negotiate a question are important criteria. Willingness to take science courses or read extensively to get the proper background is something to look for.

Budgeting and Other Financial Considerations

Costs to be considered in budgeting include the direct costs of the searches (connect time, telecommunications charges, online and offline print charges, and labor) and the indirect costs of equipment (the terminal, modem, telephone and maintenance contracts), manuals, thesauri, supplies (paper, ribbons, disks), training (vendor training, database and subject seminars, practice time, travel expenses, professional memberships), promotion (printing, online demonstrations), and overhead (facility use and modification, furniture, heat, light, and electricity) (RASD/MARS Costs and Financing Committee, 1983).

The methods of financing online services arrange themselves on a line between two endpoints: at one end, the library absorbs all the costs of computerized literature searching; at the other, the library recharges all costs including overhead. In between are almost as many variations as there are search services. One large library absorbs all direct costs up to $150. If the search goes beyond that, it is recharged to the organization's account. Brokers commonly recharge all the direct costs plus some fee to cover staffing, equipment, and other overhead. Academic libraries commonly recharge direct costs to their primary users (faculty, students, and staff) and, if they serve outsiders, tack on an extra fee for those people. Or they may subsidize even the direct costs for certain users. For example, undergraduate students may pay a $10 fee and get a search of one database plus up to 100 citations. Graduate students may get a search on their dissertation topic subsidized. A few libraries have financed online searching by cancelling subscriptions of expensive indexes and abstracts and putting the money into online searching. They do free searches until the money runs out. Even those who *could* budget for the whole amount generally feel it is better to have some kind of fee to get rid of casual inquiries or unreasonable requests (e.g., "I need everything on organic chemistry").

There is a great philosophical debate in the public and academic library literature over fee or free. One side says that online services are not basic library services and are so expensive they must be recharged. The other side argues that this restricts the information to those who can afford to pay and is in direct conflict with the library's tax-supported mission: information to all. This debate is not focused in the sci/tech world, but any library offering online services must come to grips with the cause behind the philosophy: there is a basic difference between online services and the usual kind of library materials like books, periodicals, and indexes. Once a subscription is purchased, all users of the library can access it without any additional charges. But online services are completely dependent on how much one uses them. They are thus difficult to budget for and easy to recharge.

Whether or not a particular library choses to charge the user a fee for online search services, the library must still be aware of the actual costs. When preparing the library budget, it is necessary to have specific information on the number of online searches performed and on their average length, complexity, and costs in order to base future projections. The vendor's price increases must also be taken into account in projections.

Record-Keeping

What records of online search services need to be kept? The kinds of information the librarian will want to record about the search service fall into three categories: the information necessary to complete the search transaction, administrative information, and financial information.

In keeping records of the search transaction itself, most searchers find it useful to keep a copy of their search strategy with the searcher's request, at least for a period of time. This is especially helpful if the search does not turn out as expected.

The administrative information which may be wanted about online search services includes the number of searches done each month and year, the amount of online time per search, the average number of databases searched per request, the most frequently used databases, and the turnaround time for searches.

There has been some confusion on the definition of a search (Hawkins and Brown, 1980), but it is now generally accepted that a search is a single request, no matter how many databases are used to answer the request.

Financial information includes information on the cost of the searches done for any one request (needed for recharging), information about how much has been spent on any particular vendor (in order to check it against the budgeted amount, or to check against the vendors' bills when they come in), and information about the average cost of a search (for budgeting or for telling the requestor what to expect).

Almost all libraries keep a log sheet near the terminal to record the use of online services. The log sheet can be used to record prints ordered, and is the primary source for financial information, along with the vendor's invoices.

For financial purposes, the log sheet should include date, requestor's name and organization, file searched, time spent online, logoff time (useful in contacting the vendor about prints that did not arrive or a problem in billing), number of prints ordered, cost, and the searcher's initials (if a number of searchers use the same password). A sample logsheet is included in Hawkins (1980).

Log sheets work best when the searcher simply records the data as it comes off the terminal. This means each search of a database is a separate entry. Further massaging is necessary to turn the log-sheet data into data by search. Software systems for use on microcomputers have been developed which record the transactions and help compile accounting information for recharge. An example is DIALOGLINK, a software package for DIALOG users.

Evaluation

As it is with traditional reference service, evaluation is often a neglected step in the provision of online services. However, online search services have always had a few more quantifiable aspects than traditional reference. The user receives a search result in answer to a question. That result can be judged by factors of recall, precision, user effort, and response time (Lancaster, 1977). The evaluation form designed by the MARS Evaluation Committee attempts to get information which will measure these four per-measure these four performance criteria (Blood, 1983). (MARS is the Machine Assisted Reference Services Section of the Reference and Adult Services Division of the American Library Association.)

The main reason for evaluating an online search service is to identify instances of poor search results and user dissatisfaction so that the service can be improved. But there may be other reasons to evaluate the service. There may be other goals for the evaluation. For instance, data from users might be needed to justify budgetary and staffing resources committed to the online program. Or we might need to know whether some aspect of the service is useful, or whether users are more satisfied if they are present at the terminal when their search is conducted. Warden (1981) conducted a survey that showed they are.

As we discussed in Chapter 1, Lancaster (1977) suggested that there are three forms of evaluation for information retrieval systems: effectiveness, cost-effectiveness, and cost–benefits. The MARS form is designed mainly to deal in the area of evaluating effectiveness. This is the minimum level at which we should be collecting data on our services.

Some efforts at determining cost-effectiveness and cost–benefit evaluation have been conducted. Studies comparing online and manual searching were discussed in the section on literature searching. These are mostly cost-effectiveness studies. Jensen *et al.* (1980) presented data on the dollar cost and dollar value estimates by industrial users for their online searches. Warden (1981) suggested correlating search requests with measures of productivity, such as resulting publications, patents, contracts, or prizes.

Training (Including Continuing Education)

Tenopir (1984d) found three common levels of search training: general staff orientation programs, basic search training, and searcher updates and skills improvement. The content of the general orientation is very similar to the marketing presentation discussed in the next section. The RASD/MARS Education and Training of Search Analysts Committee published a brochure entitled "Online Training Sessions: Suggested Guidelines" in 1981.

Basic search training has to be adjusted to the knowledge of the person being trained as well as the organizational setting. A trained searcher should be able to conduct an effective reference interview, determine the feasibility of searching online for information on the subject, choose the appropriate vendor system(s) and database(s) for the search, design an appropriate search strategy, execute the search efficiently, and deliver satisfactory results to the user.

Table 4.4 lists the 14 steps for training a new searcher, as determined by a committee of experienced searchers at UCLA (Kwan *et al.* 1980). The final step is continuing education.

The online world is changing so quickly that to stop learning is to fall far behind. Those databases and vendors that the searcher has mastered will be continually evolving, and the searcher must keep up with the

Table 4.4 **Fourteen Steps in Training Online Searchers**

Step 1	*An online demonstration* on the database and vendor used most often in the library. Note types of input, output, and speed of response.
Step 2	*An overview and history* of online bibliographic databases (including basic terminology for computer hardware and software such as database, vendor, telecommunications network, terminal, online, etc.), the impact of online searching on libraries, the number and variety of databases available, and the advantages and disadvantages of online searching.
Step 3	*Computer physiology and how the computer searches,* including elements of the unit record and the difference between searchable and printable elements.
Step 4	*Computer terminal mechanics,* including the location and function of keys on the terminal (e.g., correction key, carriage return, power switch, online button, duplex, etc.); loading paper and changing ribbons; telephone and modem instructions; and setting up the terminal for online searching depending on the vendor used, including appropriate phone numbers and logon procedures. If a microcomputer is used for searching, the trainee should be shown how to use the communications software.
Step 5	*The establishment of a search strategy.*
Step 6	*Initial vendor and database training.* The new searcher needs to know database selection policies; cataloging and indexing policies; types of terms searchable (and formats); searchable elements versus printable elements; database changes over time (e.g., vocabulary changes, field changes, format changes, etc.); vendor manuals that exist for the database; thesauri and other vocabulary aids that exists for the database; and hard-copy index equivalents.
Step 7	*Practice searches.* Free time from vendor training or when new databases are introduced can be used. The ONTAP files on DIALOG are inexpensive and meant for training.
Step 8	*How to conduct the reference interview.*
Step 9	*How to evaluate search results.*
Step 10	*How to interpret output.*
Step 11	*How to do recordkeeping, accounting, and other internal procedures.*
Step 12	*Where to go for help* for policy problems, search strategy problems, and mechanical problems with the terminal, network, or vendor.
Step 13	*Learn new databases and vendors* as appropriate. For each new database and vendor, the new searcher should master the same material as covered in step 6.
Step 14	*Continuing development and education.*

changes. Also, there are aspects of becoming a good searcher that become more important with experience: techniques of cost-effective searching, how to choose which vendor to use for a particular search, how to develop in-house search aids, how to introduce people to online searching, etc. Learning these techniques when they exist and developing them when they do not is part of the searcher's responsibility for continuing education.

Articles on keeping up-to-date with searching appear in a large number of sources, including DATABASE, *Library Journal*, "Messages from MARS" in *RASD Update, ONLINE, Online Newsletter, Online Review, RQ*, and *Special Libraries*. The vendors often offer update sessions along with national conferences of the professional societies such as the American Library Association, the Special Libraries Association, and the American Society for Information Science. Two conferences are held each year focusing specifically on online searching and the online industry: Online, and National Online. Exploring new databases when they become available (especially when free time is offered) is another way to keep up-to-date. A local or regional users group can provide a forum for sharing searching tips and expertise and can also be instrumental in arranging for needed training from vendors and database producers.

Marketing, Publicity, and Promotion

When online searching is first introduced as a service, it is particularly important to publicize it. All the usual marketing mechanisms are in common use: brochures, word-of-mouth, mailings to the users, etc. (Hoover, 1980). However, the online demonstration is probably the most effective, if it is done well. Crane and Pilachowski (1978), in an excellent article about the online demonstration, suggested a five-part outline. The first part is an introduction, covering definitions of terms such as database, online, interactive, and offline (they suggest getting the audience to try to define them), a description of the evolution of online searching (including the use of electronic production techniques by publishers and the major components of the database industry), and an introduction to databases (how many and what kinds are available).

The introduction is followed by a discussion of the advantages and disadvantages of online searching, a brief discussion of the production of databases, and a discussion of communication with the computer. The telephone connection is all the mechanics that needs to be mentioned, but it may be important that the audience understand that it is not the campus or corporate computer that is being searched. The final part of the demonstration is a discussion of search strategy and a demonstration of a search. Crane and Pilachowski suggest describing a search problem (preferably as close to the interest of the audience as possible, and using the library's search request form if there is one), identifying the major concepts, explaining Boolean logic, and identifying searchable fields. Then they suggest showing a sample search, discussing all the steps, and showing the end-product. Finally, they suggest explaining costs and the structure for charging, and explaining SDIs (since the capability for keeping current is a strong selling point for some users).

Equipment and Communications

Online searching requires a terminal, a telephone, and a modem. The terminal may be 300 baud (30 characters per second) or 1200 baud (120 characters per second). A few services have put in 2400-baud access, but the modems are not yet widely available, and telephone line noise is more of a problem with faster transmission speeds. The faster terminals are much to be preferred; they cut down on online time (particularly if you print online citations), so they save money; they are also less boring for the searcher. Although it is not required technically to do a search, printing capability is almost a necessity. In 1981, Lamb found that most academic libraries were using the TI 700 series of computers, but there is a clear movement toward the use of microcomputers.

Microcomputers are the most versatile terminal for online searching. They give the searcher the option of recording the search to disk and editing the search for the user before printing it. They cost in the same price range as a 1200-baud "dumb" terminal, but can also be used for many other functions (word processing, mailing list maintenance, budgeting, etc.).

Microcomputers were originally designed for stand-alone use, so in order to be used for searching, they usually need additional hardware, at least a modem and a serial interface. They also require communications software. For further information about what to look for in communications software, consult Casbon (1983), Bruman (1983), *Making the DIALOG Connection with a Personal Computer* (1983), or Saffady (1985). When the hardware already exists, the choice of software is more limited. One must choose software compatible with the serial interface, modem, display monitor, and printer controller, as well as the microcomputer. Two software communications packages which have been used successfully for online searching are Crosstalk and Smartcom (which is designed for the Hayes Smartmodem).

The cardinal rule of buying computer equipment is to test it before you buy. Actually use it to access the vendors you use before you purchase it.

General discussion of furniture for terminal use is included in Chapter 12.

Impact of Online Services on Other Areas of the Library

It was clear as soon as online services were initiated that there were impacts on other areas of the library. Users quickly receiving lists of citations on their topic of interest also want copies of at least some of the actual documents quickly. Many of the items will not be in the library, and the increased burden on interlibrary loan and acquisitions has been documented (e.g. Drake, 1978; Atherton and Christian, 1977). Document delivery, even with the advent of online ordering and automated Interlibrary Loan operations (e.g., OCLC's interlibrary loan subsystem), is still slow. In almost every case, documents not in the library must still be sent by mail. Document delivery is discussed further in Chapter 8.

Starting an Online Search Service

In order to start an online search service, one needs to consider all the management aspects of online searching simultaneously. Decisions need to be made about what services to offer, what databases and vendors to offer, where to locate the service, how to staff it, who to offer it to, and how to fund it. Search tools need to be ordered, forms designed, brochures produced, publicity started, equipment acquired, contracts with vendors signed, and searchers trained. The effects on the rest of the library must be planned for. In addition, the whole idea may have to be sold to higher management.

In spite of the number of considerations involved, an online service is a necessity in the sci/tech library of the 1980s. For further guidelines on starting an online service, consult the book by Atherton and Christian (1977) and Shroder's bibliography on starting an online service (1982).

End-User Searching

All sci/tech libraries are going to have to wrestle with the question of whether or not to allow, encourage, or promote users doing their own online searching. There are a number of trends converging to force this consideration, but it is not yet apparent what libraries will do or, indeed, what is the best long-range strategy, either for service to the user or for survival of the library.

People are using their microcomputers with modems to access services such as the Source and Compuserve. These users are developing sophistication, and some want to access the bibliographic services themselves. Indeed, DIALOG and the other online systems are being described to them in their own publications such as *Byte* and *PC World* (e.g. Hewes, 1985; Yalonis and Padgett, 1985).

The online vendors see a huge and profitable market in direct service to end users, and have already designed systems to reach that market. DIALOG has Knowledge Index and BRS has BRS/After Dark. Both of these systems offer lower rates in the evening hours, and simplified query languages and a selection of databases designed to appeal to the home market.

There is also considerable development in the area of front-end systems, gateway systems, and systems for maintaining personal databases of downloaded information. Front-end systems are software packages like Search Master, Pro-Search, and Sci-Mate Searcher, which simplify the actual search process. Gateway systems [like Easynet Business Computer Network (BCN) and the Defense Gateway Computer System] attempt to make the choice of system and the search protocol transparent to the user (Hawkins and Levy, 1986; Williams, 1986). The systems for downloaded information include SciMate, Professional Bibliographic System, Notebook II, Zyindex, SIRE, and many more. Although all the systems available in late 1985 leave something to be desired, the development is moving quickly.

The increased use of microcomputers for searching by librarians and other information intermediaries is also pushing the interest in end-user searching: the users see the librarian using the same terminal they have in their offices, and they want to know why they cannot do it themselves.

The development and innovative marketing of CAS/STN to the academic markets has also fostered end-user searching. CAS/STN let academic libraries or departments subscribe for $500 a month and allowed unlimited searching for that amount. In 1985 they changed the policy to allow academic institutions who have subscriptions to *Chemical Abstracts* to search during the off hours at 10% of the usual costs. They also conducted training for the chemists, who have always been trained to use the literature more than other scientists and engineers (see Volume 2, Chapter 15).

Libraries are also considering end-user searching because of the constantly increasing demand for online searching. Even at a modest increase like 10% a year (which is common for an established service—much larger increases are common when the service is first instituted), the need for increased staff becomes apparent. One potential way out of the bind is to teach people to do their own searching.

If libraries have not jumped wholeheartedly at the possibility of end user searching, it is because well-trained searchers are very conscious that one can search online in a very superficial way and usually get *something* on any topic. But what is obtained may not be complete, and it may not be the best available information. There may be another database that would have been better, a better search strategy, or a better choice of vocabulary. End users are more unwilling than full-time searchers to attend training and update sessions, read newsletters like *Chronolog*, and keep manuals up-to-date. Thus, many intermediaries are uncomfortable trusting users to do their own searching. They are also concerned that training users will take as much or more time as doing the search, particularly since users will search so infrequently that they will need constant re-training, and because few users continue searching after training (DesChene, 1985; Ojala, 1986).

For those deciding to encourage user searching, Hunter's 1984 article will be useful. Many pilot studies have been conducted of end-user searching. Lowry (1981) reviewed the literature, and a bibliography by Lyon brought it up to date to 1984. DesChene's 1985 article includes a good bibliography, and Janke (1984) reported on locations and results of many experiments. Flynn (1985) reported on the 3M experience, which appears to be typical. Usually only about 20% of the users continue to search after they are trained. End users also do not search as frequently, so they are less competent with the search strategy, tend to do more browsing, and spend about 50% longer online. Their searches cost more— but they also tend to be more satisfied (Warden, 1981).

Many academic libraries who wish to encourage end-user searching see it as an extension of the library's educational role. Several libraries have set up contracts with BRS/After Dark or Knowledge Index and allow students to sign up for blocks of time (such as a half-hour) in the evenings (e.g., see Halperin and Pagell, 1985). They may recharge a set fee or just budget for what it would cost if the system were used full time when it is available. One library set up an account for each student through the computer center (just like money for other kinds of computer use), and the student could use that to pay for access to bibliographic databases.

Special libraries have always had less of an educational mission. They are charged with providing information to the organization in a cost-effective way. Users who consistently use a particular database could become more expert in that one database than an intermediary who searches a broad range of databases, and thus be more cost-effective.

Many databases are marketing subsets of their databases on floppy disks or other storage media (Tenopir, 1985a). In the long term, these may offer a better solution to end-user searching in settings where the costs must be controlled.

Atkinson (1984b) points out that "The history of librarianship for the last 50 to 100 years has seen a consistent pattern of transferring tasks and procedures formerly performed by librarians to nonprofessional staff and from them to patrons" (p. 1426). Online searching is potentially one of those tasks.

Downloading and Postprocessing of Search Results

Downloading in the context of online services in libraries is the process of searching a database on a vendor and recording the search and the citations in machine-readable form rather than in print form. The citations can then be reformatted before printing, or they can be maintained as a database. There is no way for the vendor to tell if a searcher is printing a search result or downloading. Although downloading for the purpose of reformatting does not bother most database producers, the maintenance of databases with the possibility of reduced searching of the original database does. Many database producers have responded to the threat by raising online print charges. Each copyrighted database has its own rules about downloading and must be contacted directly for permission. Hopefully the industry will come up with a common policy to simplify things for the small user. In the meantime, the compiling of policies published in *Downloading/Uploading Online Databases and Catalogs* (1985) is useful, as are articles by several authors who discuss the legal framework of downloading: Warrick (1984), Hawkins (1982b), Ferguson and Ballard (1985), Garman (1986), and Inkellis (1982).

Chapter 5

Current Awareness

Current awareness is often considered a part of the information-retrieval function because it is really "anticipatory information retrieval." A request for help in keeping current, when it is expressed, is considered an information request. But there are many things that libraries do to meet the need to keep current (such as publish new acquisitions lists) which do not strictly fall within information retrieval. The importance of current awareness in sci/tech fields also suggests it deserves separate treatment. Most scientists and engineers worry that they are not doing well enough at keeping current. Although people use a large variety of sources for current awareness (personal contacts, subscriptions to major journals, attendance at conferences, etc.) and any published information may be considered too slow (as we discussed in Chapter 3), libraries have used five main ways to help people keep current: (1) displays of new items; (2) production of library publications, including new acquisitions lists, abstract bulletins, compilations of tables of contents of journals, and other alerting publications; (3) purchasing (and routing) of commercially available current-awareness publications (e.g., *Current Contents*) or abstracting and indexing tools; (4) routing of journals; and (5) setting up SDI (selective dissemination of information) services.

DISPLAYS OF NEW ITEMS

This is the simplest and least expensive current-awareness technique. It is very effective for that segment of the user population which comes into the library frequently. Considerations in planning (or evaluating) a new-item display include the following.

Contents

What should be included in the display(s)? New issues of journals and new books are common, but shouldn't all new materials, such as maps and reports (both internal and external), be included? Depending on the amount of material, there may be a need for separate display areas by type of material, but having one place that the users can go to see everything that is new in the library has its advantages from the user's point of view. However, sometimes the arrangement of the library makes separate displays more feasible.

Location

Select an attractive, clearly marked location with a place for people to sit and look at the materials. It should be close to the traffic pattern so people will see it whether or not they came to the library for the purpose of looking at it. It is useful to have a photocopy machine nearby so that the materials will not be removed from the display area for copying. If there is not a copy machine, there must be some system for getting the issues from the copy machines back to the display area.

How Long to Leave Materials on Display

There should be a clearly posted and understood time that materials will be on display. There is no data on the best length of time, but times of 1 week to 1 month are common. Many libraries simply display the latest issue until a new issue comes in, but that does not help the user who wants to systematically see only what he or she has not already looked at. A more complex system of marking the day of the week the issue arrives with a colored dot (e.g., orange for Monday, red for Tuesday, etc.) has worked for one library which has less than 2000 current subscriptions. When the serials clerk checks in all the new issues for Monday, he puts an orange dot on the cover as well as stamping the cover with the property stamp and date received. When he takes the new issues to the shelves, he first removes anything with an orange dot which is already there (these are the issues which were received the previous Monday). This relatively simple system enables users to come in once a week (e.g., every Tuesday at 10:00) and see everything that is new in the library.

How to Let People Sign Up to Receive the Items Later

There should be a card on which a user can sign up to have the book (or map, report, or journal, if it circulates) checked out to him or her later.

If someone asked to have the item ordered, his or her name should be listed before the item goes on display.

Security

Because new books and journals are so exciting to the users, there is (sigh) sometimes a problem of security with them. Some libraries lock up the new items in some different manner than other materials. Librarians are constantly in the position of protecting the materials for the many against the few. However, fear of loss should not make us lose sight of the importance of the open availability of new books for browsing in a sci/tech library.

LIBRARY PUBLICATIONS

Many of the publications produced by libraries are designed to meet the current-awareness need. The publications are of three types: (a) new acquisitions lists; (b) abstracts or other journal article alerting services; and (c) highly selective lists.

New Acquisitions Lists

The most common publication is the list of new acquisitions. The only type of material usually excluded on purpose from a new acquisitions list is new issues of journals. However, since these lists are usually produced as a by-product of the cataloging process, sometimes only items which are cataloged are included. Thus, uncataloged materials such as microfiche, reports, maps, and government documents may be excluded. For completeness, some effort should be made to include everything, whether it is cataloged or not. Herb White feels that a library bulletin should be distributed to

> . . .all appropriate professional personnel, not just the ones who ask to see it. It should most definitely be sent to general management. If there is a lack of interest at that level, the solution is not to stop sending it but to make the publication of greater interest. (Herbert White, 1984a, p. 53)

However, if a library has a more selective list of new items of interest, those who are not interested in the complete list should be spared.

Frequency of publication will depend in part on the number of items to be included. Since it is a current-awareness tool, probably it should be produced at least monthly.

Abstracts and Other Journal Alerting Services

In organizations with specialized interests and a particular need for speed (such as in corporations in particularly competitive industries), the library may produce its own abstracts of journal articles received or of patent literature.

Highly Selective Newsletter-Type Publications

Another kind of publication useful for current awareness is a short but frequently issued alerting publication which is highly selective. Only articles, reports, books, etc. which are of general interest to the organization are selected, and short annotations or abstracts are usually added. Meeting announcements, the day's stock price, etc. may also be included.

Decisions to be made about any of these current awareness publications include:

Materials to be included (everything? journals? books? meetings?).

Frequency of publication.

Format of citations (shall we include annotations, abstracts, or just citations?; and should the citations be in full form or abbreviated?).

Format of the publication (whether to arrange the citations by classification, broad subject areas, author, or something else; whether to have length limitations; whether to have a distinctive color of paper and special logo, and if so, what; and whether to publish in electronic form).

How it will be compiled (e.g., will it be produced from the cataloging records or compiled separately?).

How will users request copies of the actual documents? Some libraries include forms with the publication, and others allow users to return a marked copy of the publication to the library. Others expect the users to come into the library and track down the publication themselves.

Who to send it to. It makes sense when a publication is just starting to send at least one issue to everyone in the organization, asking them to return a form if they wish to continue to receive it. With ongoing publications, it makes sense to send out this kind of broad solicitation at least once a year.

PURCHASE OF COMMERCIALLY AVAILABLE
CURRENT-AWARENESS PUBLICATIONS

There are a number of commercially available current-awareness publications. One of the most popular, *Current Contents,* published by the Institute for Scientific Information, reproduces tables of contents of selected highly cited journals. If one of the versions of *Current Contents* or some other similar publication meets the needs of users, a subscription (or several) for routing can be an effective current awareness service. It will certainly be cheaper than reproducing tables of contents in-house.

Many of the abstracting and indexing services can also be used as a current awareness tool. Libraries can subscribe to extra copies and route them [e.g., the Information Center at Exxon Production Research (described in Chapter 2) uses *Petroleum Abstracts* this way].

ROUTING OF JOURNAL ISSUES

Anyone who has had a journal issue routed to him or her knows what the two main problems are: issues can easily be delayed by someone earlier on the routing, and if there is only one copy of the journal, there is no way to know where the issue is while it is routing. In addition, maintaining the routing lists is not a negligible clerical effort for the library. New microcomputer software (like that associated with Faxon's MicroLINX) should help with the clerical operation of producing accurate routing slips, but no one has yet solved the problem of individuals holding up the issues, or the problem of tracking the issues when they are routing. Thus, most libraries have tried to find a satisfactory substitute for routing, such as arranging attractive new-journal displays or sending out copies of the tables of contents while maintaining the actual issues in the library. Those libraries who do continue to route journals because of the high value of current information to the organization usually only route duplicates.

SDI SERVICES

The concept of selective dissemination of information is that an individual is notified of items of interest to him or her (and not notified of those not of interest). The mechanism for providing SDI can be as informal as the librarian noticing a new book or journal article and, knowing it would be of interest to a particular person, calling or sending a note (or the actual item) to that person. Some librarians have kept card files of people's interests, and these manual systems are described by Richards *et al.* (1981). But SDI was truly formalized by the computer. SDI is like a literature search which is saved and run against each update of the database. It requires the same steps of analysis as a literature search (see Chapter 4). However, because it is stored and run over and over again, it can be enhanced by an iterative process of running the profile, reviewing output, and looking for ways to remove false drops from future output, such as the use of subject categories (Sprague and Freudenreich, 1978).

All the major vendors offer an SDI service (although not on all databases), and there are numerous other vendors who offer SDI services, as well. *The Encyclopedia of Information Systems and Services* (Kruzas and Schmittroth, 1981) lists over 400 agencies under SDI Services. With these commercial services, profiles are run automatically by the vendor when the database is updated. Costs vary greatly by database and by number of hits, so they are hard to predict and thus to budget for. Kaminecki (1977) compared costs of several ways of doing SDIs (a batch system, rekeyed SDI, saved SDI, and vendor supplied SDI). The costs varied by the number of months between updates and no method was clearly superior. One sci/tech library in 1982 calculated an average cost of $150 per profile (one database, 25 hits or less each 2 weeks). This included a mix of the databases used most in the local environment (*Chemical Abstracts,* COMPENDEX, INSPEC, DOE's Energy Data Base, NTIS, and SCI-SEARCH) and an estimate of 30 minutes online time to set up and revise each profile (Pruett, 1982).

Many of the in-house SDI systems set up in the 1960s, before SDI services were available from the online vendors, survive today. The libraries purchase tapes directly from the database producers, mount them on their own computers, and run software which matches stored profiles against the tapes. For large-volume operations this can be cost effective, particularly if most of the literature of interest is in one database. In choosing whether to use commercial SDI services or to buy tapes and provide the service internally, the important factors are the availability of the information in the most useful form, hardware and software response time, security, costs, and timeliness (Archer, 1978).

SDI can also be effective when the profiles are stored, not against the world's literature, but against what has been selected for a particular library. In-house SDI profiles that can select from all the items received in the library those which match a subject profile for a particular person are a great service. An example is Bell Laboratories' control of distribution of internal reports based on notification of subject interest and stored subject profiles (Kennedy, 1978). So far, this kind of sophistication is not available on commercially available online catalog systems; those libraries which are providing it are using in-house systems.

SDI services were designed to filter the world's literature for individuals, few of whom would be able to keep up with the publications being produced. The idea was that an SDI profile on an appropriate database would select just those few new items of the huge number available which were directly of interest. However, only a relatively small proportion of scientists and engineers are using it, compared to the many which were expected when SDI was designed (Mauerhoff, 1974). There are probably a number of reasons for the relative lack of use, including the apparent lack of timeliness (journals themselves are quicker), that the information is not evaluated (a citation may be on the subject, but it may not be any *good*), and that people feel they are well informed or that they have too much to read already. Interestingly, SDI has been used increasingly for business and management, and the rate of requests for copies of the journal articles in a study by M. Jackson (1978) was found to be higher than the average rate of request for sci/tech journals.

There is evidence that SDI is being used for background information and to give a general sense of what is being published on a topic. This is particularly useful for a scientist or engineer expert in one area who is temporarily working in another area and does not want to lose touch with what is going on in the original area. Several studies have found that retrieval of background citations is one of the ways people use SDI listings (Cole, 1981, p. 444).

Warden found that the General Electric Whitney Library Current Awareness Service had "alerted users to other persons working in the same field, provided new research leads, prevented the duplication of research already conducted, and resulted in more free time for reading or research" (Warden, 1978, p. 459).

SDI can also be used by libraries for collection development. Seba and Forrest (1978) used the results of SDIs to evaluate the journal collection and increase its relevancy. Cancelling journals which did not seem relevant helped offset the cost of the SDI service. It is also possible to store profiles on databases which include books, to enable the library to be notified of new books which match a subject profile (Quinn, 1985). For example, an energy library might want to be notified of new dictionaries and directories which appear in the Energy Data Base.

We may see more use of electronic mail as a method for disseminating SDI output. For example, an in-house SDI could run daily against newly cataloged items and the references be sent via electronic mail direct to the person.

For further information, Lavendal (1981) gives an overview of SDI services in sci/tech libraries.

CHOICE OF A CURRENT-AWARENESS METHOD

The choice of a current awareness method will depend on the cost and on the needs of the organization for timeliness and coverage. No one method is a clear winner. In her master's thesis, McVicker (1979) compared online SDI with an in-house current awareness bulletin (produced manually). The results indicated that SDI would provide better coverage than the manual service, but it would be slower. Chaloner and de Klerk (1980) reported that Carnegie-Mellon University Libraries conducted a 6-month study of two current awareness services: the provision of computer updates from bibliographic search services, and the distribution of selected contents pages of in-house journals. The value and desired frequency of computer updates varied. Distribution of contents pages was extremely popular. Blick *et al.* (1982) compared running computer SDI profiles on commercial online databases with manual scanning of journal sources. SDI retrieved 43% of items found by manual scanning, whereas manual scanning retrieved 61% of items found by SDI. They found that costs were of the same order and that manual scanning was the better alternative as a stand-alone current-awareness system. Although manual scanning is superior, however, it depends on in-house staff and, at least in the corporate setting, money is usually more available than staff (Herbert White, 1984a, p. 53).

Since no technique is clearly superior to any other in all circumstances and since the function is so important to sci/tech users, we must continually strive to improve techniques of current awareness, and to make as many methods available as possible.

Chapter 6

Collection Development

> . . . by making your collection smaller you can actually provide more and better service.
>
> Daniel Gore (1976, p. 172)

Collection development is the third of the five primary functions of sci/tech libraries. It is closely tied to the information-retrieval function, since the primary source for answers to the user's questions is the library's own collection, and since feedback from the information-retrieval staff about what questions are being asked is helpful to selectors. Academic and special sci/tech libraries display different attitudes toward collection development. Often in academic libraries the collection-building aspect takes precedence over the information-retrieval function, whereas in special libraries active information retrieval compensates for a more limited collection.

The current-awareness function is also dependent on collection development, since new acquisitions lists and new-book displays depend on how well collection development is done, both in terms of selection quality and timeliness.

Collection development also has a close relationship with the collection control function, which will be discussed in Chapter 7. Ideally, the collection control sequence should begin at the time a particular item is ordered. Thus, acquisitions is the first step in collection control. Also, collection development is dependent on collection control for access. That is, if a book is in the collection but not cataloged usefully, it is almost as if it is not there.

Collection development is also closely related to document delivery, and there is a need for information exchange between the two functions. The information about what had to be borrowed should get back to the selectors.

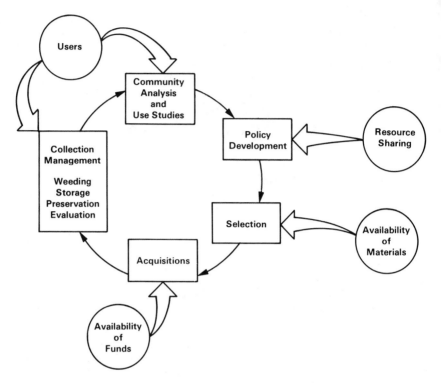

Fig. 6.1. The collection development process and environment.

Most of the terms referring to collection development in the library literature are not well defined. People seem to use the terms "collection development," "selection," "bibliography," "acquisitions," and, more recently, "collection management" interchangeably. In this book, we use collection development to include the whole process of analyzing user needs, developing policies, selecting and acquiring materials, and collection management. Collection development implies a planning process. So we discuss "collection development policies" rather than "selection policies" or "acquisitions policies," because they are planning documents and should encompass the whole process. "Selection" is the process of choosing materials which meet the policies, and "acquisitions" is the process of acquiring the materials. "Collection management" encompasses all those functions that happen after the collection exists, such as storage, weeding, preservation, and evaluation, although current writers use the term to encompass the whole process. To some extent the terminology reflects the changes in emphasis over the last 30 years, from acquisitions in the 1960s, to policy-making in the 1970s, to collection management in the 1980s (Mosher, 1982; Cooper, 1983).

We will consider the collection development steps in a logical order: community analysis and use studies, collection development policy development (including resource sharing), selection, acquisitions, collection management (including weeding, storage, preservation, and evaluation) and finally the management aspects (organization, staffing, and budgeting). However, the process is actually circular and affected constantly by the community, the resources available, and the publishers of materials, as shown in Fig. 6.1.

COMMUNITY ANALYSIS AND USE STUDIES

In Chapter 3 we discussed the way scientists and engineers use the literature and the types of literature they use. We found that any sci/tech library needs to be concerned with special forms of materials other than monographs (journals, technical reports, conference preprints and publications, data compilations, patents, etc.). We also saw that timeliness is an overriding concern. The particular subject or setting can introduce variations. In Volume 2, Part II, we focus on the needs of users in particular subject areas, which should be helpful for collection development in those subject areas.

For any subject, the first step in rational collection development is to determine who the library's users are, and this will depend on the library's mission. In an academic library, the mission is usually to serve the students and faculty of the institution. In a special library, it is usually to serve the employees of the corporation, or sometimes a special segment of the employees such as the scientists and engineers. In almost all sci/tech libraries, the users are more well-defined than they are in public libraries, where many of the community analysis studies have taken place.

Once we know who the prime users are, we gather data about their interests and information needs. First, we look for previously gathered data. There may be some data available in written form on the number of employees with particular subject backgrounds (e.g., physicists, chemists, geologists, mechanical engineers). There may be annual reports of activities which list the faculty individually with their subject interests and publications. The catalog of courses indicates what courses they teach. We can gather publications of the user group and look at what kinds of materials they are citing as an indication of what materials they are using. For example, an analysis of theses produced can provide excellent in-materials they are citing as an indication of what materials they are using. For example, an analysis of theses produced can provide excellent information about what is being used in the local environment (Kriz, 1978; Crissinger, 1981).

If the written materials are inadequate, we can conduct a survey, either by sending questionnaires or by interviewing people. Interviewing people has the great advantage ge of establishing rapport and allowing them to ask questions about library services. The survey should focus on the subject areas in which they do research or teach, and should be tailored to the setting. Interviews of heads of divisions in an industrial setting might include the following questions, the first set focusing on the research activities of the division, and the second set focusing on the division's information needs and sources.

Possible questions to ask about research include: Describe the work of your division. What work do you expect to be doing in the future? Is the work of your division stable or unstable? Describe related work in other divisions. Describe where other related work is going on outside the company (e.g., which universities, other laboratories, etc.). What are the funding sources for your division's work? What professional organizations are most closely concerned with your work? What profession do you identify with? What about the other members of your division? At what conferences do you present your work? How do you keep current? What journals do you and other members of your division read most often? Who are the most active library/literature users in your division? Can you suggest how I might keep up with what is going on in your division? Are there department or division reports? Meetings? Informal seminars?

Possible questions to ask about information needs and sources include: How would you evaluate the library's collection in your area of interest? Are there areas of the collection in which we are weak? Areas that need weeding? Journal titles that we should or should not have? Are there types of materials that we should keep forever and that you rely on us to have? What other sources for documents do you rely on (e.g., office collections, personal collections)?

Other questions about specific library services or plans for the future could also be included. All questions should be geared to the setting and to what the library really needs to know. An interview like this usually takes about an hour, and may well produce literature-search or SDI requests which require follow-up time.

The data collected in an analysis of the community can be organized on a microcomputer (e.g., Borovansky and Machovec, 1985) or in a manual file. The analysis of the community should be used to help formulate the collection development policy (discussed in the next section), as well as in specific decisions about what to weed, store, preserve, etc.

Determining the Core Collection

One of the most striking patterns in the data of almost every form of information use is Bradford's law of scattering (also referred to as the Bradford distribution, the Bradford–Zipf phenomenon, or the 80–20 rule). The Bradford distribution was originally described as a feature of citation data for journals, and has since been found to hold true in a large number of subject areas, in book circulation studies and in all phases of information generation, dissemination and use (Subramanyam, 1980b, pp. 349–351).

In a typical citation analysis, the researcher takes a body of literature, groups the citations by journal title, and ranks the journal titles by the number of citations each produces. A curve like the one in Fig. 6.2 results

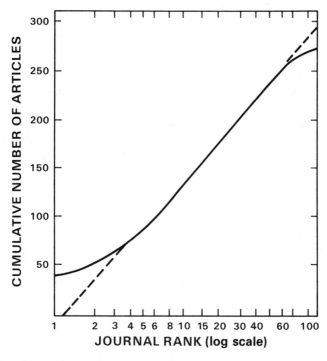

Fig. 6.2. An example of the Bradford distribution. [From Subramanyam, 1980b, p. 350. Reprinted by permission of the publisher.]

when the cumulative number of citations is plotted against the log of the rank of the journal. As an example, when Subramanyam (1976) did this for computer science, the top two journals produced nearly one-third of the citations, the top seven journals produced nearly one-half of the citations, and the top 19 journals produced about three-fourths of the citations. The pattern persists, and the journals with lower ranks receive fewer and fewer citations (Subramanyam, 1980b, p. 353).

The implications for library management are that by subscribing to a relatively small number of journals, the library can satisfy 80% of the demand, but the last 20% of the materials are scattered in a large number of sources, each of which contributes only one or two items. The highly used or core material should not be stored or weeded, and may require special loan policies.

Strict Bradford distributions on books are not usually done; they require the complete circulation history, something few libraries have. However, Trueswell (1966, 1976) found a mathematical way to use LCD (last circulation date before the current circulation) to develop a distribution for books which could be used to determine a core collection. The procedure for doing this is as follows:

A. To determine the last circulation date (LCD) that will satisfy any percentage of demand:

1. Take circulation records of material currently on loan and determine the last circulation date prior to the current loan.

2. Arrange the records in order by LCD.

3. Determine what percentage were last borrowed 6 months before, 1 year before, 2 years before, etc. This is plotted as the top curve in Fig. 6.3.

4. Decide what percentage of requests you want to satisfy (e.g., 90%) and look at the data in Fig. 6.3 to see what LCD would have produced that (e.g., 60 months would satisfy 90% in this example).

5. The core collection of books then consists of everything which has not circulated in the last 60 months and can be expected to satisfy 90% of the demand.

B. To determine what percentage of the total collection is the core collection:

1. Do a similar study of LCDs of randomly selected materials in the stacks (Trueswell selected the second book on each shelf). The results are plotted as the lower curve of Fig. 6.3.

2. Use the data from the two curves in Fig. 6.3 to plot percentage of circulation satisfied versus percentage of holdings satisfying that percentage of circulation. (For example, in Fig. 6.3, 75% of the circulation (see point A) came from 15% of the holdings (point B), and this makes

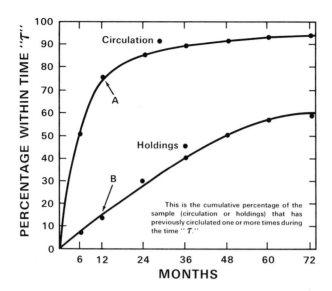

Fig. 6.3. Percentage of circulation versus time. At point A about 75% of the items have circulated in the last 12 months. Only 15% of the materials in the stacks (point B) have circulated in the same amount of time. Points A and B plot as point C on Fig. 6.4. [From Richard W. Trueswell, "Growing Libraries: Who Needs Them? A Statistical Basis for the No-Growth Collection," in Daniel Gore, *Farewell to Alexandria: Solutions to space, growth, and performance problems of libraries* (Greenwood Press, Westport, CT, 1976), Figures 2, 3 and 4. Copyright© 1976 by Daniel Gore. Reprinted by permission of the publisher.]

one point (point C) on Figure 6.4. Figure 6.4 can also be used to see that in this particular library only 20% of the holdings is needed to satisfy about 80% of the circulation demands. About 50% of the collection would satisfy 90% of the needs.

These analyses can be done within particular subject areas (and there is quite a bit of variation by subject area), as well as for a general collection. Most libraries, under pressure to weed or store, pull a last circulation date out of the air ("well, if it hasn't circulated in 10 years, let's put it in storage"). The advantage to the Trueswell approach is that it gives us a number for the cutoff date that is related to the percentage of circulation we expect to satisfy.

Collecting Data on Use

In addition to analysis of the community and determinations of core collections, evidence about what is needed in the collection can be gathered

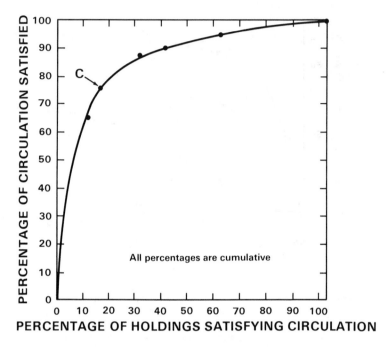

Fig. 6.4. Percentage of circulation satisfied versus percentage of holdings satisfying circulation. Point C is point A from Fig. 6.3 plotted on the Y axis and point B on the X axis. [From Richard W. Trueswell, "Growing Libraries: Who Needs Them? A Statistical Basis for the No-Growth Collection," in Daniel Gore, *Farewell to Alexandria: Solutions to space, growth, and performance problems of libraries* (Greenwood Press, Westport, Ct, 1976), Figures 2, 3, and 4. Copyright© 1976 by Daniel Gore. Reprinted by permission of the publisher.]

from a variety of use data, including circulation data, in-house data, citation analyses, and availability studies. Availability studies are discussed in the section on evaluation.

Circulation Data

Most collection development experts agree that past use is the best indicator of future use, and that titles may be weeded or stored if they have never circulated, or have not circulated for a long period of time (Fussler and Simon, 1969; Slote, 1982; Mosher, 1980a, pp. 174–175).

There are a number of ways circulation use can be recorded. Automated circulation systems should record the number of times (and when) each item circulates. Then the system can be programmed to deliver a list of everything which has not circulated since a particular date, and the weeders can work from the list of candidate items rather than the items themselves. (Still, periodicals rarely circulate, so automated circulation systems do not help establish use of periodicals).

In libraries which are not automated, or if the automated circulation system does not provide this information (and, unfortunately, many systems on the market do not), then some record should be kept on the physical item. Unfortunately, libraries have discarded data that would have been useful without realizing its importance. For example, date due slips are discarded from the back of the books when they are filled. If none of this information has been kept, short-term studies can be conducted to collect use data.

A study by Simmons (1970) showed how a backfile of automated circulation records could be used effectively. The two studies he reported resulted in the purchase of over 2000 additional copies of heavily used books. However, the analysis was not simple, and the final decisions to buy extra copies could not rely completely on the analysis. Other factors which were considered included the number of different borrowers who used the book; the number of days a book was off the shelf and thus unavailable for loan; the number of holds and call-ins placed on a book by different borrowers; the loan history of copies of the title held by other branches of the library system; the availability of newer editions of the work; the presence of the title on one or more reading lists; and other factors known only to librarians who knew the book collections and students' reading habits well (Simmons, 1970, p. 63).

Circulation data have also been related to academic departments (which have been classified by LC) and to number of students in each department by Jenks (1976) to identify areas of the collection which are under- or overutilized. To identify overused or underused collections, all we need to know is the percentage of the collection occupied by any LC class and the percentage of total circulation accounted for by that same class in any time period.

> To take a simple example, suppose that mathematics occupies 12% of the total collection of a science library and that geology occupies 9% of the same collection. Probabilistically, we would expect that mathematics should receive 12% of the use and geology 9% of the use. We might find, however, that geology accounts for 15% of the current circulation and mathematics for only 6%. Mathematics is an underused class and geology an overused class in this library. (Lancaster, 1982a, p. 19)

However, what this means for collection development is sometimes difficult to determine. Many other studies are summarized in the book by Kohl (1985).

In-House Use

In-house use is more work to measure than circulation, since there is no record of the transaction. For the book collection, McGrath (1971) showed that in-house use of books is proportional to circulation in any one library. For example, if computer science circulates five times more than chemistry, it can also be expected to be used five times more often within the library. If this is true, it means we can do a one-time study to establish for each subject area what the ratio is of in-house use to circulation and then rely on the proportion without having to measure in-house use (Lancaster, 1977, pp. 24–25).

For the periodical collection, however, there may be no circulation records at all, so some kind of study has to be done. The study can be of short duration and request some user cooperation in returning slips or in not reshelving materials (which are then counted before returning them to the shelf). Colored stick-on dots on the spine or stamps inside the back cover of periodicals can also be used to indicate in-house use of noncirculating materials (e.g., Shaw, 1978).

Citation Analysis

The usefulness of citation studies is based on the assumption that highly cited materials are used more often. This assumption depends on the source the researcher uses to count citations. For a particular library, a source which is specific to the users of the library (e.g., dissertations and faculty publications in a university and internal reports in a special library) is better than an overall citation analysis (such as ISI's *Journal Citation Reports*). For example, Kriz (1978) used analyses of the citations in theses produced at his institution to show that books were more important than journals to the particular users of his library. This led to cancellation of journals in order to maintain the monograph budget (at a time when everyone else was borrowing from the monograph funds to maintain the serials).

Ranked lists of journals developed based on citation data are useful for determining the core collection (as we discussed previously). When cost data is added, a ranked list can be used to determine the cost of collecting all the journals relevant to a given subject and what fraction of the total coverage would be available at any specified cost. Ranked lists can also be used to determine the best distribution of journals between a branch and central library, and to make storage decisions. And citation data can also be used to establish the need for expensive serials and monographic series (Subramanyam, 1980b, pp. 362–363)

Citation studies have also been used to try to find a correlation between age and use, particularly in the science and technology journal literature. These have found that use reaches a peak about 2 years after publication and then tapers off (Brookes, 1970; Buckland, 1972; Line, 1974; Morse and Elston, 1969; Cole, 1965; Rao, 1973) and that articles that are highly cited immediately after publication continued to be cited frequently during the next 12 years (Line, 1974). Several authors have proposed the concept of "half-life" of periodicals (Burton and Kebler, 1960; Line, 1970; MacRae,

1969), but the variation in half-life between disciplines and among libraries is large so the concept has not proved as useful as it might have (Bourne, 1965). Nevertheless, citation studies can be used at least as general predictors of future use (Subramanyam, 1980b, p. 365).

THE WRITTEN COLLECTION DEVELOPMENT POLICY

A written statement of collection development policy is useful primarily for communication. In large libraries, it provides a necessary vehicle for communication among the numerous selectors and the collection coordinator, if there is one. In all libraries, even small ones, it is a vehicle for communicating the library's mission to its users, and to new selectors when the responsibility changes. In all libraries, it also guides the selector in day-to-day decisions of what to buy and what to discard. To libraries engaging in any kind of shared resources (and what library in this day does not?), it provides a means for communicating between libraries.

Unfortunately, all these people have different needs from the policy, and for that reason it is difficult to structure a policy that will satisfy everyone's needs. In a large university system, the collection development coordinator needs to know that the chemistry library collects the QDs. The chemistry and biology librarians have to know which aspects of biochemistry each is responsible for. The chemistry selector, trying on a day-to-day basis to select new materials for the collection or to choose materials for storage, needs to know what levels, formats, etc. to collect in much more detailed areas.

The guidelines for the formulation of collection development policies formulated by the American Library Association, Resources and Technical Services Division, Collection Development Committee (1979), although focused on large academic libraries, can be useful for other libraries as well if the subject analysis is carried to enough detail.

The guidelines recommend that a collection development policy statement should include two parts: (1) an overall introductory statement of general institutional objectives and limitations, and (2) a detailed analysis by subject field, arranged by classification.

The general introduction should include a statement of the mission of the library as a whole; identification of the programs and clienteles it is established to serve; definition of the kinds of user needs; the implications of cooperative collecting agreements, local, regional, or national; and any statement of level of collecting, languages, forms of materials collected, or chronological periods or geographical areas collected which are true for the collection as a whole.

The detailed subject analysis should be arranged by the classification which the library uses. Using an established subject classification helps to define the subjects in a standard way. If the classification used is Library of Congress (LC), there is the added advantage of being able to compare other library collections which were measured in the Library of Congress National Shelflist Count project (Ortopan, 1985; Branin *et al.*, 1985).

Whatever scheme is used, the decision on level has to be made. The Collection Development Committee Guidelines recommend as a minimum refinement the 500 categories used in *Titles Classified by the Library of Congress Classification: National Shelflist Count* (Berkeley, General Library, University of California, 1977). However, they admit that many libraries will prefer to analyze their collections in more detail. For science libraries, the 500 subdivisions do not provide enough detail to be useful, at least not for selection decisions, and sci/tech librarians will probably choose to go to a more detailed level.

For each class chosen, three levels of collecting should be indicated, using the definitions of levels included in the Guidelines:

> A. Comprehensive level. A collection in which a library endeavors, so far as is reasonably possible, to include all significant works of recorded knowledge (publications, manuscripts, other forms) for a necessarily defined and limited field. This level of collecting intensity is that which maintains a "special collection"; the aim, if not the achievement, is exhaustiveness.
>
> B. Research level. A collection which includes the major published source materials required for dissertations and independent research, including materials containing research reporting, new findings, scientific experimental results, and other information useful to researchers. It also includes all important reference works and a wide selection of specialized monographs, as well as an extensive collection of journals and major indexing and abstracting services in the field.
>
> C. Study level. A collection which supports undergraduate or graduate course work, or sustained independent study; that is, which is adequate to maintain knowledge of a subject required for limited or generalized purposes, of less than research intensity. It includes a wide range of basic monographs, complete collections of the works of important writers, selections from the works of secondary writers, a selection of representative journals, and the reference tools and fundamental bibliographical apparatus pertaining to the subject.
>
> D. Basic level. A highly selective collection which serves to introduce and define the subject and to indicate the varieties of information available elsewhere. It includes major dictionaries and encyclopedias, selected editions of important works, historical surveys, important bibliographies and a few major periodicals in the field.
>
> E. Minimal level. A subject area in which few selections are made beyond very basic works. (American Library Association, Resources and Technical Services Division, Collection Development Committee, 1979)

These levels should be indicated for three different aspects of the collection: the strength or level of the existing collections in that class, the actual level of collection development being sustained in the current budgetary year, and the level of collection development deemed appropriate for the support of the institution's goals.

There may be some difficulty in applying these definitions in practice. For example, if a library collects to the "research level" in engineering, what exactly does that mean? Can we predict what it will be buying?

For each class, the following aspects should also be considered and recorded:

Languages collected or excluded.
Chronological periods or geographic areas collected or excluded.
Forms collected or excluded.
Any other exceptions or considerations relating to the specific class.
Shared resource obligations for the specific class.

Although these guidelines are most appropriate for large academic research libraries, they are recommended here because there is an advantage to a standardized approach, particularly in resource sharing. Even if this pattern is not followed, however, every library should have a written statement of collecting policy which describes the goals and decisions that have been made.

RESOURCE SHARING

Resource sharing can take many forms, including:

An agreement by one organization to maintain a subscription to a particular journal and to loan (or provide photocopies) to others on request.

An agreement to maintain a whole subject area (e.g., spectroscopy).

An arrangement to purchase some item or collection that the whole group needs with pooled money.

A complete shared network. This includes discussing purchases in advance, dividing subject responsibilities, and maintaining a shared bibliographic network and a document delivery service.

Kent has pointed out five requirements which are necessary for resource sharing to work: understanding of the use of the collection; the bibliographic apparatus to permit appropriate access; an efficient document delivery system; coordinated purchasing of materials; and understanding of the philosophy of remote access in the user community (Kent and Galvin, 1977, p. 13).

Understanding the Use of the Collection

Use studies and all efforts toward developing written collection development policies are helpful in understanding the use of the collection. This understanding of use is critical, because, as Kaiser says, "It is obvious that only little-used, or less frequently used materials can be shared" (Kaiser, 1980, p. 145).

The Library of Congress National Shelf List Count and the Collection Analysis Projects sponsored by the Association of Research Libraries are examples of useful methodology toward the understanding of the collections (Kaiser, 1980, pp. 141–144). A more recent project is the National Collections Inventory Project (NCIP) sponsored by the Association of Research Libraries, which involves having a large number of libraries complete the Research Libraries Group (RLG) Conspectus online (Farrell, 1986). One sci/tech subject (chemistry) was included in the 1981–1982 test, and the full inventory will include the other sci/tech subjects. The result will be shared collection development policies in a common format, all available online.

Bibliographic Apparatus to Permit Appropriate Access

Shared online files have the greatest effectiveness in resource sharing. Each participating library can see what the others have. Examples of networks which share bibliographic information are OCLC, RLIN, WLN, UTLAS, and Bell Labs. Some of these will be discussed further in the chapters on collection control and document delivery. In libraries which are not online, the difficulty of knowing what the other library has is much greater, but copies of printed book catalogs, microfiche catalogs, and/or periodicals lists can be exchanged. Up-to-date accurate information is critical if resource sharing is to succeed.

An Efficient System for Document Delivery

The two important things about document delivery from the collection development point of view are: (1) as a borrower, if you have chosen not to buy something because it is available elsewhere, you expect to be able to physically get the item in a reasonably quick manner; and (2) as a lender, if you have accepted the responsibility to develop shared resources, part of the responsibility is to provide the materials when requested. Systems that exist go from mail (the national ILL network) to telefax to daily courier service. Often the document delivery service is supported by an electronic ordering component (e.g., OCLC). Document delivery is covered in more detail in Chapter 8.

Coordinated Purchasing

The coordination of purchasing is more difficult than the sharing of existing services. Partly this is an information problem, since acquisitions records are not shared as frequently as catalogs. But it is also a political problem. It is difficult for one university library to have another tell it that it should or should not buy a particular item.

Patron Understanding

Although many users are appreciative that we can borrow from elsewhere to get them what they need, they are not always so understanding when an item they want is checked out to another organization through interlibrary loan.

Everyone believes in resource sharing in the ideal, but in reality it does not always work too well. The materials in demand in one place tend to be in demand elsewhere as well. Also, political realities may set in and budgets may get tighter. Consider the case of a branch library in a university which has agreed to maintain a little-used subscription for the local network. What happens when the budget gets tight and the librarian tells the faculty they have to cancel something they want more than that subscription?

SELECTION

Once we know what subjects and forms of materials our users need, we can select materials for the collection. Selection has two main steps. The first is to gather information about what is available in the publishing world which might satisfy our collection policies. The second is to judge what is available against our criteria for selection. I am assuming in this section that the selector is the librarian. Variations on who has the real responsibility for selecting will be discussed in the section on management aspects.

Selection may be on an item-by-item basis, but whenever possible, selectors try to make decisions which will bring large amounts of relevant material into the collection without individual decisions. For example, they may set up standing orders for all American Chemical Society publications or subscribe to all microfiche produced by NTIS in particular COSATI categories.

Selection processes vary by the type of material being selected (e.g., books, serials, technical reports, etc.). Selection of formats other than books and serials is considered in Volume 2, Part I.

For all forms, it is important to be aware of what is being published and by whom. Selectors need to get on mailing lists for the major publishers, both trade and society, in the areas in which they want to build collections.

In the last decade, the ever-hungry periodicals budget took large bites of the budgets for materials in many sci/tech libraries. Funds for buying monographs in some libraries became almost nonexistent in the struggle to maintain the periodicals collections. Cooper (1983) discusses the importance of the monograph to scientists and engineers, suggesting that this trend should be reversed. There is also evidence (Kriz, 1978) that monographs are particularly important in certain subdisciplines of science and engineering.

Monographs can be selected individually or through mass buying programs.

Individual Selection

Critical, timely reviews by subject experts are of great help in selecting science books. Unfortunately, these are not as available or accessible as we would like. Chen found that the average overall mean time lag for the appearance of a review in general science was 12.2 months (with a range of 5.5–25.3 months), and in engineering, 9.7 months (with a range of 6.5–20.1 months) (Chen, 1978, p. 366).

However, even when reviews appear as much as 2 years after a book's publication, they can cover critical items that might have been missed, and the books advertised and reviewed in journals are often the ones the users will be asking for.

Most selectors read the technical journals and look for advertisements and reviews. But which technical journals? Chen's 1976 book *Biomedical Scientific and Technical Book Reviewing* analyzed 168 reviewing journals in terms of their authoritativeness, quantitative coverage of reviews, timeliness, duplication of reviews among journals, and which American and British book publishers get the most reviews.

She identified a core list of journals which review books. In general science, the top five journals (in terms of quantitative review coverage) were *American Scientist, Nature, New Scientist, Science,* and *Scientific American.* Of the subject oriented journals, the following rated highly: *American Mathematical Monthly, American Statistical Association Journal, Observatory, Sky and Telescope, Applied Optics, Contemporary Physics, Physics Today, Chemistry and Industry, Chemistry in Britain, Journal of Chemical Education, Journal of the American Chemical Society, Bulletin of the American Association of Petroleum Geologists, Earth*

Science, Engineer, Aeronautical Journal, Chemical Engineer, Civil Engineering, IEEE Spectrum, Journal of Material Science, Mechanical Engineering, Journal of Metals, and *Nuclear Science and Engineering.* Chen's book should be consulted for the full list of reviewing media for each subject. The distribution of book reviews obeys Bradford's law of scattering: by scanning a relatively few core journals, the selector should be able to see the large majority of book reviews in any subject.

Library science has quite a few publications which include short reviews designed to aid in the selection process. These selection tools include *Choice, Publisher's Weekly,* and *Library Journal* (which has an annual "Best Sci-Tech Books of the Year" feature), all highly rated selection tools for general libraries. There are also some specialized selection aids aimed particularly at science materials, such as *AAAS Science Books and Films.* Unfortunately, none of these sources are very useful at the *research* level in science. They are primarily geared to general libraries, and do not cover the very specialized materials which are usually the ones sci/tech libraries are collecting. For sci/tech reference materials, *American Reference Books Annual* and *Scientific & Technical Libraries* should be scanned. In addition, any new guides to the literature which are published in the subjects of interest should be checked for lists of reference books which might be needed.

In 1970, Sadow wished for a scientific book review index which was comprehensive, prompt, and included critical reviews by qualified reviewers (Sadow, 1970, p. 196). Unfortunately, the need still exists.

In searching for a review of a particular book, Gale Research's *Book Review Index* and Wilson's *Book Review Digest* can be useful, although their coverage of science is less than 10% of their total coverage, according to Chen (1978, p. 362).

Technical Book Review Index (TBRI), begun in 1935 by the Special Libraries Association and currently published by JAAD Publishing Company, quotes excerpts from reviews appearing in scientific, technical, medical, and trade journals. About 3000 reviews are excerpted annually. Unfortunately, the reviews appear in the scientific journals slowly, and it is even longer before they are excerpted in *TBRI.* Also, *TBRI* at its best only covered about one-fourth of the scientific books which were published annually.

Two other helpful compilations are *New Technical Books,* published monthly by the Research Libraries of the New York Public Library, and the *Aslib Book List,* which is also a monthly list of new books in English in science and engineering. Both lists include annotations which are mostly descriptive, but often contain recommendations of reader or library suitability (Chen, 1978, p. 363).

Other sources of information useful for selection besides reviews include online databases, accessions lists from other similar libraries, publishers' flyers ("blurbs"), book dealers, and the users.

The abstracting and indexing services such as *Science Citation Index, Engineering Index, Applied Science and Technology Index, Mathematical Reviews, Applied Mechanics Reviews, Bioresearch Index,* and *Chemical Abstracts* include short but critical book reviews which are not frequently used for selection by librarians (Chen, 1978, p. 362). Selectors could store SDI profiles on appropriate databases and be alerted to new materials in their area of interest. For monographs, *Books in Print* and the LC MARC files are possibilities.

Because of the need for timeliness, sci/tech librarians often select from publishers' blurbs rather than waiting for reviews. Many contain detailed descriptions of the contents, biographical information on the author, and other information of real value in making a selection decision.

> They have the advantage of being timely, are usually easy to handle, and they can be discarded easily when no longer useful. They are widely used by academic, special and large public libraries, because when honestly written they provide better information than any other source, with the exception of reviews. (Hamlin, 1980, pp. 198–199)

Sci/tech libraries should get on mailing lists for flyers from Academic, Elsevier-North Holland, McGraw-Hill, Springer Verlag, and Wiley-Interscience at least.

Many dealers and jobbers provide notification of available titles in the form of cards, multiple-order forms, or lists. They usually only have the basic bibliographic information.

Because of the specialized nature of sci/tech materials, the users are one of the best sources of information. They can have various roles in selection. They may provide most of the recommendations for purchase, and thus be an information source, or they may be involved primarily in the judgment phase. This is discussed further under management aspects. Although they may make recommendations for the library to purchase materials, few faculty these days have time to consider in a broad way what the library should be buying. Some users buy for themselves what is truly important to them, and only recommend to the library those items which they want to have available, but do not want to buy themselves. The knowledge of what they are ordering for themselves is an excellent aid for collection development. If at all possible, the library should have some access to these personal orders. In many special libraries this is accomplished by the library being responsible for ordering all books for the corporation. This is reasonable, since the library knows the sources and can do the job efficiently. It also has the advantage that orders for personal copies of books may be reviewed by the library's selectors with an eye to whether they should also be purchased for the library collection.

Criteria for Evaluating Monographs

The second step in selection, after information-gathering, is judging whether or not to order a particular monograph. This step is also somewhat different in a sci/tech library than it is in general libraries, where many collection development policies recommend waiting for reviews. In sci/tech libraries the need for timeliness means that often the decision to buy must be made before a review appears.

Monographs may be selected individually more often than other types of materials. In selecting individual monographs, the following list of criteria is appropriate:

1. Is it on a subject important to the users?
2. Is the level of treatment appropriate?
3. What is the reputation of the publisher, author, editor in the field?
4. What is in the collection already? Does it duplicate materials already there?
5. What is the price? Is it appropriate for the budget for that subject or type of material?

Although the problem does not come up too often in research libraries, some question can arise about the selection of pseudoscience materials. Normally, consideration of the publisher, author, and editor will keep out unverified materials. Pollet (1982) discusses this aspect of selection further in her article "Criteria for Science Book Selection in Academic Libraries," reassuring us that excluding pseudoscience materials is not censorship.

Mechanics of Selection

Selection is a very repetitive process. Publishers' blurbs and catalogs (sometimes in duplicate and triplicate) arrive and the librarian does not know by looking at them whether the library already has something, or has already ordered it, or should receive it on standing order or an approval plan.

The selector needs a systematic way of dealing with all the paper. Probably the most efficient method is for the selector to mark what should be in the library and a clerical staff person to check the catalog to see if it is there already. If not, the item is ordered. However, some selectors prefer to do the initial checking themselves in order to stay familiar with what is already in the collection. They may at the same time look at other similar items in the collection before making the final decision whether to buy or not. Online catalogs facilitate this kind of checking.

Mass Selection Programs

Selecting monographs individually can be an inefficient way to select. In order to streamline the selection process, libraries sometimes set up standing orders, approval plans, or blanket orders.

When a series is identified that the library wants, a standing order can be placed with the publisher or the dealer. As new volumes in the series are published, they are sent to the library automatically. Standing orders are particularly helpful for new editions of reference tools and publishers' series that the library is sure to want and needs as soon as they are published.

Approval plans and blanket orders are less frequently used than standing orders in sci/tech libraries. An approval plan can be set up with either a publisher (for just that publisher's books) or with a jobber (for books from a variety of publishers). A profile of the library's interests in terms of subjects, geographic areas, publishers, languages, prices, formats, and intensity of collecting is turned in to the publisher or dealer, and they then send to the library a copy of each new book which matches the profile. This profile is critical to the success of the plan (Stueart, 1980, p. 209).

Since approval plans allow the library to return anything which does not match the profile, the decision to keep or return can be made with the book in hand, a great advantage in selection. Stueart reports that a miss rate of greater than 10% (returns or additional orders) is generally considered unacceptable (Stueart, 1980, p. 210).

The problem with approval plans from the point of view of selection is that one is never sure what is going to come on an approval plan. The publisher may not think a book which you would have ordered is in your subject area. The jobber may not include all the publishers you expect. The blurbs announcing new books come out months ahead of publication, and selectors must keep some kind of file to be checked against what arrives on approval to be sure that everything that would have been ordered has come. Many libraries who have tried approval plans have gone back to just placing the orders, since they were keeping track of them anyway.

The problem is particularly acute in regard to series standing orders (reprint series, publishers' monograph series, society publication series or annuals, irregularly published handbooks, and directories), which must be clearly defined in an approval plan. Will the library cancel its standing orders and have them be part of the approval plan, or will it exclude all series (except perhaps the first volume of a new series) from the approval plan (Stueart, 1980, p. 211)?

The advantages of approval plans include broad coverage, timeliness,

selection with the book in hand, extra time to spend on fugitive material
if the routine things come in without title-by-title selection, and person-
alized service from the dealer (acquisitions can be streamlined by using
the dealer's computer facilities for title selection, accounting, and keeping
the library informed). The only disadvantage is that it may be more trouble
than it is worth, especially in a specialized library.

Blanket orders are orders for everything that a publisher puts out. A
blanket order does not generally include return privileges. Blanket orders
can be very useful for societies or other publishers whose publications
are always right in the subject area of interest. For example, an academic
chemistry library might have a blanket order for everything the American
Chemical Society publishes. Or a library catering to the nuclear industry
might have a blanket order for everything that the American Nuclear Society produces.

Not all publishers accept standing orders or blanket orders or provide
approval plans. (You would think they would jump at the chance to have
guaranteed buyers, but they don't always). And all these programs have
to be monitored to see that the library is getting what it intended to receive.
They also complicate the budgeting process, because one cannot predict
accurately how much will come in on any of these kind of orders in one
period of time. When money is tight, they reduce flexibility. However,
they are alternatives to individual monograph selection and should be
carefully considered in each setting.

Retrospective Selection

Most of the above discussion was focused on selection of current monographs in a particular subject area. However, there are several circumstances in which a selector will need to consider older materials for addition
to the collection: the library may be moving into a new subject area, the
selection may not have been done well previously (for whatever reason),
or a new library may be developing from scratch.

In these cases, one must find out what the core literature in the subject
area is and acquire it. One of the best starting places is with the guides
to the literature, if one or more exists for the subject of interest. One can
also choose current monographs or reference tools in the subject and consult their bibliographies. The bibliographies at the end of articles of interest
in sci/tech encyclopedias like the *McGraw-Hill Encyclopedia of Science
and Technology* or Kirk and Othmer's *Encyclopedia of Chemical Technology* can be starting places. Experts in the subject will know the classic
works in the field, and if the selector is not an expert, he or she should
consider bringing one in. It can be useful to visit another library which

collects in the subject and browse their catalog and shelves. There may be a list of "best books" in particular subjects compiled by the professional societies. Many of these methods are similar to those used for collection evaluation (discussed later).

Serials

As we noted in Chapter 3, serials are a very important part of the sci/tech library collection. Selection of serials is different from selection of monographs in several ways. First, when a serial is selected, one is committing to a continuous expenditure. Although a serial can be cancelled, it is a more difficult decision once there is a backfile established. For a serial that exists already, the selector must consider the backfile as well as the current issues. Also, the subject coverage may not be as precisely defined as in a single monograph: if a journal has only one article of interest in each issue, is it worth subscribing? When should a journal be duplicated? How do you decide which branch should have it if there are several? Does the *Journal of Chemical Physics* belong in Chemistry or Physics?

In the 1970s much attention was given to evaluation of journal collections as inflation hit serials and libraries were forced to consider their subscriptions carefully. These studies list the following criteria for selecting a journal:

1. What is its subject? Does it fit within the scope of the collection?

2. Where is it abstracted and indexed? If it is abstracted or indexed in the prime index or database for the subject area, it is likely to be cited and asked for. This is often not known for a brand new journal.

3. Who is the publisher, editor, editorial board? Are they known and reputable?

4. How often and where is it cited? For established journals, ISI's *Journal Citation Reports* can be useful.

5. What is the price versus apparent quality and quantity?

6. Is it available elsewhere? Does some other branch in the library system or some other nearby library have it?

7. What is the past use? In evaluating existing subscriptions for weeding, *use* is a prime consideration, but normally with a new journal, there is no evidence of use. In questionable cases, it is useful to enter a trial subscription and keep track of use and collect comments from users for a year.

8. Is the format appropriate? The availability of microfiche or microfilm backfiles may be a consideration in libraries where space is a problem.

Models for serials selection are considered in Volume 2, Chapter 5.

Other Special Formats

Not much has been written about selection of technical reports. Like monographs, they can be selected individually or en masse. When individual reports are selected, it is usually because they appear in a bibliography, a literature search, or an SDI profile. When the report's subject matches the interests of the library, it is ordered for the collection. The great bulk of reports, however, are received in most libraries more systematically. If a particular agency is doing research that is relevant, the most timely way to get the reports is to get on the initial distribution lists. This way they will come direct from the agency. The government agencies sponsoring large numbers of technical reports also maintain distribution lists and have various methods for distributing reports. The selection of technical reports is considered further in Volume 2, Chapter 12, and the selection of other special formats is considered in the other chapters in Volume 2, Part I.

ACQUISITIONS

Acquisitions is the process of actually acquiring the materials. It differs from the document delivery function in that the materials being ordered are to become a permanent part of the library collection (or, in the case of materials ordered for the staff, of the organization's permanent collection). However, both functions search the sources outside the library in response to a request for a particular item. The new document delivery sources which are appearing, particularly online, have become an additional source for acquisitions, and with their increasing use the line between these two functions is blurring. It is particularly blurred when the library orders copies of items for the end user—what is the difference between that and delivering a photocopy of an article which the library obtained for him or her through ILL?

Acquisitions encompasses checking to see if the item requested is already in the collection or on order, finding the necessary ordering information if it is not on the request, choosing the source, placing the order, following up on the order, receiving the material when it arrives, paying the bill (or at least seeing to it that it is paid), and accounting for funds (that is, producing accounts of amounts expended by fund).

In general, the *process* of acquisitions in sci/tech libraries is similar to that in general libraries. Although we will summarize important points about acquisitions, more detail can be found in Ford (1973) or Grieder (1978). We will concentrate on an overview of how the process relates to the other functions and how it is different in sci/tech libraries. The primary differences include the need for timeliness and the existence of special sources which deal especially in sci/tech materials, and the need to acquire special formats such as technical reports.

The need for timeliness means that the sci/tech library should be able to order materials itself and not go through a purchasing department or a centralized ordering department. In a special library setting, often the case can be made that the library is more familiar with sources than are people who deal primarily in other commodities. If there is a centralized purchasing operation, the library should try to have one person designated to work with library materials, and should foster a good relationship with the person. However, the library should at the very least have the ability to prepay, to go direct for rush items, and to set up deposit accounts. Many government publishers and small societies require prepayment, and it can hold up an order interminably if the library cannot write checks. The need for timeliness also means that all steps of the acquisitions process should constantly be evaluated in terms of their efficiency. Price can be less important than speed, especially in special libraries.

Many sci/tech libraries use a jobber for books from trade publishers and go direct for publications from professional societies and other non-trade sources. It is always a disadvantage to have to put *all* orders through a jobber. However, using a jobber for routine orders cuts down on the number of invoices and the number of checks to write. This can be of great advantage in some settings. In some cases, using a jobber also speeds up service: the publisher is likely to give a jobber better service because of the volume of business. In other cases, it will slow things down since jobbers sometimes wait to place orders until they have collected a large number of them in order to get a better discount from the publisher.

Choosing a jobber is an individual matter. Having a local jobber results in advantages in communication. If a good local one is not available, national jobbers are listed in the *Bowker Annual* and usually have booths at the national library conferences.

The American Library Association has published *Guidelines for Handling Library Orders for In-Print Monographic Publications* (1984). This has a good discussion of the expectations library dealers and libraries should have of each other. It does not, however, deal with online ordering. Progress is being made by the BISAC committee of the American Association of Publishers Books Industry Study Group toward an order format that will allow online orders to be transmitted (BISAC and SISAC: Computer-to-computer ordering, 1984). A number of book dealers have accepted orders for their own materials in machine-readable form since about 1974. The Bowker Book Acquisition System (BAS), available on

BRS, is the first online ordering system which will accept orders for materials from any vendor. BAS sends orders from Bowker to the different vendors via telecommunications, magnetic tape, or print for vendors not equipped for electronic order receipt.

Boss (Boss and Marcum, 1981; Boss and McQueen, 1982) and Rush (1984) cover in detail what to look for in an automated acquisitions system. Libraries considering automated acquisitions have to address a number of issues, including the relationship of automated acquisitions to other automated library functions, cost, reliability, how easy the system is to adopt and how easy to cancel, and priorities among system features. The most frequently requested features are:

Database access. This could be an in-house database, the database of a bibliographic utility, the database of a book jobber, or all three.

Name/address file. Publishers' addresses should be easier to maintain online than manually.

Purchase order writing.

Online ordering (the ability to transmit orders online to the desired vendor). Online ordering to a single wholesaler has been available for several years, and there is a book industry standards committee (BISAC) working on a format for sending electronic ordering messages.

In-process file. Information about in-process items should be searchable in real time.

Claiming. Automatic claims should be sent after a time set by the library and variable for each vendor.

Receiving/paying. In an automated system, when material is checked in the accounts should be automatically adjusted. If the price has changed, the price field should be automatically adjusted. The receipt function should also allow for all parts of the receiving process, including returning materials, routing items, and authorizing payment of invoices.

Funds accounting. Reports on the status of funds should be produced periodically, showing encumbered and expended totals for each fund.

Management information. Types of management information that can be provided by an automated acquisitions system include summary reports on expenditures, funds balances, vendor performance reports, and statistical reports on the numbers of items and titles acquired.

Vendor monitoring. A vendor file should be available, which can record claim cycles, performance statistics, discount information, and vendor statements (Boss and McQueen, 1982, pp. 21–27).

Most of the challenges in sci/tech library acquisitions are in the area of acquiring the special format materials (technical reports, patents, standards, maps, etc.) These are discussed in Volume 2, Part I.

COLLECTION MANAGEMENT

After materials have been selected and acquired and have become part of the library's collection, there are several related aspects of collection development which come into play. The aspects of weeding, storage, preservation, and evaluation have become much discussed in the past decade as libraries (especially academic libraries) have become concerned with space and more resigned to their inability to buy and store everything. However, they are not new topics. In 1899, Harvard's President wrote:

> One who watches the rapid accumulation of books in any large library must long for some means of dividing the books that are used from those that are not used, and for a more compact mode than the iron stacks supply of storing the books that are not used. . . The devising of these desirable means of discrimination and of compact storage seems to be the next problem before librarians. (Brough, 1953, p. 124)

In the almost 100 years since this speech, use has been found to be the best "means of discrimination", but it is rarely applied without additional criteria.

The role of academic and government sci/tech libraries in collection management is different from that of special libraries. Special libraries have always weeded heavily. They are usually limited in space and require quick accessibility (and smaller collections are more accessible). They also have been able to rely on the academic and government sci/tech libraries to provide the 10–25% of requests for materials not in the collection. Academic research libraries have had size of collection as one requirement in standards, and accreditation and their reputations as "great" libraries seem to be related to the number of obscure items that only they have. Also, they have a different role: if the academic and government research libraries do not preserve the little-used materials, who will?

Whether the purpose is to select materials for weeding or for storage, the first step is to determine the core collection based on last circulation date, as we described earlier. The core collection should be considered for shortened loan periods, duplication of heavily used titles, and preservation. The remaining items should be considered for weeding or storage.

Weeding

Because of the factors just mentioned, academic research libraries rarely weed based on past use alone. Mosher recommends that they always involve subject experts in the review, either during the item-by-item review or as a final step (Mosher, 1980a, pp. 177–178). However, this review is probably more important for political reasons than to improve the quality of the weeding. Opello and Murdock, for example, reported that in a weeding study of the NOAA library in Boulder, there was "almost total agreement between scientists' opinions and the objective criteria" (1976, p. 453).

Typical criteria for weeding include:

1. Use.
2. Suitability of the subject.
3. Quality of the author, publisher, or sponsor.
4. Language.
5. Format.
6. Level or quality of treatment.
7. Availability of multiple copies.
8. Availability elsewhere (through resource sharing locally or through a national center like the Center for Research Libraries).
9. Physical condition.
10. Currency (noncurrent textbooks and old editions replaced by current revisions are candidates for weeding).
11. Number of better books on the subject on the shelves.

However, even something that seems a simple decision for a science library, like discarding older editions of standard works in favor of the newer versions, cannot be considered "safe" if the library has any responsibility toward history of science. If someone is tracing the development of a scientific concept, a set of older editions of textbooks can be treasures for determining when the concept appeared in the standard works and how it changed over time. Those libraries who are trying to protect against losing historically interesting material might also wish to apply the criteria developed by Columbia University in the 1960s discussed by Cooper (1968a, p. 349). The criteria included preserving books by Columbia University authors, European publications with dates before 1800, American publications with dates before 1900, and consecutive editions of more than three when the library had a copy of each.

Books which appear to be candidates for discard can be removed to a separate area for subject expert review before the final discard, or they can have a book slip attached to them soliciting feedback from any users and be left on the shelf for some period of time (Mosher suggests a minimum of one year) (Mosher, 1980a, pp. 177–178)

Storage

Decisions for storing items should clearly be made on the basis of use, although other methods have been proposed and used. For example, some libraries make mass decisions (e.g., move all of QD 255 to storage), and others store by age of the materials. However, the Yale study found it was impossible to isolate entire groups by either subject or form (Ash, 1963), and the persistence of older books which continue to be well used makes the age criteria suspect. However, in a library which has reclassified the most-used part of the collection into LC, what remains in Dewey may be the less-used portion and thus a good mass candidate for storage.

Sometimes the choice is made to store all of a classification or everything before a certain date because of cost. Determining circulation data can be costly [Stueart (1985) estimates two and a half times more costly than weeding by publication date], and there is a high cost involved in changing catalog records. Libraries sometimes leave the records as is and just tell users "All the QDs are in storage" or "Everything before 1960 is in storage." The records must be changed if individual items are selected. In a cost analysis which does not take into account the cost to the user of a delay in service for frequently used materials, age can come out the least costly alternative. But, as Andrews (1968) shows, when a reasonable cost to the user of a delay is incorporated, circulation data prove to be the better choice. As more and more catalogs become online catalogs, the cost of changing location records should approximate the cost of a loan transaction (Gore, 1976, p. 179) and storing less-used materials individually rather than by broad category will become less costly.

Stayner (1983) developed a model which provides storage and retrieval costs for different kinds of storage, including compact shelving, microform and remote storage.

Serials are often considered for storage because they occupy so much space and can be moved as a block to storage, changing only one record for a large number of volumes. Journals should be considered as sets, and individual volumes should not be separated from their sets.

Materials for storage can be identified just like materials for weeding— by an item-by-item review with various criteria in mind, using the book-slip method or review by subject experts as a final input, if the setting requires it. Mosher finds that the book-slip method is used more often for storage than weeding, because the intellectual decision that the item is a candidate for storage need not be followed right away by the clerical procedure of changing the records (Mosher, 1980a, pp. 177–178).

Storage decisions are related to larger issues of policy and procedure. How will the materials be stored (compact shelving, by size, by regular classification number)? How will the records be changed to show the item is in storage? How will the materials be transferred, and once transferred, how delivered on demand? Will the materials still be "owned" by the contributing library or be transferred to the storage collection? Regional storage centers will have to decide whether to store only one copy, and if so, which library's will be discarded. If ownership does not transfer, anything that circulates from storage should be returned to the main collection, although in practice, some discretion is usually used. For example, if one volume of a journal circulates, returning it to the main collection would separate it from the set. And if a particular item was recalled from storage for an outside user, perhaps there is no need to return it to the main collection.

Weeding, storage, and evaluation are closely interrelated. Indeed, many of the studies which say "evaluation" in their titles do not really evaluate the collection in terms of how good it is; they are really studies to determine which books or journals to weed or store.

Evaluation of the Collection

> How good are our library collections? This perfectly reasonable question has proved extraordinarily difficult to answer with any meaning. The literature of librarianship does not even agree on the meaning of the term good with regard to collections, to say nothing of providing us with ready techniques for analyzing or evaluating their quality. (Mosher, 1980b, p. 527.)

Nevertheless, we must try to evaluate our collections, even if we cannot precisely define "good," in order to try to determine whether the collection is meeting its objectives, how well it is serving its users, in what ways or areas it is deficient, and what remains to be done to develop it (Mosher, 1980b, pp. 529–530).

There are various methods of evaluating collections, none of which is perfect, and none of which has approached true evaluation. The methods include subjective evaluations by subject specialists; checking against a list, either one prepared especially for purposes of the evaluation or a preexisting list such as the holdings of another library presumed to be especially strong in a particular subject; evaluating the collection in terms of its use; numeric counts; formulas; interlibrary-loan analysis; and an analysis of machine-readable cataloging data. Herbert White suggests that in the special library setting, a collection which meets 50–75% of the needs will appear reasonable to corporate management (1984b, p. 49).

Subject Specialists

The use of subject specialists is the least quantitative method, but in some collections the most useful. Someone familiar with the classic works and basic tools in a subject can easily browse a collection and the catalog and give an opinion on the quality of the collection. But the opinion is quite subjective, the choice of expert is critical, and it is difficult to use this kind of evaluation to show improvement.

Although the examination of the collection by a subject expert is not very quantitative, it can be useful in pointing out inadequacies, which can be followed up by more quantitative studies, as can a simple examination of the collection with the collection development policy in hand.

Standard Lists

The study by Burns (1968) is an example of the use of standard lists for collection evaluation. In evaluating science and technology collections at the University of Idaho library, a large number of kinds of lists were checked, including standard checklists, guides to the literature, holdings of libraries known for excellence, lists of journals covered in indexing and abstracting services, checklists of recommended books and journals published by professional societies, references cited in terminal bibliographies, and lists of professional society publications.

Checking the collection against lists and bibliographies has the same kind of problem as selecting a subject expert: the choice of list is essential and often difficult. Also, it is difficult to use the technique to show improvement. The evaluator could choose a number of current research monographs in the subject, check their bibliographies against the collection, and report on percentage available. However, the obvious next step is to purchase the missing items, assuming they are judged relevant. Then what tool do you use to evaluate the collection next time? The problem is that any list you can devise for collection evaluation in a research collection could also be used as a collection development list, which negates its value for collection evaluation the next time.

Evaluating the Collection in Terms of Its Use

Use data has the most potential to show change quantitatively. We have already discussed using circulation information to identify little-used portions of the collection and to identify a core collection. Availability studies are another form of use study which has the potential to provide a quantitative measure of the collection's availability. There are two types of availability tests: document delivery tests, and shelf availability studies. The document delivery test uses a list of citations compiled to simulate the needs of the library community. The shelf availability study uses the actual needs of users.

The document delivery test has been used by Orr *et al.* (1968) in medical libraries and by DeProspo *et al.* (1973) in public libraries. Orr starts with a citation pool of 300 citations (drawn at random from about 2000 citations), checks each citation in one day against the library catalogs and shelves, and records the outcome of each search. Each outcome is then assigned a speed code ranging from 10 (less than 10 minutes) to 10,000 (more than a week). From these scores a mean speed code is derived, and from this a capability index (CI) for the library is produced. Although the capability index would be most useful in comparing libraries, it could also be used to compare the same library at different times, and thus to show change.

DeProspo *et al.* express the results of their test as three probabilities: Pr(O), the probability of ownership; Pr(B), the probability of availability of books owned; and Pr(A), the probability of availability. Pr(A) = Pr(O) × Pr(B). These correspond to Gore's "holdings rate" (the percentage of books wanted by a patron and recorded in the catalog), "availability rate" (the percentage of wanted books which are available on the shelves), and "performance rate" (the percentage of all books a patron wants which are available to him on the shelves). As with DeProspo's probabilities, the performance rate is the holdings rate times the availability rate (Gore, 1976).

From the patron's standpoint, it is the performance rate [or the probability of availability, Pr(A)] which matters, since it represents the probability that the patron will find what he or she came to the library for. To improve this ratio we can increase either of the two factors, but if the holdings rate is already high, increasing it will not have much effect. For example, if a library has a 90% holdings rate and a 50% availability rate, it has a 45% performance rate. Improving the holdings even to 100% only improves the performance rate to 50%. But purchasing extra copies of needed materials, moving at least 20% of the unused collection to storage (or weeding it), and shortening loan periods to bring the availability even to 75% will bring the performance rate to 67% and thus have more effect than purchasing large numbers of new items.

However, if document availability studies show a low percentage of holdings (and no one has really defined low), more effect might come from improving the collection by studying what the users want.

The main problem in conducting a document delivery test is the creation of the list which simulates the user's needs. Line (1973) used a random sample of users to create a list. Each user was given a stack of cards and asked to fill one out every time he encountered a reference to a document he needed in his work. The users also indicated what steps they took to obtain the document (including use of the library if this source is selected) and whether or not the search was successful. All the cards can be considered a list representing the latent needs of the library's community, which is thus an excellent list to be checked against the holdings of the library or for use in an availability test.

A list could also be created from samples of journal articles, bibliographies in important monographs, samples of index and abstract entries, etc. But the value of the test is dependent on the quality of the list.

Shelf-availability studies do not require a list. They have been used by Buckland *et al.* (1970), Urquhart and Schofield (1971, 1972), and Kantor (1976a). In this kind of study, users are asked to report their failures to find what they came to the library for, either on failure slips or by way of interviews. In both cases, all failures are followed up immediately to determine whether they are collection failures, catalog-use failures, failure to find an item actually present on the shelves, or failures due to an absence of items from the shelves (in use in the library, in circulation, missing, waiting to be reshelved, at bindery, etc.). From this kind of study we can establish the probability that a user looking for a particular item will find that item at the time it is needed as well as identify the major causes of failure in the known item search situation (Lancaster, 1977b, pp. 37–38).

Of the two types of studies, the document delivery tests are more useful for comparison of libraries. They also allow for "latent" needs better than a study based on failure analysis of present users. Unfortunately, the creation of appropriate lists remains a problem. It would be useful if someone would make up a list of citations that could be expected to be asked for from a sci/tech library. That list could then be self-administered, like the DeProspo test is in public libraries, and sci/tech libraries could compare themselves to other sci/tech libraries, both in terms of collection and document availability. The sample should cover books, periodical articles, reports, standards, technical reports, etc.: everything that users might ask for.

Other Methods of Evaluation

Counts of the number of volumes a library has are normally collected for statistical purposes and are sometimes included in library standards. The simple count is useful for comparisons with other libraries and as a rudimentary check on adequacy. In addition, the number of volumes a library has in various subject classifications can be counted by measuring the shelf list (at about 100 volumes per inch), by measuring the shelves (at about 10 volumes per foot), or from online cataloging or circulation records.

These studies can point out areas which need more or less collection development effort, if the data is interpreted properly.

Clapp and Jordan (1965) and others have devised formulas for determining the minimal adequacy of an academic library's collection to provide support for user needs. Most formulas rely on the assumption that there is a relationship between collection size and adequacy. They attempt to relate collection size to factors such as the numbers of faculty and students in various categories and the numbers of undergraduate or graduate subjects or degree programs. These are useful for establishing basic target collection sizes for libraries in new colleges and universities.

Interlibrary-loan analysis has also been used to study collection adequacy. ILL requests can be counted by subject or program, publication date, language, and format (document, book, serial, etc.) to

> . . .identify areas of collection weakness or unmet user need, whether there is unmet demand for current or out-of-print materials, whether there is need for attention to publications of a particular region, and if the library is giving inadequate attention to a particular format. Like any data, the results are susceptible to varying interpretation; certainly the library's missions and goals and the collection development policy statement should be used in interpreting the findings. (Mosher, 1980b, pp. 536–537)

An example of the use of ILL statistics used to study collection development in several medical libraries was recorded by Byrd *et al.* (1982).

The data available from machine-readable cataloging can also be useful. It is possible to perform collection analyses on library catalog data contained on OCLC archival tapes. Many other studies have been done with analysis by LC classification: for instance, a comparison of the percentage of new acquisitions by LC class compared with a percentage of the entire collection.

Probably the best collection evaluation comes from a combination of several approaches such as a subjective analysis followed with a more quantitative study of particular aspects of the collection. A yearly document availability study is useful, if only to show change from year to year. Examples of studies using a combined approach are reported in the articles by Campbell (1975), who analyzed the science periodicals at Wolverhampton Polytechnic Library using the factors of internal use within the library, of interlibrary loan, and of citation (within local publications), and by Wenger and Childress (1977), who did a journal evaluation study of two NOAA libraries, using data collected from a use study, circulation and interlibrary loan statistics, a core list (including subscription costs), questionnaire returns, and library and patron input.

Summaries of collection evaluation problems and methods include Chapter 5 in *The Measurement and Evaluation of Library Services* (Lancaster, 1977), and the articles by Bonn (1974) and Mosher (1979).

PRESERVATION

The item-by-item review of the collection can also be used to identify candidates for preservation. Science librarians have shown less concern with preservation than their colleagues in the humanities and social sciences, because of their concentration on current materials. However, many scientists value first editions of the landmark works in their fields as much as humanists do those in their fields, and will expect the library to preserve them as physical entities, not just their content. And they will expect their librarians to provide advice about books, paper, binding, etc. Also, every library has a concern with simple repairs, binding, proper environmental controls, and disaster planning, which are considered part of the topic of preservation.

Acidity, heat, light, dust, moisture, air pollutants (especially sulfur dioxide and nitrogen oxides), salts, alkalis, and biological agents (including fungi, bacteria, insects, and rodents) are the major enemies of physical books (Wessel, 1972; Paper and its preservation, 1983). To defeat these enemies, controlling the environment of the books is critical. Books should be kept as cool as possible (preferably 60–65°F) and at constant relative humidity of about 50% plus or minus 5%. Airborne pollutants should be removed. Books should be protected from light, especially the shorter wavelengths (e.g., ultraviolet). They should be stored upright on the shelf (not on the fore edge), with bookends (or shelf fullness) adjusted to hold the books straight (but not too tightly). They should be carefully handled, especially during photocopying. Boxes and encapsulation can be used to protect valuable items (Banks, 1972, p. 191).

If these prevention techniques fail or are applied too late, however, what to do next depends on whether the artifact as well as the information needs to be preserved. If the artifact needs to be preserved, probably it will have to be turned over to a conservator. A trained conservator (and there are less than 100 certified paper conservators in the entire country) can do wonders on an individual item, but the work is very labor-intensive, and thus expensive (e.g., several hundred dollars to deacidify one book). The main approaches are to clean and deacidify the paper chemically. A test at the Library of Congress of the DEZ (diethyl zinc) deacidification process has the potential for getting deacidification down to about $5 per volume, but currently no inexpensive mass deacidification process is available.

When the information content of materials needs to be preserved, but the physical item has no intrinsic value, microfilming or xerography is an alternative (see Darling, 1974).

More often than the need for massive deacidification programs, the science librarian will need everyday repairs. What is the best way to mend a ripped page? Mount a map? Repair a binding? Banks recommends Japanese paper for repairing tears and encapsulation (rather than lamination) for maps. Rebinding, though difficult from the conservator's point of view because it is irreversible (the binders cut the edges of the paper), is a logical solution to many problems where the artifact is not overly valuable.

The literature of conservation is reviewed by Lowell in his 1982 article "Sources of conservation information for the librarian." Other useful sources, which should be in every librarian's collection, are *Conservation of Library Materials* (Cunha and Cunha, 1971), *Preserving Library Materials* (Swartzburg, 1980), *Library Binding Manual* (Tauber, 1972), and *Cleaning and Preserving Bindings and Related Materials* (Horton, 1967).

MANAGEMENT ASPECTS

Organization

The organization of collection development in any setting can be evaluated by how well it provides for communication and gets the needed information to the responsible person. The selector must have information about user needs (for all the users and potential users, not just a select few), about funds allocated and spent, about available materials, and about how to evaluate materials against the library's policies. The person who allocates budgets must have information about costs and forecasted costs for the various areas (whether they are split by subject or format), about user needs, and about priorities for the organization and the library. Any good collection development organization must provide for communication among selectors in different subject areas or formats, between the selectors and users, and between the selectors and those who set the budgets.

Perhaps the most usual arrangement in large university libraries is to have a collection development officer who budgets and coordinates, various selectors for subject areas and/or formats (e.g., serials and microforms are often broken out for budgetary purposes), and faculty committees to advise on the various subjects (Hamlin, 1980; Dudley, 1980). Faculty may propose items for selection, may help evaluate, or may make the final selection, although the trend is toward an advice mode rather than a controlling one. In 1978, Baatz found that in 75% of science branches in the ARL libraries he surveyed selection was done by librarians. However, up until the 1960s, most selection in American university libraries was done by faculty.

Staffing

Staffing varies by the setting and the goals of the library, but it is clear that for collection development, the more knowledge the selector has of the subject area and of the publishers, authors, and bibliographic apparatus of the subject, the better.

Bogardus (1960) described the minimum qualifications of a subject literature specialist as:

> . . .a knowledge of the authors and publishers who are doing the best work and writing in a given field, the associations and organizations which are fostering quality work in the area, the present interest and directions or research within the field, a recognition of the "classics" in the subject literature and at least a speaking acquaintance with the terminology and subject content. Naturally, this implies, above all, that the librarian knows the bibliographic tools available.

Large libraries may hire a true subject specialist or bibliographer, but this is more common in the humanities and social sciences than it is in the sciences.

In smaller libraries, collection development is rarely someone's full-time job, but usually part of the assignment for some professional, either the head librarian, a reference librarian, or the head of acquisitions. As Dudley says, in describing branch libraries in universities,

> . . .most of them are so small that all the selection done by the library staff . . . is done by the head librarian, who is also the only librarian. In the larger libraries where a differentiation of function is possible and indeed necessary, collection development may still be the responsibility of the head librarian, or it may be assigned to the head of acquisitions, the head of serials, the head of reference, or any other staff member as an additional assignment. Rarely, if ever, does a branch library have a full-time selector. (Dudley, 1980, p. 26).

Budgeting

Levels of budgeting vary from the allocation of collection development funds among campuses (D. Smith, 1984), to division of funds in a large university among formats and/or subject areas, to allocation within a particular subject area for formats or smaller subject areas, to the selection and purchase of a particular item.

Every sci/tech library will need to forecast what will be needed in the next budget cycle, either for all of science and technology or for a specific subject area like chemistry. It is usually useful to consider serials, books, maps, reports, and any other categories of importance separately.

The following sources of information are useful:

1. Past data, including the number of items purchased for the past few years and the amount spent in various categories (e.g., serials, monographs, reports, standing orders, etc.). Serials, because they are a continuing expense, are often budgeted (at least in large libraries) by taking what was spent last year and adding a figure for inflation. If there is not

that much money available, individual titles have to be cancelled. It is very short-sighted to cancel serials impulsively, however, because it is costly to get them later. Monographs can also be forecast by getting a figure for the number of books purchased in the preceding year, figuring out what the inflation factor is for books in your subject area, and budgeting for that amount. This standard technique is only useful in fairly stable environments, however.

2. Notes of any unusual purchases necessary for the upcoming period. These might include the 5-year cumulation of *Chemical Abstracts,* new subject areas that need retrospective development, or large microform purchases in order to gain space.

3. Forecasts of book output and of average costs of serials, books, etc. for the upcoming budget period. Sources for these include *Publishers Weekly,* the *Bowker Annual,* and an annual article in *Library Journal* on serial prices. Dealers can also provide this kind of information. Unfortunately, the standard sources do not always break down subject areas into the detail we would like. Further information about useful sources can be found in the articles by Clack and Williams (1983), S. Williams (1984), and Lynden (1977).

Often the forecast is submitted to a collection development officer or other high-level manager to integrate into a larger budget. There are various ways of allocating funds among units, and unless the methods take into account the high cost of sci/tech materials, inequities can develop. If a sci/tech library is chronically underfunded, it is useful to examine the premises upon which the money is allocated. It may be helpful to gather data on the number of students (undergraduate and graduate), faculty, or other categories of the user community and to calculate the volumes purchased per user, comparing the ratio for the sci/tech areas versus other subject areas. It may also be useful to investigate other models for allocation, and if so the articles by Sweetman and Wiedemann (1980), Schad (1978), Sampson (1978), M. Martin (1980), and T. Pierce (1978) are recommended (Sanders, 1983).

Within the sci/tech library it is not unreasonable to have the proportion of funds going to serials as high as 90% (Koenig, 1984). Mount (1985) found in the 16 sci/tech libraries he studied that the average breakdown of collection funds was 24% to books, 67% to periodicals, and 9% to binding.

New Technology and Collection Development

Online developments in acquisitions systems, information retrieval, networks, the catalog, and circulation files have all affected collection development. Mainly they have affected either the amount of information which is available (for acquisitions, verification, selection, collection evaluation), or the speed with which items can be acquired. So far, the actual collection itself has not been put online.

But this is changing. In the early 1980s, the American Chemical Society made its journals available in full text, searchable and printable online. Increasingly, the discussion is about access rather than purchase.

Chapter 7

Collection Control

Collection control is the fourth primary function of a sci/tech library. The purpose of collection control is to allow us to find the items in our collection that we want when we want them. The usual mechanism for collection control is to put some kind of number on the physical item and produce a catalog or index which refers to the item number. The items are shelved in order by the number. Access by other features such as author, title, and subject are provided by the catalog or index. Although this is a simplistic description, it is useful to keep it in mind as we discuss the details and exceptions later in this chapter.

The traditional library functions which fall into collection control are cataloging, classification, indexing, and abstracting (these are aspects of *bibliographic* control) and circulation and shelving (these are aspects of *physical* control). Bibliographic control and physical control interrelate in the shelf arrangement. The bibliographic record in a library catalog must indicate the item's location in some manner.

The advent of automation in libraries has disturbed all the traditional divisions in collection control. In an online catalog which displays both the cataloging record and the circulation record, where is the division between bibliographic and physical control? What is the difference between indexing a technical report in the NTIS database and cataloging it for the library catalog? Indeed, when the catalog is automated, what is the distinction between the automated system and collection control? Unfortunately, most of the literature of collection control deals with a single aspect, and few authors have attempted to deal with the principles of bibliographic records for all purposes and in both printed and automated formats. The notable exception is Ronald Hagler, in his book *The Bibliographic Record* (1982). He considers the various aspects of bibliographic records without ignoring the details, but also without losing sight of the big picture.

Automation has also caused a reexamination of traditional areas. Many aspects, especially access points, authority control and subject analysis are being reexamined after a long period of stability. Considerable change can be expected in these areas.

It is useful to examine the relationship of collection control to the other four primary functions. Collection development is the starting place for collection control, since ordinarily the acquisitions record created in the collection development process contains the basics of the cataloging record. But collection development is more dependent on collection control than cataloging is on the acquisitions record. No collection development decision can be made without knowing what is already in the collection. And there is no point in doing collection development superbly if your bibliographic and physical control is so poor you cannot find what is there. Also, as we discussed in Chapter 6, an automated catalog can be helpful in providing statistics which can be used to evaluate the collection development function.

Collection control is also essential to information retrieval, document delivery, and current awareness. A useful, up-to-date catalog with appropriate access points (including effective subject headings) is necessary for locating materials in answer to questions, as is a good physical arrangement and circulation control. The first step in locating a document of interest is to check the collection. What is not there has to be delivered from elsewhere. And the library's ability to perform the current awareness function is dependent on collection control because accessions lists and internal SDI services are often produced from the cataloging data.

Collection control is a function which may vary widely among sci/tech libraries. Historical decisions about automation or networking cause variety in collection control. And although the principles are basically the same (to allow the user to find what is in the collection), decisions about choices of access points, arrangement of the catalog, circulation policies, etc. can vary because of the setting.

In our hypothetical science branch of a university, often the collection control function and sometimes the major decisions about collection control are handled by a centralized technical processing unit rather than in the branch. Universities are also commonly committed to resource sharing, networks, and cataloging to meet national standards. Much of the challenge in collection control in a science branch is in convincing the centralized technical processing unit of the importance of timeliness of cataloging for science materials, of the importance of series entries (and that many science users prefer analyzed series to monographs classed separately), and of the importance of special types of materials like technical reports, maps, etc., which are unappealing to many catalogers.

Communication with the centralized technical processing staff is critical, and a branch science librarian needs to know enough about national standards and local practice to argue effectively in cases where the standards are not acceptable. Rarely does a science branch library have the opportunity to design or select its own automated catalog. This is usually done for the library system as a whole. But there may be an opportunity to influence the choice or to make sure it can accommodate science materials. Usually if a branch is developing its own in-house system, separate from a centralized cataloging unit, it is for material which does not fit the overall system (e.g., technical reports or maps). Although some libraries catalog technical reports as monographs, the system for technical reports developed at Branner Earth Sciences Library (described in Chapter 2) is a typical example of a special system. Another characteristic of collection control in the academic setting is that often academic libraries have such large collections that retrospective conversion of the entire catalog is not an option. Split catalogs are permanent fixtures in many academic libraries, forcing librarians to educate users to look in the "old" card catalog and the "new" computer output microform (COM) or automated catalog.

In contrast, in the corporate setting, collection control is usually within the control of the sci/tech library. Although sometimes corporate libraries are restricted to particular online catalogs in order to be compatible with other libraries in a corporation, one of the most common questions from the new special library is "what system should I use?" The choice of system is less restricted than in any other setting, and the problem is narrowing the choices. The established corporate sci/tech library faces a slightly different problem. Corporate sci/tech libraries were the first to have automated catalogs and indexes. Their automated files were often developed in the early 1960s before most of today's national standards were developed. They are now considering whether MARC-based systems would provide enough advantages to make it worth changing.

The growth of shared cataloging networks and bibliographic utilities is causing many special libraries to consider changing to the national standards. There are four major bibliographic utilities: Online Computer Library Center, Inc. (OCLC), 6565 Frantz Rd., Dublin, Ohio 43017; Research Libraries Group Research Libraries Information Network (RLIN), Jordan Quadrangle, Stanford, California 94305; Western Library Network (WLN), Washington State Library AJ-11, Olympia, Washington 98504; and University of Toronto Library Automation Systems (UTLAS), 80 Bloor St. West, Second Floor, Toronto, Ontario, Canada M5S2V1. Although each has additional services to offer its members, each built its membership base as a shared cataloging source. Each utility mounts some of the LC MARC tapes on its computer and allows member libraries to search those records as well as records contributed by other member libraries. Members ordinarily are asked to catalog in the system any item that they do not find (so that it can be shared). Each system provides catalog cards for those libraries which want cards, and also archival tapes—the records of what the library has cataloged in the system in machine readable form. Lately, these utilities have been working with vendors of online catalogs to provide interfaces so that records can be cataloged on the utility and immediately transferred in machine-readable form to the library's online catalog. This cuts down the wait for the archival tape, and makes cataloging more efficient.

There have been several factors keeping sci/tech libraries in the corporate sector from joining networks. First has been the privacy issue: most corporate libraries do not want the world to know what they have. Second is the make-up of the networks themselves (the IRS rules have restricted the percentage of for-profit membership they could have) (Hill, 1985). Third has been the cost factor: the hit rate (i.e., the percentage of items cataloged which would be found in the database) can be low for libraries with very specialized subject areas.

However, as the databases of these utilities and the other services they offer both grow, corporate libraries are increasingly asking new questions: Should the library switch to national cataloging standards? Is there a cost advantage to joining a network that would make converting a nonstandard file cost effective? If the library joins a network, can the "hits" be downloaded into the existing system and can company privacy be protected?

Thus corporate sci/tech librarians are having to learn more about AACR2, MARC, OCLC, RLIN, WLN, UTLAS, and lots of other acronyms in order to decide whether the developments in the broader library world are cost-effective for them.

For a corporate sci/tech library, collection control is almost synonymous with automation. The collection is normally small enough (and money available enough) that a retrospective conversion process is possible (that is, all the records can be converted to machine-readable form).

Sci/tech libraries in other settings (e.g., government, professional societies, nonprofit organizations) fall somewhere in between these two examples. Government libraries often have the same need to meet national standards that academic libraries do, but may have more control over the choice of system that will be used.

Once a library has chosen a particular form of catalog, automated system, network, or set of standards, its interests narrow to issues related to those choices. This chapter and the related one on automation (Chapter 11) focus on basic principles rather than on specific systems, and will not be as useful to those with established systems, at least not until the choice has to be reconsidered for some reason. Contact with the vendor and with other users of the same systems or standards is the most effective information source for those with particular systems already in place.

BIBLIOGRAPHIC CONTROL

Bibliographic records are used in three products: library catalogs, indexes and abstracts, and bibliographies. Each product has standards associated with it. For library cataloging, these include the *Anglo American Cataloguing Rules,* Second Edition (AACR2), and MARC (Machine Readable Cataloging), a standard for the format of machine-readable records. The line between the two is blurred because MARC, the primary standard for transmission, is used to transmit Library of Congress (LC) cataloging (as well as other cataloging). The tapes which contain LC's cataloging (done to AACR2 standards) are called the MARC tapes, so MARC is commonly confused with a cataloging standard. Standards are vital to any library creating a catalog entry for a traditional library catalog, participating in an online shared cataloging network, or creating records which will be shared with other libraries. Although they were designed for general libraries, they are applicable in sci/tech libraries as well.

There are also standards for indexing and abstracting, a common activity in sci/tech libraries. These include COSATI (see Volume 2, Chapter 12), the American National Standards Institute (ANSI) standards for indexing and abstracting [e.g., *Basic Criteria for Indexes,* Z39.4-1968(R1974) and *Abbreviations of Titles of Periodicals,* Z39.5-1969(R1974)], and international (ISO) standards. However, these are applied much less consistently than AACR2 and MARC are applied to cataloging, and a library might choose to follow a pattern established by one of the abstracting and indexing services, or it might use different standards for the descriptive record, the abbreviations of periodicals, or the subject terminology.

Many different "standards" for references used in bibliographies, publications, or at the end of articles also exist. They include the *American National Standard for Bibliographic References* (ANSI Z39.29-1977), style manuals such as the *Chicago Style Manual,* and requirements of individual professional societies and even of individual journals. All of these vary in the details of the elements to be included, the arrangement of the elements and the punctuation.

Although in the past these different uses (catalogs, indexes, and bibliographies) for bibliographic records were distinct, they are converging because of online searching, online library catalogs, word processors, and microcomputers. Librarians want to take references from the shared cataloging networks and put them in their online library catalogs. Other librarians want to take records from the technical reports databases (e.g., NTIS) and do the same thing. Library users want to take references in machine-readable form from a subject-oriented database or from an online library catalog and put them in their word processors to use in bibliographies. Even though the basic data is the same, in each of these situations the bibliographic reference varies: there are different elements, in different order, with different punctuation.

To demonstrate the variety of bibliographic records, Figs. 7.1–7.8 are examples of various citations as they appear in different indexes or catalogs. The library catalogs are represented by a fully tagged MARC AACR2 record from the RLIN database (Fig. 7.1), and the same entry displayed in DOBIS/Leuven, an IBM-supported online integrated library system (Figs. 7.2 and 7.3). The abstracting and indexing systems are represented by a record from DOE's Energy Data Base as it appears in the

```
ID: DCLC8025862-B          RTYP: c      ST: p     FRN:    NLR:023333   MS: c   EL:  AD: 09-16-81
CC: 9110      BLT: am     DCF: i    CSC:      MOD:      SNR:         ATC:          UD: 01-01-01
CP: dcu        L: eng     INT:      GPC:      BIO:      FIC: 0       CON: b
PC: s          PD: 1981/            REP:      CPI:0     FSI: 0       ILC: a    MEI: 1    II:1
MMD:           OR:     POL:      DM:      RR:          COL:         EML:      GEN:      BSE:
010       8025862//r85
040       $dCStRLIN
050 0     TK9145$b.K58 1981
082 0     621.48$219
100 10    Knief, Ronald Allen,$d1944-
245 10    Nuclear energy technology :$btheory and practice of commercial nuclear
          power /$cRonald Allen Knief.
260 0     Washington :$bHemisphere Pub. Corp. ;$aNew York :$bMcGraw-Hill,$cc1981.
300       xv, 605 p. :$bill. ;$c24 cm.
490 0     McGraw-Hill series in nuclear engineering
504       Bibliography: p. 571-587.
500       Includes indexes.
650 0     Nuclear engineering.
650 0     Nuclear energy.
```

Fig. 7.1. Fully tagged MARC AACR2 record for a monograph from RLIN. [Courtesy of Research Libraries Group.]

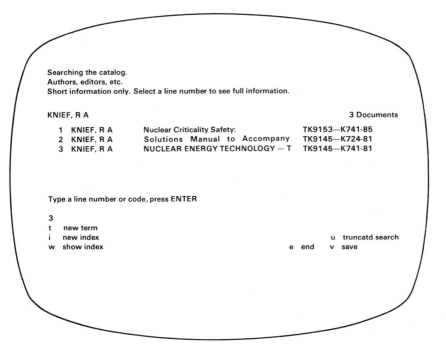

Searching the catalog.
Authors, editors, etc.
Short information only. Select a line number to see full information.

KNIEF, R A   3 Documents

  1 KNIEF, R A   Nuclear Criticality Safety: TK9153—K741-85
  2 KNIEF, R A   Solutions Manual to Accompany   TK9145—K724-81
  3 KNIEF, R A   NUCLEAR ENERGY TECHNOLOGY — T TK9145—K741-81

Type a line number or code, press ENTER

3
t   new term
i   new index   u   truncatd search
w   show index   e   end   v   save

Fig. 7.2. The short information display for the same record in a DOBIS online public access catalog. The user who wants more information types "3" and gets the screen shown in Fig. 7.3.

DOE/RECON system (Fig. 7.4), in NTIS (File 6) on DIALOG (Fig. 7.5), and in hard copy in *Energy Research Abstracts* (Fig. 7.6). Figure 7.7 illustrates a bibliographic reference in the format for *Science* magazine, and Fig. 7.8 shows the same reference in ANSI format.

Although much of the variety seems unjustified and is unfortunate, there are some real reasons for differences among bibliographic references. The differences among the abstracting and indexing services exist in part because there is no motivation to share records and thus no reward for standardizing (Wood, 1982b, p. 349). Also, differences in purpose result in differences in the content and display of the reference. A bibliographic record which included all items which might be necessary in one application (for example, the detailed physical description necessary in describing a rare manuscript or the in-depth subject analysis of a technical report) would be unnecessary in others, and thus unnecessarily costly.

Because of the variety of types of records, the topic of bibliographic control is complex, and there are a number of ways to approach it.

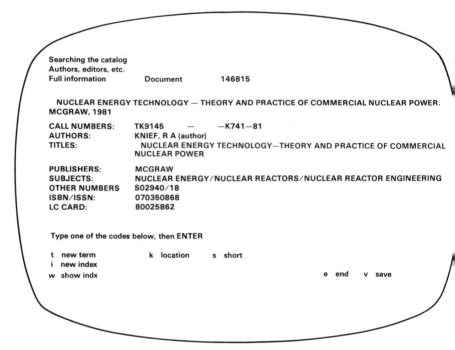

Searching the catalog
Authors, editors, etc.
Full information   Document   146815

NUCLEAR ENERGY TECHNOLOGY — THEORY AND PRACTICE OF COMMERCIAL NUCLEAR POWER.
MCGRAW, 1981

CALL NUMBERS: TK9145   —   —K741—81
AUTHORS: KNIEF, R A (author)
TITLES: NUCLEAR ENERGY TECHNOLOGY—THEORY AND PRACTICE OF COMMERCIAL
                   NUCLEAR POWER

PUBLISHERS: MCGRAW
SUBJECTS: NUCLEAR ENERGY/NUCLEAR REACTORS/NUCLEAR REACTOR ENGINEERING
OTHER NUMBERS   S02940/18
ISBN/ISSN: 070350868
LC CARD: 80025862

Type one of the codes below, then ENTER

t   new term   k   location   s   short
i   new index
w   show indx   e   end   v   save

Fig. 7.3. The full information display for the same record in a DOBIS online public access catalog. Note that both Fig. 7.2 and this figure are easier to understand than the fully-tagged MARC record in Fig. 7.1.

We could consider separately the traditional library functions of cataloging, classification, indexing, and abstracting. The distinction between indexing and cataloging is not always obvious. "Cataloging" has tended to be used to refer to book cataloging and to cataloging of serials as a whole, and "indexing" to indexing of journals (as in the abstracting and indexing services), or of the contents of books (as in the index in the back of the book), or of individual papers from a conference or of an edited collection of papers. The difference would appear to be that an index provides access to the *contents* of a physical item, where cataloging focuses on the item as a *whole*. However, indexing is also commonly used to refer to indexing of nontraditional materials (e.g., technical reports, laboratory notebooks, slides, etc.) which are commonly treated as whole items, and there are mechanisms within cataloging to handle parts of the whole (e.g., analytical entries). Another distinction that is commonly made is that in an index, the user consults an access point (e.g., "catalysis") and is referred to another place to see the full entry. In contrast, library

DIS 2/5/000001-000001/1
<ACCESSION NO.> 83C0142139
<REPORT NO,PAGE> CONF-8305103—1;DE83014717
<TITLE(MONO)> Developments in eddy-current computation with eddynet
<EDITOR OR COMP> Turner, L.R.; Lari, R.J.
<CORPORATE AUTH> Argonne National Lab., IL (USA)
<TYPE> R
<PAGE NO> 4
<AVAILABILITY> NTIS, PC A02/MF A01.
<ORDER NUMBER> DE83014717
<CONTRACT NO> Contract W-31-109-ENG-38
<CONF TITLE> COMPUMAG conference on the computation of electromagnetic fields
<CONF PLACE> Genoa, Italy
<CONF DATE> 30 May 1983
<DATE> 1983
<CATEGORIES> EDB-420500
<PRIMARY CAT> EDB-420500(ENGINEERING; MATERIALS TESTING.)
<ABSTRACT> The eddy-current computer code EDDYNET has been modified to incor-
porate 2pi/n symmetry, specification of current-crossing boundaries, a polar mesh generator,
calculation of current density and temperature rise, calculation of forces and torques, a
mesh with multiple sheets, and fields and fluxes from filamentary current rings. It has been
successfully applied to find the forces and temperature rise of a conducting flat plate in a
pulsed magnetic field and the currents in a hollow conducting torus.
<DESCRIPTORS> *EDDY CURRENT TESTING—computer codes; *EDDY CURRENT
TESTING—data analysis;CURRENT DENSITY;E CODES;FELIX FACILITY;MESH
GENERATION;MODIFICATIONS

Fig. 7.4. A record from the Energy Data Base as displayed on the Department of Energy's
RECON system.

card catalogs have full information at each entry, and thus some difference
could be seen between the two. But library catalogs in microfiche and
online often use indexes, so this distinction has become blurred.

There has always been more confusion about the difference between
cataloging and indexing in sci/tech libraries than in general libraries. Most
sci/tech libraries have created indexes to some subset of material that is
not handled properly by traditional library cataloging. Even those sci/tech
libraries which do not *create* indexes use them. Index records in online
databases like INSPEC, COMPENDEX, and NTIS and their hardcopy
equivalents actually provide part of the collection control function for sci/
tech libraries by indexing the contents of their collections of journals and
technical reports. Although indexes and abstracts, like catalogs, rely on
descriptive analysis and subject analysis, for historical reasons they do
not follow the standards for library catalogs and there is much less con-
formity among them. Indexes have vocabulary control and syndetic structure (that is, the cross-reference structure); catalogs have subject headings

PRINTS   User:U0016585 21Oct85   PRINT 1/5/1

DIALOG File 6: NTIS - 64-85/ISS21 (Copr. NTIS)

1015998   DE83014717
   Developments in Eddy-Current Computation with Eddynet
   Turner, L. R. ; Lari, R. J.
   Argonne National Lab., IL.
   Corp. Source Codes: 001960000; 0448000
   Sponsor: Department of Energy, Washington, DC.
   Report No.: CONF-8305103-1
   1983   4p
   COMPUMAG conference on the computation of electromagnetic
fields, Genoa, Italy, 30 May 1983.
   Languages: English   Document Type: Conference proceeding
   NTIS Prices: PC A02/MF A01 Journal Announcement: GRAI8401;
NSA0800
   Country of Publication: United States
   Contract No.: W-31-109-ENG-38
   The eddy-current computer code EDDYNET has been modified to
incorporate 2 pi /n symmetry, specification of
current-crossing boundaries, a polar mesh generator,
calculation of current density and temperature rise,
calculation of forces and torques, a mesh with multiple
sheets, and fields and fluxes from filamentary current rings.
It has been successfully applied to find the forces and
temperature rise of a conducting flat plate in a pulsed
magnetic field and the currents in a hollow conducting torus.
(ERA citation 08:044440)
   Descriptors: *Eddy Current Testing; Data Analysis; Computer
Codes; Current Density; Mesh Generation; Modifications; E
Codes; FELIX Facility
   Identifiers: ERDA/420500; NTISDE
   Section Headings: 14B (Methods and Equipment--Laboratories,
Test Facilities, and Test Equipment); 9B (Electronics and
Electrical Engineering--Computers); 94J (Industrial and
Mechanical Engineering--Nondestructive Testing); 41G
(Manufacturing Technology--Quality Control and Reliability)

Fig. 7.5. The same record from the NTIS database as it appears in an offline print from
DIALOG.

(one kind of structured vocabulary control) and authority control (a concept which includes syndetic structure). Thus, dividing the discussion of
bibliographic control between traditional cataloging and indexing is somewhat artificial, especially in sci/tech libraries, most of which have to consider control at least of journals and technical reports, and many kinds
of other formats as well.

A second way of dividing the discussion of bibliographic control is to
consider the content of a bibliographic record as including two parts: (1)
descriptive analysis (i.e., description of the physical item) and (2) subject
analysis (i.e., analysis of the content, which includes subject headings and
classification). This division is used in traditional cataloging and is particularly useful in discussing collection control in sci/tech libraries because
descriptive standards are generally accepted while subject standards often
are not. AACR2 and MARC are acceptable as descriptive standards in

⌐/185).
⌐eb 1983).

⌐nal natural convective
⌐u of arbitrary cross section
two configurations: (i) concen-
⌐nuli formed by an inner hexagonal
⌐ular cylinder. Also embodied in the
⌐ predict local as well as mean heat trans-
⌐ conditions of the form T'x/sup m/ can be
⌐u for the Rayleigh number (Ra/sub R/) varied
⌐randtl number (Pr) varied from 0.7 to 3100, and
⌐ to 2.0. Even with these large variations, the present
⌐neory collapses all the experimental data for the annu-
⌐tries to a single line. The physical problem appears to be
⌐tely specified by a single equation when the following is
⌐wn: thermal boundary condition (i.e., m), the fluid (i.e., Pr), the
⌐aspect ratio, the Rayleigh number, and the geometry. This work
demonstrates that the present theory is applicable to annuli of arbitrary cross section, and therefore the theory will be extended to include curvature effects and axisymmetric geometry.

⌐cal

⌐ form the
⌐in the sense
⌐discussing sev-
⌐ing the behavior
⌐ouple," we discuss
⌐sonant tubes to pro-
⌐nd, for restricted heat
⌐e results are analyzed
⌐oacoustic theory based on
⌐coustic engine are general-
irreversible heat engines of
a special case. Finally the re-
intrinsically irreversible engines
⌐ the efficiency of such engines
geometry or configuration rather

⌐ transfer in fibrous insulations: Part
⌐ing, T.W.; Tien, C.L. (Department of
⌐ring, University of Kentucky, Lexing-
⌐06). *Journal of Heat Transfer;* **105:** No. 1,
⌐.
⌐rpose of this work is to develop models for predicting
heat flux in lightweight fibrous insulations (LWFI). The
transport process is modeled by the two-flux solution and
⌐ar anisotropic scattering solution of the equation of transfer.
⌐adiative properties of LWFI consistent with these solutions
⌐ been determined based on extinction of electromagnetic radi-
⌐on by the fibers. Their dependence on the physical characteris-
⌐cs of fibrous insulations has been investigated. It has been found
that the radiant heat flux can be minimized by making the mean
radius of the fibers close to that which yields the maximum extinction coefficient. The results obtained in this study are useful to
those concerned with the design and application of LWFI.

4205 Materials Testing

REFER ALSO TO CITATION(S) 43753, 44120, 44446

44440   **(CONF-8305103—1) Developments in eddy-current** ◄
computation with eddynet. Turner, L.R.; Lari, R.J. (Argonne
National Lab., IL (USA)). 1983. Contract W-31-109-ENG-
38. 4p. NTIS, PC A02/MF A01. Order Number
DE83014717.
From COMPUMAG conference on the computation of electromagnetic fields; Genoa, Italy (30 May 1983).
The eddy-current computer code EDDYNET has been
modified to incorporate $2\pi/n$ symmetry, specification of current-
crossing boundaries, a polar mesh generator, calculation of current
density and temperature rise, calculation of forces and torques, a
mesh with multiple sheets, and fields and fluxes from filamentary
current rings. It has been successfully applied to find the forces and
temperature rise of a conducting flat plate in a pulsed magnetic
field and the currents in a hollow conducting torus.

44441   **(CONF-8305103—2) Survey of eddy-current programs.** Lari, R.J.; Turner, L.R. (Argonne National Lab., IL
(USA)). 1983. Contract W-31-109-ENG-38. 5p. NTIS, PC
A02/MF A01. Order Number DE83014734.
From COMPUMAG conference on the computation of electromagnetic fields; Genoa, Italy (30 May 1983).
Portions are illegible in microfiche products.
This paper describes the results of a literature search and
questionnaire survey of eddy-current computer programs. The capabilities of each program are compared in the first four tables and
the references listed. Results of a recent questionnaire survey of researchers working in this area are given.

Fig. 7.6. The same record as it appears in the hard copy of *Energy Research Abstracts.*
Note that descriptors are not listed.

most sci/tech libraries. However, standards for subject analysis and classification are less widely accepted in the sci/tech setting. Sci/tech libraries,
particularly those outside of academia, have often found alternatives to
Library of Congress Subject Headings (LCSH), the most widely used
general subject headings list, and to the two most-used classification
schedules (the Library of Congress classification and the Dewey Decimal
classification system).

Subject access is receiving increased attention in the 1980s as a result
of online public access catalog use studies which emphasize its importance
and people's problems with it. The topic of subject access includes not
only the topic of subject analysis (how to analyze a document, assign
subject headings, etc.) but also the user and his or her process of searching.

ular-
...s move
...larization
, around the
causing oscilla-
...nitted, whose fre-
...cession frequency.
...oms get out of phase
...sions with the walls of
...taining the atoms or other
...t make their precession fre-
...s vary slightly, so that the polar-
...ation also decays with a characteristic
time, the transverse relaxation time (T_2
in nuclear magnetic resonance parlance).

... depen-
might signal
...rred frame of
...esearchers com-
of their beryllium-9
...f a hydrogen maser.
...s in a container are ran-
..., so any effect causing a
quency with orientation
...ge out. An essential feature
...lium-9 clock, therefore, is the
field of the Penning trap. In the
...c field, the same laser light as
used for the cooling places the
...pped ions in a specific quantum state,
...so that the spin angular momentum vec-

With this technique, the Washington investigators were able to rule out any changes in the precession frequency of mercury-201 as the earth rotated with respect to prospective preferred frames of reference in the universe down to 3 microhertz.—ARTHUR L. ROBINSON

References

1. A. L. Robinson, *Science* **222**, 1316 (1983).
2. A. Brillet and J. L. Hall, *Phys. Rev. Lett.* **42**, 549 (1979).
3. G. F. Smoot, M. V. Gorenstein, R. A. Muller, *ibid.* **39**, 898 (1977).
4. J. J. Bollinger, J. D. Prestage, W. M. Itano, D. J. Wineland, *ibid.* **54**, 1000 (1985).
5. J. D. Prestage, J. J. Bollinger, W. M. Itano, D. J. Wineland, *ibid.*, p. 2387.

Fig. 7.7 A bibliographic reference in *Science* magazine format (August 23, 1985, p. 747). Notice the title of the article is not included. [Copyright 1985 by the AAAS; reprinted by permission.]

A third way of dividing bibliographic control is to consider the format of the record separately from access to it. Access covers how the records are arranged in files, catalogs, indexes, etc., and also how they can be searched [e.g., by author, title, subject, keywords, report numbers, publishers, dates, International Standard Book Numbers (ISBNs), etc.]. This is a convenient approach in that the MARC standard for the format of records can then be used in various kinds of local systems which provide different access points and possibly different displays. However, decisions on the content of the record mean some decisions about access are already made: for example, you cannot have access by report number or geographic coordinates unless they are included in the record.

---

1. Robinson, A.L. New test of variable gravitational constant. Science; 1983; 222(4630): 1316-1317.

Fig. 7.8. The same reference in ANSI Standard format.

The final approach to considering bibliographic control requires us to step back and look again at the purpose of collection control as we stated it in the beginning of this introduction. *The purpose of collection control is to allow us and our users to locate the materials in the collection that are of interest.* In designing a system to do this, we have to consider the purpose of the index or catalog, what the final product will be, what will be included in the bibliographic description of the individual items, and how the index or catalog can be produced most efficiently.

The arrangement of the remainder of this chapter encompasses some aspects of all the many ways of dividing the discussion of bibliographic control.

THE PURPOSE OF LIBRARY CATALOGS AND INDEXES

Charles Cutter, in 1867, defined the purpose of the library catalog. The same purpose was adopted as part of the Paris Principles (International Conference on Cataloguing Principles, Paris, 1961 (1963), a document which still provides the basis for library cataloging. Cutter said the catalog should:

1. Enable a person to find a book about which one of the following is known: the author, the title, the subject.
2. Show what the library has: by a given author, on a given subject, and in a given kind of literature.
3. Assist in the choice of a book as to its edition—bibliographically—and as to its character—literary or topical (Cutter, 1904).

A similar purpose statement for a sci/tech library might read that the catalog should:

1. Enable a person to find a book, technical report, map, patent, standard or specification, laboratory notebook, etc. about which the following is known: (a) any author; (b) any corporate body; (c) any title; (d) the subject (including geographic area); (e) any report number, patent number, standard and specification number, or control number.
2. Show what the library has by a given personal author or corporate body and on a given subject.
3. Assist in the choice of a book as to its relevance to the subject at hand.

It is a useful exercise to think through and record the purpose of a library's own catalog and any indexes that it creates. The purposes of indexes may vary widely, but they should also be clearly defined.

By asking that the catalog be able to show what the library has by a given personal author or corporate body or on a given subject, we require that authors and subjects always be entered the same way in the catalog—so that they will be grouped together. This requires uniform headings for authors, corporate bodies, and subjects, and the use of uniform headings requires authority control (also called cross references, or syndetic structure).

FORMS OF CATALOGS AND INDEXES

There are four physical forms of library catalogs: the card catalog, the book catalog, the microform catalog, and the online catalog. Indexes also can appear physically on cards, in printed book catalog form, on microform, and online. Each format has its advantages and disadvantages. The online catalog's potential for increased access and the overwhelming support of users for online catalogs (Matthews, 1984) are making online the choice of most libraries.

Card Catalogs

The card catalog has been the standard form of library catalog for most of the twentieth century. With a card catalog, a card set is prepared for each item. Each card is usually a "unit record" and contains the full bibliographic information for the item. Added entries (for title, subject, added authors, etc.) are typed across the top of the unit card and filed separately, so there is access by author, title, and subject (by the first word of the controlled subject heading only) and full bibliographic information at each access point. Card catalogs are easily updatable and there is only one place for the user to look (unless a filing backlog develops).

Originally, all card sets were typed. Later, means of duplicating card sets were developed so the headings could just be typed on a unit card. The availability of card sets from the Library of Congress increased the popularity of card catalogs. Today card sets can also be ordered from various sources (such as OCLC, RLIN, WLN, UTLAS and other commercial suppliers). Card sets now come with all the added entries overprinted, and can be delivered in filing order. They can also be produced on computers and word processors. Details of how to compile a card catalog are covered by Strauss et al. (1972).

There are some limitations to the card catalog. It is usually limited to one location. Increased access points (for example, a card for each word of the title) require additional cards and the effort to type or highlight the filing word and to file them, so there is a tendency to limit access points.

The card catalog can only be approached one way, no matter what the experience level of the user is. Also, changing the cards to indicate a new location for materials moved to storage adds greatly to the costs of storing or relocating materials. And changing headings is also difficult since old cards have to be pulled manually and corrected and refiled. As the catalog grows, the costs of filing and maintenance grow.

Book Catalogs

A book catalog is simply a printing of the card catalog with many entries to a page. Access is the same as for the card catalog: by author, title, and subject. If printed from machine-readable records rather than from photographs of cards, additional access points can be added. It can be available in numerous locations, but it is instantly out-of-date. Supplements must be printed (or the entire catalog reprinted), and studies show (e.g., Blackburn, 1979) that people rarely look in supplements. Lately printing and paper costs have increased so that book catalogs are not very common. Costs of book catalogs are discussed in detail by Malinconico (1977).

Microform Catalogs

Microform catalogs are currently quite common. Generally these are produced as computer output microform (COM), which can be either microfiche or microfilm. The records are in machine-readable form and the computer can produce microform with no print intermediary.

In a COM catalog, access points in addition to the traditional author, title, and subject can be added for little extra expense. For example, keyword access to titles and subjects can be provided. Costs for COM catalogs vary with the size of the database, the frequency of compilations, and the number of copies produced. Additional copies can be produced for small marginal costs.

Microform catalogs also require special equipment to read them. The equipment can be a significant initial expense, and all equipment needs maintenance. Readability can also be a problem.

Like book catalogs, microform catalogs may require supplements if new cumulations are not produced frequently. Thus, there is a delay while the supplement is being produced, and once it arrives, it may not be used. Studies by Blackburn (1979) at the University of Toronto Library and by Dwyer (1979) at the University of Oregon showed a typical pattern of low use of supplements. At Toronto, use of the supplement was less than 3% of total use and mostly reflected staff use. Microform catalogs, like book catalogs, are also always out-of-date.

Usually the backup catalog for an online catalog is microform because it can easily be produced from the online file. Also, many libraries which use one of the bibliographic utilities (e.g., OCLC, RLIN, WLN, or UT-LAS) have chosen to close their card catalogs and produce COM catalogs from archival tapes instead. There was a particular flurry of this activity around early 1980 with the adoption of AACR2 by the Library of Congress. The form of a large number of headings changed with the adoption of AACR2, creating a serious problem of incompatibility with the old headings. But other reasons can also cause the switch to microform: the need for distributed catalogs, increasing costs of maintenance of card catalogs, and the attractiveness of increased access points.

Online Catalogs

> The design of the public catalogs currently being developed does not—and should not—offer merely improvements in efficiency over earlier catalog designs. Like the jet airplane, which cannot be considered simply as a more efficient glider, the online catalog is a far more powerful instrument than any of its predecessors. It will profoundly change the way people go about the business of living, and its purpose must be expressed in societal concepts rather than in narrow bibliographic terms. (Kilgour, 1984, p. 320)

Online catalogs have several advantages over the other forms of catalogs. They are normally updated quickly (the exception is if the library has to wait for a new tape from a bibliographic utility). With no need for supplements, there is only one place for the user to look (unless everything is not online, of course). They can be distributed widely at little extra cost (although there are increased costs for terminals, electricity, and computer use). Numerous extra access points and search commands can be available: keyword access to the title and subject headings is common; publisher, date, and ISBN (International Standard Book Number) are available access points in some systems; and Boolean logic is often available to aid searching. Some systems can show the location and status of an item (e.g., circulation data) as well as whether the library owns it. Some systems have global authority control. This means that if you change a heading in the authority file (e.g., from aeroplane to airplane), the system will automatically change the heading on all bibliographic records which use it. Some systems help correct user errors with spelling checkers, automatic truncation, and automatic reverse of order of authors' names if nothing is found. A few systems such as CITE allow the user to enter queries in natural language. The program then performs selective term stemming and displays ranked lists of potentially useful keywords (Doszkocs, 1983). Some systems offer several different interfaces, such as menu driven modules for the novice user and command driven modules for the experienced user.

The disadvantages to the online catalog are that the availability of the catalog is dependent on the reliability of the system, and that the costs can include a high initial capital investment and ongoing maintenance costs of the system.

There are more and more online catalogs available in the marketplace, and their capabilities are changing too quickly to evaluate particular systems here, even though that is the very data that every library manager thinking of automating wants. Actually acquiring an online catalog requires close examination of what is needed in the local environment as well as what is available at the moment, and what it costs. Some of the aspects that are relevant to consideration of the online catalog include ease of input and modification, response time, display capability, "user friendliness," capability of handling MARC records, and the availability of a flexible record length, desired access points, Boolean search capability, authority control, an online thesaurus, help screens, printing capability, and an interface with circulation and acquisitions systems.

When the standard for a common command language proposed in March, 1986 by the National Information Standards Organization (NISO-Z39) Committee G is approved, the availability of this language will also be a consideration in choosing a system (Wall, 1986).

The term "user friendly" is often misused, but it means a system which does not require the user to have much training or expertise in order to interact successfully with it. Matthews and Williams (1984) suggested a way of rating computer systems from user-oriented through user-crabby to user-hostile and user-vicious. There are probably online library catalogs which would meet each description.

The characteristics of specific systems were covered by Matthews (1985a, 1986), Fayen (1983b, 1984), Seal (1984), Hildreth (1982) and Salmon (1983). Hildreth's review in the *Annual Review of Information Science and Technology* (1985) is also useful. Further guidelines and sources are included in Chapter 11.

Retrospective Conversion

In order to create an online catalog, the records must be in machine-readable form. Retrospective conversion is the process of converting manual catalog records to machine-readable form. There are vendors who will do this for a fee, including Carrollton Press (the REMARC system), Autographics, Blackwell North American, OCLC, and many of the regional networks such as AMIGOS and SOLINET (Hewitt, 1984, p. 209). Adler and Baber (1984) collected fourteen case studies of retrospective conversions of card catalogs to machine-readable form. Each case covers time and budget constraints, coming to terms with money, administration and personnel. Peters and Butler (1984) provided a cost model for retrospective conversion alternatives, and Epstein (1983a) discussed the implication of creating copy-specific records. There is also an ARL Spec Kit (number 65) on retrospective conversion available from the Association of Research Libraries.

CATALOG-USE STUDIES

There are many choices to be made in creating a catalog or index in a sci/tech library. As with other library activities, decisions should focus on the library user, but the information about the user is limited.

Catalog use, like the information transaction, is a complex behavior. The user approaches the catalog with all kinds of attributes which can influence his or her success or failure (background, experience, training, expectations, perseverance, intelligence). The kind of information he or she has can affect the success of a known-item search (how complete is it? is it correct?) or of a subject search (what is the subject the person is looking for? does he or she know other things it might be called?). How the person chooses to conduct the search is also a factor (e.g., many users search by author, but title would often be quicker and more reliable). Also, factors in the catalog itself can affect the success or failure of the search. (Does the subject vocabulary allow entry by the terms the user will use? How many see references are there? What access points are included? Is it a large catalog or a small one? How are things filed?) In online catalogs the system interface can also be a factor. (What is the command structure? Is the system easy to use? Do the displays provide the right information?)

Not enough is known about all these factors. Partly this is because catalog use is complex, partly because it is hard to study effectively. The usual way to study catalog use in depth is the interview method. An interviewer walks up to someone using the catalog and asks if he or she can record the user's behavior as the person is doing the search. This method was used in the classic card catalog use studies, which are summarized in Lancaster (1977) and Markey (1984b, 1985). The difficulty with this approach is that it is unclear how much the interview modifies the result—it is hardly an unobtrusive method. When, in order to be more unobtrusive, the interview is conducted after the search is complete, the accuracy of the person's memory of what he or she did is in question. Questionnaires are even more dependent on user cooperation and memory, but have been used successfully in the series of online catalog use studies sponsored by the Council on Library Resources (CLR) in 1981 and 1982 and summarized in *Using Online Catalogs: a Nationwide Survey* (Matthews *et al.*, 1983). Another technique of studying users is the focussed group interview, used successfully by Markey (1983a,b, 1984a).

The online catalog offers us for the first time the ability to record unobtrusively what users actually do, so even better information about the use of library catalogs should be forthcoming. The computer can record each transaction (or particular kinds of transactions, such as those that result in no hits) to see what the user really does. Transaction logs could be particularly helpful in improving subject access. Every time the searcher puts in a subject word and gets no hits, the computer should record the entry vocabulary, which could then be examined and the subject headings either changed or enhanced with cross references. Transaction logs have been used to study MELVYL at the University of California (Radke *et al.*, 1982), the LCS system at Ohio State (Borgman, 1983), the NOTIS system at Northwestern (A. Taylor, 1984; Dickson, 1984), and five online catalogs as part of the CLR studies (Tolle, 1983a).

Failure analyses also provide some data useful in considering catalog use. A user who tries the catalog and does not find what he or she is looking for fills out a slip. Then a staff member tries to figure out why he or she found nothing. This kind of study can point out collection deficiencies and problems caused by processing delays, as well as changes that should be made in the catalog or in user training (Lancaster, 1977, p. 59).

Few user studies have singled out sci/tech libraries for attention. Although the American Library Association (Mostecky, 1958) study included one sci/tech library (MIT Engineering Library), the analysis did not separate the behavior of the sci/tech library users from the rest. Four of the studies Markey cites are sci/tech libraries, or could be considered to be: McGeever (cited by Montague, 1967, p. 96) studied the John Crerar library; Thurlow (cited by Palmer 1972, p. 37) studied the biology, zoology and agriculture library at Columbia University; Tagliocozzo (1970, IV, p. 29) studied the Yale University medical library; and Maltby and Sweeney (1972, p. 194) studied several polytechnic libraries (Markey, 1984b, pp. 76–77). Sci/tech research libraries could be expected to be closer to the pattern of academic libraries than that of public libraries, and that seems to be true, based on these few studies.

Let's consider what the numerous catalog-use studies tell us about users and how they use catalogs, and how the catalog itself affects use. The original studies should be read by anyone involved in catalog design; we have only summarized the most generally useful results.

1. About 41% of library users do not use the catalog. (Matthews, 1985a, p. 8).

2. Users have a positive attitude to the online catalog and prefer the online catalog to the card catalog. (Matthews, 1985a; Bishop, 1983; Dowlin, 1980). Even nonusers perceive the online catalog to be easy to use (Matthews, 1985a).

3. Most people have little knowledge of the structure of the catalog. (Mostecky, 1958; Maltby, 1971, 1973).

4. People do not remember complete bibliographic information, and people with bibliographic information written down are no more likely to have it correct than a user who has it in his or her head. The University of Chicago "Requirements Study for Future Catalogs" (1968) was particularly helpful in understanding what people remember about books. In this study, 104 volunteers were exposed to a collection of 180 psychology books. Each person chose five books of interest, examined them, and wrote a brief comment on some aspect of each book that particularly interested him or her. Several weeks later, the participants were given their own comments on the books and tested to determine which characteristics of the books they remembered. Only 10 out of 440 responses provided a complete and correct bibliographic citation. In 25% of the cases, enough information was given to make it relatively easy to locate the item in the University of Chicago Harper Library catalog, although 71% provided enough information to locate the book eventually (Lancaster, 1977, pp. 40–41). Titles are remembered better than authors. People also remember nonstandard clues such as imprint, color, size, etc., which could be useful in narrowing the search if they were included in catalogs (Cooper, 1970).

5. Title information is also more accurate. For example, Ayres et al. (1968) conducted a study of 450 known-item requests at the Atomic Weapons Research Establishment, Aldermaston, England. Of the 450, 90.4% of titles were correct, and an additional 2.9%, although incorrect, were traceable in the catalog. Authors were correct in 74.7% of the cases and traceable in an additional 14%. Untraceable authors (11.3%) exceeded untraceable titles (6.7%).

A title search is also more efficient. For example, Grathwol (1971) found that the search length for an author's name which was in the same form used in the catalog would have been 7–8 cards. When the citation only had initials, the average search length was 26 cards. When there was no initial, the search length could have been as long as 139 cards. In contrast, most titles constituted a file of one and the average search length was only 1.32 cards.

Although titles are more accurate, more efficient, and remembered better than authors, most users approach the card catalog by author. For example, the Yale study found 62% approached by author and 28.5% by title (Lancaster, 1977, p. 38).

6. The proportion of users searching in card catalogs for known items as opposed to subjects was 48% in the ALA study (Mostecky, 1958), which included both public and academic libraries. The academic libraries ranged from 65% in the University of Michigan study (Tagliocozzo *et al.*, 1970; Tagliocozzo and Kochen, 1970) to 73% in the Yale study (Lipetz, 1970, 1972). Faculty were much more likely to be seeking a known item than were undergraduates. The few sci/tech libraries which were studied followed the academic pattern. Known-item searches ranged from 63 to 66%, and subject searches from 32 to 33% (Markey, 1984b, pp. 76–77). Subject searching may be more common in online catalogs than it was in card catalogs. The more recent online catalog studies have revealed that users are searching for subject from 34 to 65% of the time (Markey, 1984b, p. 78).

7. Success rates for known item searches ranged from a low of 66% in the ALA study to a high of 84% in both the Yale study and a study of the University of Michigan General Library by Palmer (1972).

8. We know about some common mistakes that people make in online catalogs. Dickson (1984) used transaction logs from 1 month's transactions in the NOTIS system to analyze the types of errors which users made in searching for titles and authors. Although the study was system-specific, it is interesting that 37% of the title searches and 23% of the author searches were zero-hit. About 16% of the no-hit searches under title were searches under articles (a, an, the). Entering forenames first accounted for about 18% of the zero-hit author searches. Another 10% were mistakes in the forenames beyond the initial letter, such as entering the wrong second initial, not spacing or punctuating, and misspelling the forename (Dickson, 1984, pp. 36–37).

9. The average number of access points provided in the catalog influences the success rate. For example, in searching for the inexact titles from the University of Chicago data, Hinckley (1968) found that keyword access to titles would have improved success rates to 80% from 30% with first-word access only. Multiple access and redundancy also greatly increase the chances of success when searching under incomplete or inaccurate bibliographic information.

10. People have more trouble with subject searching than with known item searching. The primary problems include (a) matching subject vocabulary with the controlled subject vocabulary and (b) reducing or limiting highly posted subject sets.

a. Studies of BACS, SULIRS, LUIS, and MELVYL showed that from 35% to 57.5% of subject searches resulted in no retrievals (Tolle, 1983a; Kern-Simirenko, 1983; Markey, 1984b, p. 83). This is consistent with the ALA and U.K. studies that found that about half of all users could think up a subject heading or entry word that would get them an entry or see reference in the catalog on first try (Meyer, 1977). And even 50% is probably an understatement of the subject matching problem, since there is no way to know how often the results are irrelevant when there *is* a match.

b. Highly posted sets occurred in 14% of all MELVYL searches (Larson, 1983) and 8% of all SULIRS searches at Syracuse University (Tolle, 1983a), at times when both online catalogs only contained a small number of these library's holdings. The problem can be expected to increase as online catalogs become larger.

11. Users were asked to choose the features they would most like to see in an improved OPAC (Online Public Access Catalog). The six top-ranked additional OPAC features were:

a. Ability to view a list of words related to the search words.
b. Ability to search a book's table of contents, summary, or index.
c. Ability to know whether or not a book is checked out.
d. Ability to print search results.
e. Searching by any word or words in a subject heading.
f. Providing step-by-step instructions (Markey, 1984b, p. 84).

12. Not one of the 189 searchers consulted the *Library of Congress Subject Headings* or other subject heading guide in the course of a subject search in the Subject Access Project sponsored by OCLC (Markey, 1983b).

13. Users do not normally record more than the call number. In nearly two-thirds of the subject searches in the Subject Access Project, the only information the users recorded was the call number. Full bibliographic information was recorded in only 7% of the searches (Matthews, 1983c).

14. Online catalogs have a number of factors which were not of concern in card-catalog-use studies. For example, for card catalogs the descriptive elements, their arrangement on the card, and the access points to the file were all standardized: the LC card format and access by author(s), title (first word), and subject heading (also first word). The user–system interface was also standard and simple: the user uses the file by flipping through the cards. Format, access, and the user interface are all more varied in online catalogs. Also, users want printing capability and an interface with circulation, neither of which were a possibility in card catalogs. In addition, users have a hard time knowing just what they can find in the online catalog. This was probably a problem with the card catalog, too, but it has been pointed out by the online use studies.

a. Users find brief displays satisfactory. Palmer (1972) found that a computer catalog which included an abbreviated citation of only five elements—author, title, call number, subject headings, and date of publication—would satisfy approximately 84% of user requirements. The online catalog use studies show that when a brief display includes call number, author, and title, less than 5% of patrons go to the full display (Matthews *et al.,* 1983).

b. On video screens, users prefer clearly labeled elements to the card-catalog layout. Fryser and Stirling (1984) tested users' preferences for the standard Library of Congress format, a "table of contents" layout, and a format with vertically arranged and underlined field labels. Both the variations were preferred to the Library of Congress format. This is consistent with research by Tullis (1981), which was not library-specific, but found that both a structured display format and a graphic display format resulted in significantly shortened response times than the narrative format.

c. The call number should be displayed prominently and away from other numbers and characters which could be confused with call numbers (Markey, 1984b, p. 129).

d. Mnemonic symbols are preferred over numbers in menu-driven systems. (That is, users prefer to enter A or AU for author than 1.) Users also had difficulty with codes and abbreviations.

e. People read upper- and lower-case letters more easily than only upper-case print. But they read labels faster that are all upper-case (Tinker, 1963; Poulton, 1968; Galitz, 1980, 1981).

f. Users want the human–computer interface to lead them to the next step. Users at all five of the OCLC Subject Access Project sites had trouble entering commands during the ongoing search. They did not like to use help screens. They wanted to be prompted with the appropriate option at the appropriate time.

g. "No matter how well designed the learning materials are, e.g. command charts at the terminal, exercises with sample searches, or programmed instruction, users prefer to plunge directly into their searches" (Hildreth, 1985, p. 263). When confronted with a problem, the average user will first read the screen (looking for the answer), next try online help if it is available, then ask someone standing nearby if he or she knows how to do it, then consult written materials, and only as a last resort will consult a library staff person.

h. Users had considerable trouble knowing exactly what they could expect to find in the catalog.

> Focussed-group participants at nearly every library expressed surprise when they found out that their library's computer catalog did not contain the entire contents of *Reader's Guide,* especially patrons at libraries where "all the library's holdings" were in the computer catalog. (Markey, 1984b, p. 138)

The top-ranked type of material to be added to OPACs was "journals or magazine titles" (Markey, 1984b, p. 86).

Although these use studies have not yet been run in sci/tech libraries, one would expect that because of the importance of the journal literature to sci/tech users, that the desire for access to journals would be even stronger. The only online catalog that currently contains journal articles as well as books is PaperChase, the catalog of Beth Israel Hospital in Boston. PaperChase is an end-user-oriented system. Users can search the medical literature by author, article title word, medical subject heading (MeSH), or journal title (Cochrane, 1982; Horowitz and Bleich, 1981). They can enter their query in natural language, and PaperChase establishes an initial match with the index vocabulary, then monitors the user's search strategy, asking for information to refine the search or offering suggestions for improving the search (such as suggesting MeSH terms to use). Beth Israel Hospital thus has a user-friendly catalog which truly represents the contents of its collection—journal articles as well as books. A medical library has an advantage over other sci/tech libraries in developing a catalog like this: the availability of Medline tapes which probably index all the journals available in the library. In a more general sci/tech library, probably no one abstracting and indexing service covers all the journals held, so the difficulty and cost of creating a catalog like this would be much larger. However, it is also probably what our users really need.

DESCRIPTIVE ANALYSIS

Library Catalogs

The two main standards for traditional library descriptive cataloging are AACR2 (the *Anglo American Cataloguing Rules,* 1978) and the MARC formats. AACR2 covers the content of the record and MARC is a standard for transmission of machine-readable bibliographic records. It is used for content designation and to maintain the records on a local system as well as for transmission. MARC was originally an acronym for Machine Readable Cataloging. It is sometimes confused with a cataloging standard because the Library of Congress cataloging (based on AACR2) is the primary product made available on MARC tapes.

Sci/tech libraries have a few differences from general libraries with regard to descriptive analysis. One difference is the relative unimportance of some things that general-library catalogers spend a great deal of energy on (such as uniform titles for Shakespeare's plays, pseudonyms for personal authors, and period and form subdivisions for literature). On the other hand, some things such as timeliness, specificity and accuracy of subject headings, and the handling of series entries and corporate-name entries are critical in sci/tech libraries.

In spite of these differences, most sci/tech librarians have found that AACR2 and the MARC formats can be used at least for the book collections. However, not all sci/tech libraries use them. One reason is that they are complex standards, and in a small library, a sci/tech librarian may feel the usefulness of meeting standards is not worth dealing with the complexity. However, this could be short-sighted. Not using the standards may make it impossible in the future to share cataloging or to use many of the sources of cataloging data. Also, most commercial online catalogs are being developed to use the MARC formats, and changing systems will be easier if the database is standardized. It should be mentioned, however, that there is variety even among records which meet MARC standards. LC, OCLC, RLIN, and many local catalogs adapt the tapes to their own internal formats, and some conversion process is usually needed from these to the MARC communication format.

Another reason these standards are not used in all sci/tech libraries is that the standards are relatively new. Many corporate libraries developed their computerized files long enough ago that the MARC formats were not available. These libraries are trying to decide whether the benefits of sharing are worth the cost of converting their files. In any case, libraries making the choice *now* should clearly choose MARC if at all possible.

The AACR2 rules on the description of library materials incorporate parts of the International Standard Bibliographical Descriptions (ISBD), which define punctuation and arrangement of bibliographic records (International Federation of Library Associations, 1977). The purposes of ISBD are to make it easier to recognize data elements whatever the language of their content, to facilitate the application of computer processes to the manipulation of bibliographic data, and to standardize national practices in the content and arrangement of the bibliographic record. Gorman (1981) summarized the basics of using AACR2 in Fig. 7.9.

In addition to discussing in detail the elements of a bibliographic description, AACR2 covers rules on the choice of access points, main entry, headings and uniform titles, and cross references (needed for authority control). Although AACR2 provides a structure, there is still judgement involved in cataloging, particularly in choice of entry and choice of access points.

Usually the author's name is the choice as the main entry for a work by a personal author, following a long cataloging tradition, but it depends on the type of publication. As more and more catalogs become online catalogs, the choice of main entry is becoming less important in many applications, since the user has equal access by any access point which has been made searchable. However, there are implications for collocation of records and filing arrangements if the "main entry" is dropped.

AACR2 follows the principle of uniform headings, which requires us

Rule 1. Describe the item you have in hand. Record the following details in this order and with this punctuation:

Title: subtitle/author's name as given; names of other persons or bodies named on the title page, label, container, title frame, etc.—Edition (abbreviated).—Place of publication: Publisher, Year of publication.

Number of pages, volumes, discs, reels, objects, etc ; Dimensions of the object (metric).—(Name of series)

Descriptive notes

Examples of descriptions

i) His last bow: some reminiscences of Sherlock Holmes / A. Conan Doyle.
   —London: Murray, 1917.
   305 p. ; 20 cm.—(Murray's fiction library)

ii) A white sport coat and a pink crustacean / Jimmy Buffett.—New York: ABC, 1973
   1 sound disc; 12 in.
   Backing by the Coral Reefer Band.

iii) Little Ernie's big day / by Norma Eustace ; designed by Doris Manier.—2nd ed.—Chicago: Little Folks, 1980.
   1 filmstrip; 35 mm.—(Big day filmstrips)

If the item is a serial (periodical, etc.) add the numbering of the first issue before the place of publication and leave the date and number of volumes "open" as in this example: Circulation systems review.—Vol. 1, no. 1–   . New Orleans: Borax Press, 1980–
   v.; 25 cm.

Rule 2. Make as many copies of the description as are necessary and add to each the name of the author and of other persons or bodies associated with the work.

Rule 2A. Give the names of people in their best known form.
   Wodehouse, P.G.
   Buffett, Jimmy
   Harris, Emmylou
   Seuss, Dr.

Rule 2B. Give the names of corporate bodies in their best known form.
   Yale University
   Coral Reefer Band
   Newberry Library

If a corporate body is part of another body, give it as a subheading *only if* it has an indistinct (blah) name.
   United States. Department of the Interior
   F. W. Woolworth Company. Personnel Division
   University of Michigan Library

Fig. 7.9. A summary of the AACR2. [From Michael Gorman's "The Most Concise AACR2." *American Libraries,* Sept. 1981, © 1981 ALA, reprinted with permission of the American Library Association.]

to enter all the works by a particular author under a uniform heading regardless of how many names or how many forms of a name an author has used (Chan, 1981, p. 99). This groups all of one author's works together. The principle of uniform heading requires the establishment of name authority records. Name authority files generally include the heading to be used in catalog entries, the sources upon which decisions were made in establishing the heading, and tracings for cross references to this heading (Taylor, 1984; Avram, 1984). Name authority files are similar to subject authority files, which we will discuss under subject analysis. Sci/tech libraries which do not conform to AACR2 do not usually worry about authority control of personal authors, except to decide whether the authors' names will be entered in full or just with initials. Pseudonyms are not as much of a problem—most scientists are trying to be clearly identified with their work.

However, corporate names are more problematical than personal names, and they are vital to sci/tech libraries because of their large collections of association publications, conference proceedings, government reports, and other publications by corporate bodies. The problem is particularly acute because corporate bodies change names frequently. As Hagler says,

> When a person changes name, there is no doubt that it is the same person before and after the change, and consequently that only one name should represent that person in a controlled vocabulary. However, when a corporate body changes its name, it is often because the body has become something different and/or wishes to project a new image. Many name changes result from a change of purpose or constitution, form a merger with another body or a split from one, etc. (Hagler, 1982, p. 200)

When a name changes there are two choices in how to handle it in a catalog:

1. Preserve the continuity of the body by having a single heading to represent the body before and after the change. If this is done, we still have to decide whether to change all the old headings to the new, or to continue to use the old for the new material. Old cataloging rules required the use of the most recent name of a corporate body, so when Los Alamos Scientific Laboratories became Los Alamos National Laboratories, all the old headings should have been revised. When a body did more than change its name (for example split or merged in some complex way), catalogers had to judge whether to retain the old heading.

2. Treat any change in a body's name as if a different corporate body has come into existence. This means we establish headings for both names and link them by cross references. Under this arrangement, which is what is required by AACR2, both ``Los Alamos Scientific Laboratories'' and ``Los Alamos National Laboratories'' are acceptable headings. One problem with this approach is what to do with a book about the history of Los Alamos which covers the lab under both names. The other problem is that this approach tends to scatter closely related works even when the changes are relatively minor.

AACR2 also has rules for the *form* of the corporate entry. The general rule is to enter a subordinate body under the lowest element in the hierarchy which has a name, omitting intervening elements unless the omission results in two or more bodies with the same heading. If it does, then the name of the lowest element in the hierarchy that will distinguish between the bodies is interposed. For example, the National Bureau of Standards is established as "United States. National Bureau of Standards" without Department of Commerce in between. A body created or controlled by a government is normally entered under its own name (such as U.S. Atomic Energy Commission), except for those with names implying administrative subordination (such as Department of Commerce) and those serving an executive, legislative, or judicial function (Chan, 1981, pp. 112–113). There are also special rules for entries of subordinate or related bodies, for government bodies, and for conferences. For example, headings for conferences are established in the form of: "*Name of conference* (number if any: date: place or institution)" (Chan, 1981, p. 112).

Because of the emphasis on uniform headings and titles, "see" and "see also" references need to be made from other possible headings and titles. Another set of rules, the COSATI guidelines, is widely used for the establishment of corporate entries for technical reports (see Volume 2, Chapter 12).

Transmission of Machine-Readable Bibliographic Records

Most *non*bibliographic machine-readable records use fixed-field data. Fixed-field means that a certain number of characters is allowed for each field of the record (e.g., 4 for year, 8 for purchase order number, 10 for ISBN, etc.). Each field also has a fixed place on the record. However, fixed-field records have been found to be inappropriate for bibliographic data. Many (indeed, most) of the elements of bibliographic data have unpredictable lengths (e.g., title, call number, author), and the results when truncated to some arbitrary length are unacceptable to most librarians. It is also unsatisfactory to leave a large enough space in the fixed field to accommodate the longest possible title because this wastes space on the magnetic tape or other storage media.

The MARC formats are variable-length formats. They use content designators to let the computer know where a field begins or ends. Content designators can be delimiters which act as location markers at beginning or the end of a subfield, a field, or a record, or they can be identifiers (including tags, indicators, and subfield codes) which identify specific types of data.

There are several variable-length formats for bibliographic information. The framework for the MARC formats is ISO 2709, a standard adopted in 1973. It has three parts: a label (or leader), a directory, and the data fields (Fig. 7.10).

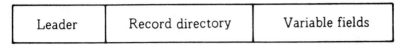

| Leader | Record directory | Variable fields |

Fig. 7.10 ISO 2709 and MARC structure.

The label ensures that every program written to accommodate any ISO 2709 format can identify to a computer the structure and relationships of the entry which follows the label. The directory includes an entry for each field; each entry includes a tag which designates the field, the length of the field, and its starting position. The data fields themselves are separated by a field delimiter, and the entire record is separated from the next record on the tape by a record delimiter.

MARC follows the ISO 2709 framework. The three-part structure is illustrated in Fig. 7.11, a view of how a record looks conceptually on magnetic tape with all the tags and delimiters.

The tags which identify the variable data fields are three-digit numbers and the first digit of the tag identifies the variable data. The groups include:

0xx = Variable control fields
1xx = Main entry
2xx = Titles and title paragraph (title, edition, imprint)
3xx = Physical description
4xx = Series statement
5xx = Notes
6xx = Subject added entries
7xx = Added entries other than subject, series
8xx = Series added entries
9xx = Reserved for local implementation.

For comparison, Fig. 7.12 is the same MARC record as displayed fully tagged in the RLIN database.

MARC actually has a number of variations. There are different MARCs in many countries, OCLC and RLIN each have their own version, and there is UNIMARC, adopted by IFLA and used for transmitting records between countries. Details of the variations are available in W. Crawford (1984).

Indexing and Abstracting

Many materials will not fit into traditional library cataloging. In the past, special libraries have been particularly adept at creating indexes of special materials. These range from pictures to laboratory notebooks to internal reports to maps. Studies show there is an advantage to having

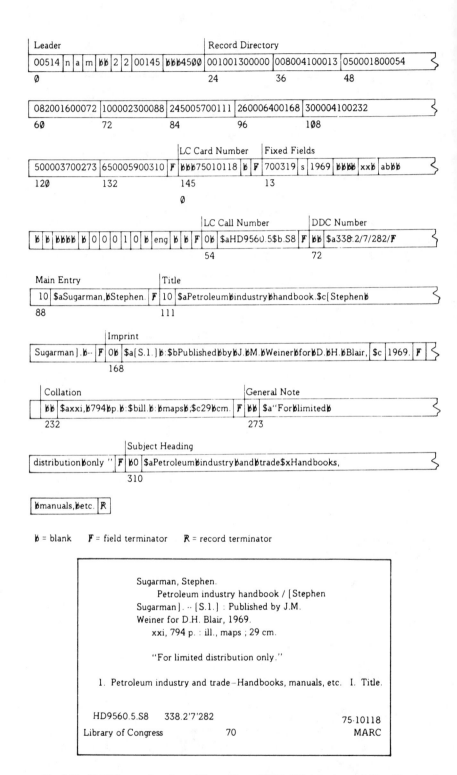

Fig. 7.11. MARC record on tape. [From Chan, (1981). "Cataloging and Classification," McGraw-Hill, 1981, p. 322. Reproduced with permission.]

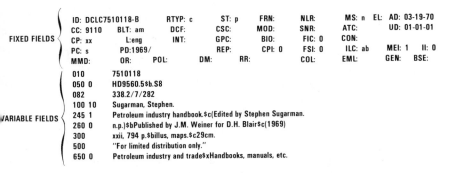

FIXED FIELDS

ID: DCLC7510118-B	RTYP: c	ST: p	FRN:	NLR:	MS: n	EL:	AD: 03-19-70
CC: 9110	BLT: am	DCF:	CSC:	MOD:	SNR:	ATC:	UD: 01-01-01
CP: xx	L:eng	INT:	GPC:	BIO:	FIC: 0	CON:	
PC: s	PD:1969/		REP:	CPI: 0	FSI: 0	ILC: ab	MEI: 1 II: 0
MMD:	OR: POL:	DM:	RR:		COL:	EML:	GEN: BSE:

VARIABLE FIELDS

010	7510118
050 0	HD9560.5$b.S8
082	338.2/7/282
100 10	Sugarman, Stephen.
245 1	Petroleum industry handbook.$c(Edited by Stephen Sugarman.
260 0	n.p.)$bPublished by J.M. Weiner for D.H. Blair$c(1969)
300	xxii, 794 p.$billus, maps.$c29cm.
500	"For limited distribution only."
650 0	Petroleum industry and trade$xHandbooks, manuals, etc.

Fig. 7.12. Fully-tagged MARC record displayed in RLIN. [Courtesy of Research Libraries Group.]

all material in one index with similar subject approaches [for example, as maps were added to the main catalog at the Colorado School of Mines, usage increased from 2031 used in 1978–1979 to 10,087 used in 1982–1983 (Phinney, 1983)]. But when the levels of indexing or what users need from the file differs, separate files are frequently maintained.

There are no standards like AACR2 or MARC for indexing and abstracting, except for the UNISIST Reference Manual (UNESCO, 1974), which, like MARC, is a communication format designed as a medium for the exchange of data from one organization to another, but, unlike MARC, is not widely used. For this reason, and because there is no economic reason for the abstracting and indexing services to standardize, there is considerably more variety in the descriptive records among the indexing services than there is among different libraries, most of whom are dependent on LC for cataloging.

This also makes the indexing and abstracting services less useful as a source of cataloging data, although progress is being made. As we mentioned, PaperChase uses Medline tapes to provide an online catalog of books and journals available in the library. The Department of Energy's Office of Scientific and Technical Information is working to create a database of Energy Data Base records which are in MARC format (see Volume 2, Chapter 12).

Sources of Bibliographic Records

Long ago, libraries realized the foolishness of everyone in the country doing the same cataloging operation on a particular book. Many mechanisms have been developed to share cataloging data: the Library of Congress has produced proof slips of new cataloging and published the *National Union Catalog*. Books from most major trade publishers include on the verso of the title page Cataloging in Publication (CIP) data prepared by the Library of Congress. The production of the MARC tapes provided the opportunity for online shared cataloging networks to develop. MARC tapes are purchased by the various networks (OCLC, RLIN, WLN and UTLAS are the largest) and shared among the members, who also share their original cataloging. The MARC tapes are also available through DIALOG and BRS. In addition, they are used by some commercial vendors (e.g., BroDart and Blackwell North American) as bases for their own systems. Other vendors provide LC cataloging in microform (e.g., MiniMARC and MARCFiche), and some are offering optical or video disk systems (e.g., Bibliofile). So there are many sources of cataloging data (McQueen and Boss, 1985), and most libraries with any large amount of book cataloging are using one or more of these sources. No commercial enterprise has competed with the Library of Congress to provide the original cataloging. This is because LC prices the MARC tapes only to recover the costs of making them available in machine-readable format, not to recover the entire costs of producing them (i.e., the intellectual effort). No commercial organization can compete with that, so they all supply LC-based cataloging. Hence the importance of AACR2 and MARC as standards: they are LC's standards, and virtually all cataloging available in the marketplace is based on what LC does.

The choice of where to get cataloging data, whether to join a shared catalog network, etc. is a complex decision, and is rarely based on cataloging alone. The need for holdings information for interlibrary loan is often a consideration for joining a shared cataloging network. There have been many studies comparing the four major bibliographic utilities. The usual parameters compared include size of database, access points, ability to retrieve needed information, system and terminal reliability, fiscal reliability, and available functions (all four provide shared cataloging, but some also provide retrospective conversion, online authority files, online acquisitions, online interlibrary loan, serials control, circulation control, or "search only" access). Libraries often do studies of hit rate to determine what proportion of their cataloging will be found in a database of interest.

SUBJECT ANALYSIS

Subject analysis is the process of analyzing the subject content of materials and assigning appropriate subject headings or other controlled vocabulary so that the items can later be retrieved by the user.

The intellectual side of subject analysis is similar to the process for information retrieval in that the initial analysis of the document (like the analysis of the information request) should be independent of the subject heading system employed. The first step is to examine the work and determine its subject content. Guides to the subject content (as possible substitutes for reading the whole thing) are the title, abstract, table of contents, chapter headings, preface, introduction, and accompanying descriptive materials. Sometimes external sources, such as bibliographies, catalogs, and other reference sources, may have to be consulted. An international standard for determining document subjects and choosing indexing terms, ISO/DIS 5963, is in draft form (New international standard for determining document subjects and choosing indexing terms, 1985).

The second step is to identify the main subject or subjects or principal concepts and their relationships to each other.

The third step is to represent the subject and concept according to a particular subject heading system or classification scheme.

A number of ways exist of using computers to assist in various phases of the subject analysis. Usually these techniques depend on counting words which occur in the abstract or the text itself, and using words which occur frequently as index terms or as guides for the indexer in choosing assigned vocabulary.

The Choice of a Subject Scheme

There *are* indexes in which vocabulary is not controlled. Index producers are attracted by lack of subject control, because of the speed with which the index can be produced. For example, consider the *Science Citation Index Permuterm Index*. No intellectual effort is put into indexing the journal articles by any controlled vocabulary. Words from the titles of articles are simply pulled out and combined with all other words from the title. And indeed, SCISearch has consistently the quickest turnaround time of any of the major journal indexes. The problem is that the speed gained at the beginning is often lost in the search process. The activity of bringing together all articles on one topic falls on the searcher rather than the indexer. The searcher has the burden of thinking up possible synonyms and variant spellings, instead of the indexer already having done it.

The main drawback of such an index is that it complicates searching in several ways. Searching becomes more difficult and uncertain, since this type of indexing spreads similar entries over many synonymous terms, ignores misspellings, and confuses any general–specific term relationships that certainly exist in the implied indexing language being unconsciously used. Terms for entering the index list are confusing, and a user has to search from one to another in order to get all the possible approaches to the subject of interest and must second-guess the author as to what terms to use. Since one of the qualities of a good index is that it quickly focuses on those entry terms that express the needs of the users and connects the index language to their way of thinking, uncontrolled concordance-type indexes are time-consuming and place a burden on the searcher. The speed and economy achieved in creating such indexes are paid for by the user in time and effort. Even in a computer-controlled retrieval system, the user may have to query the system over and over because of the limitation of no vocabulary control. The computer only allows the user to find the wrong things faster. (Cleveland and Cleveland, 1983, p. 30)

All the experts recommend controlled vocabulary to save the user time. Controlled vocabulary requires authority control (or syndetic structure)— the recording of which terms are approved and of cross references from synonyms, broader terms, narrower terms, related terms, and other terms that the user might use which are not used in the vocabulary.

There are a number of standard choices for a subject scheme. The *Library of Congress Subject Headings* (1980) (LCSH) is in use in a large number of sci/tech libraries, particularly those which are part of large general libraries. The application of LCSH is discussed in Chan (1981) and Curley and Varlejs (1984). Figure 7.13 shows part of the LCSH. Note that LCSH uses see, see also, and, in the authority file, x for "make a see reference from" and xx for "make a see also reference from."

The main advantages to using LCSH is that the subject analysis is available as part of LC cataloging, which is widely available from suppliers and that there is an organization keeping it up-to-date. However, only an average of 1.8 subject headings per item is added by LC (Mischo in Cochrane, 1984e), and this may not be enough for a sci/tech library. Also, lack of currency in the quickly changing sci/tech subject areas has been a traditional problem with using LCSH in sci/tech libraries (although recently sci/tech subject headings have appeared to be more current).

There are other problems with LCSH, and these are problems for general libraries as well as sci/tech libraries. These have to do with the basic rules for LCSH and the lack of consistency with which the rules are applied. For instance, one of the most basic rules for alphabetical catalogs is that of specific entry, which dates back to Cutter's *Rules for a Dictionary Catalog* (1904). This rule is that a book is to be entered under the most specific term that describes its entire contents. The idea is that a subject heading applied to a book should have the same scope of coverage as the contents of the book, and should be neither broader or narrower. For example, a book on PASCAL should not have a heading of computer programming languages. However, if the specific heading is not available, LC will apply the broader term. So users following the rule of specific entry are not necessarily successful.

As Bates says,

If there is only one heading assigned to a book—which, as has already been noted, is the case over half the time with LC—then a subject search *must* use that heading if that book is to be retrieved. Furthermore, in the LC system, synonyms are supposed

Fig. 7.13. Part of a page from the *Library of Congress Subject Headings*. Note the use of "s" to mean "see," "sa" to mean "see also," "x" to mean "make a see reference from," and "xx" to mean "make a see also reference from." For example, if we look under "Creep of materials" we should find a cross reference that says "see Materials—Creep," as indicated by the entry "x Creep of materials." Compare to Fig. 7.14. To find information in LSCH on creep of material, one must consult Materials—Creep. To find all information on mechanical properties, one must search a number of headings.

to be controlled. It can be argued that this does not always happen, but to the extent that it does, other terms are going to be on *different* subjects, related subjects perhaps, but not the same—desired—subject. Thus to truly home in on one's topic of interest, it is necessary to show perfect subject term choice. (Bates, 1977, p. 166)

Another rule that is applied inconsistently is the rule of direct entry. This means that PASCAL is entered under PASCAL, not Computer programming languages—Pascal. Topical subdivisions are a feature of an alphabetico-classed catalog, not the dictionary catalog which LCSH is supposed to be. Although subdivisions by form, geography, and period are allowed, topical subdivisions were not supposed to be (Haykin, 1951), but the introduction to LCSH says they are being used frequently (Bates, 1977).

Several authors have summarized the weaknesses of LCSH (Cochrane, 1984a; Mischo, 1982; Kirtland and Cochrane, 1982). Bates (1977) suggested that LCSH might be unsatisfactory for the specialist in any field—that its choice of headings in economics and psychology indicated that it thought all users would be laypersons.

Many sci/tech libraries which are branches in large academic libraries will use LCSH in spite of these weaknesses. However, in sci/tech libraries not tied to LCSH by some higher organizational structure, other subject heading schemes are often more attractive. Even if a library chooses a subject analysis scheme other than LCSH, it can still take advantage of available cataloging from networks and commercial vendors for the descriptive cataloging. Local subject headings can be added in the MARC record (various libraries have used the 653, 690, and 691 fields).

Many sci/tech libraries have chosen to use thesauri developed for the abstracting and indexing services in the area of their interest as controlled vocabulary (e.g., an energy library might choose to use the *Energy Data Base Subject Thesaurus*). An example of part of the *Energy Data Base Thesaurus* is given in Fig. 7.14. Note that the cross reference structure in thesauri is more specific about relationships than the cross reference structure of cataloging. Thesauri usually specify related terms (RT), broader terms (BT), narrower terms (NT), and terms instead of which the heading is used (UF). Many users find the thesaurus-type notation easier to follow than x, xx, sa (Cochrane, 1984, p. 147). Attempts to redo LCSH into the thesaurus structure have not been successful.

In general, the abstracting and indexing services do a better job of using specific, up-to-date language in the sci/tech fields than LC does. Sometimes libraries take these thesauri as a base and then add new terms, with all their related terms, and gradually the thesaurus grows to be a new product. Corporate libraries almost always want to index by project names or subject headings of local importance, so whatever subject heading scheme is chosen should be able to incorporate additions of local terms.

MEᵤ.
DA   Lᵣ.
BT1 Impeᵤ

MECHANICAL k.
DA   June 1977
BT1 Artificial Organ.
BT1 Kidneys
BT2 Organs
BT3 Body
BT1 Prostheses
BT2 Medical Supplies
RT   Dialysis
RT   Urinary Tract

MECHANICAL POLISHING [01]
BT1 Polishing
BT2 Surface Finishing

➡ **MECHANICAL PROPERTIES [01]**
UF   *Mechanical Effects*
UF   *Properties (Mechanical)*
NT1 Brittleness
NT1 Compressibility
NT1 Compression Strength
➡ NT1 Creep
NT1 Dilatancy
NT1 Fatigue
NT2   Corrosion Fatigue
NT2   Thermal Fatigue
NT1 Flexural Strength
NT1 Fracture Properties
NT1 Hardness
NT2   Microhardness
NT1 Impact Strength
NT1 Plasticity
NT1 Poisson Ratio
NT1 Shear Properties
NT1 Tensile Properties
NT2   Ductility
NT2   Elasticity
NT3   Photoelasticity
NT3   Thermoelasticity
NT2   Flexibility
NT1 Ultimate Strength
NT1 Wear Resistance
NT1 Yield Strength
NT1 Young Modulus
RT   Destructive Testing
RT   Physical Metallurgy
RT   Stresses

MECHANICAL STRUCTURES [01]
UF   *Columns (Mechanical)*
UF   *Structures (Mechanics)*
NT1 Bellows
NT1 Bridges
NT1 Domed Structures
NT1 Honeycomb Structures
NT1 Intake Structures
NT1 Outlet Structures
NT1 Supports
NT2   Foundations
NT2   Fuel Racks
NT2   Powered Supports
NT1 Towers
NT2   Cooling Towers
NT3   Mechanical Draft Cooling
       Towers
NT3   Natural Draft Cooling Towers
NT2   Power Transmission Towers
RT   Buildings
RT   Civil Engineering
RT   Construction
RT   Ratcheting
RT   Response Functions

Nᵣ
NT₁
NT1
NT2 ⌐.
NT2 Elecᵣ
NT2 Hydrauᵣ.
NT2 Hydrodynaᵢ.
NT3   Electrohydrouᵣ
NT3   Magnetohydrodyᵢ.
NT2 Magnetogasdynamics
NT2 Pneumatics
NT1 Fracture Mechanics
NT1 Quantum Mechanics
NT1 Rock Mechanics
NT1 Soil Mechanics
NT1 Statistical Mechanics
RT   Anharmonic Oscillators
RT   Canonical Transformations
RT   Center-of-Mass System
RT   Equations of Motion
RT   Galilei Transformations
RT   Hamilton-Jacobi Equations
RT   Harmonic Oscillators
RT   Kinetics
RT   Laboratory System
RT   Lagrange Equations
RT   Lagrangian Function
RT   Moment of Inertia
RT   Physical Metallurgy
RT   Surface Forces
RT   Virial Theorem

MEDEC PROCESS
DA   August 1980
BT1 Radioactive Waste Processing
BT2 Waste Processing
BT3 Processing
BT3 Waste Management
BT4 Management
RT   LMFBR Type Reactors
RT   Sodium
DEF A process for removal of
     elemental sodium from LMFBR
     radioactive wastes.

MEDIASTINUM [01]
BT1 Chest
BT2 Body Areas
BT3 Body
RT   Aorta
RT   Esophagus

Nᵣ
Nᵣ
NᵢI
RT
RT   ᵢ

MEDICINE
UF   *Internᵤ.*
NT1 Balneologᵣ
NT1 Dentistry
NT1 Gynecology
NT1 Hematology
NT1 Industrial Medicine
NT1 Neurology
NT1 Nuclear Medicine
NT2   Radiology
NT3   Biomedical Radiography
NT4   Fluoroscopy
NT4   Ionographic Imaging
NT4   Osteodensitometry
NT4   Renography
NT3   Radiotherapy
NT4   Afterloading
NT4   Neutron Therapy
NT5   Neutron Capture Therapy
NT1 Ophthalmology
NT1 Pediatrics
NT1 Preventive Medicine
NT1 Surgery
NT2   Adrenalectomy

Fig. 7.14. Part of a page from the *Energy Data Base Subject Thesaurus* (U.S. Department of Energy, 1984). Note the use of "UF" for "used for," "BT" for "broader term," "NT" for "narrower term," and "RT" for "related term." This structure is more exact about the relationships between terms than is the LCSH structure. Partly because the relationships are specified exactly, EDB can allow the searching of hierarchies. For example, searching EDB by Mechanical Properties will also retrieve all materials indexed with Creep, Fatigue, or any other narrower term listed under Mechanical Properties, and the terms can also be searched separately.

A large collection of subject heading lists, classification schemes, and thesauri, begun by the Special Libraries Association in 1924, is currently available at the University of Toronto. For information about the collection and interlibrary loan and photocopying services contact Subject Analysis Systems Collection, Faculty of Library Science Library, Room 404, 1140 St. George St., Toronto, Ontario, Canada M5S 1A1; telephone (416) 978-7060.

Some examples of the kinds of thesauri and subject heading lists which are available and could be used as bases for library indexes are listed in Table 7.1.

For some collections, no existing thesaurus or subject headings list will be satisfactory. In this case, the library may choose to develop its own. Although some guidelines exist (e.g., American National Standard Guidelines for Thesaurus Structure, Construction and Use, 1980; Heald, 1966; UNESCO, 1973; Bernier, 1976; Blagden, 1968), the development of a new list of subject terms with the recording of all the cross-references is not a trivial task. Like cataloging and indexing projects themselves, it should not be undertaken lightly. But if it is undertaken, it should be done well and thoroughly and the results should be published so others can use it. Depositing a copy at the University of Toronto is a useful mechanism for making it available to others, too. Dirlam (1986) investigated sources and

---

TABLE 7.1.   **Selected Subject Thesauri Which Could Be Used for Subject Analysis**

Aluminum Association. Committee on Technical Information. (1980). *Thesaurus of Aluminum Technology,* 3d ed. The Aluminum Association, Washington, D.C.

COSATI Subject Heading List. (1964). Committee on Scientific and Technical Information of the Federal Council for Science & Technology, Washington, D.C.

Fish and Wildlife Reference Service Thesaurus. (1981). 2d ed. Denver Public Library Fish and Wildlife Reference Service, Denver, Colo.

Food Science and Technology Abstracts Thesaurus. (1981). 2d ed. International Food Information Service, Reading, England.

Medical Subject Headings (1983). National Library of Medicine, Bethesda, Md.

Metals Information Thesaurus of Metallurgical Terms. (1981). 5th ed. American Society for Metals, Metals Park, Ohio.

NASA Thesaurus. (1982). National Aeronautics and Space Administration, Scientific and Technical Information Branch, NASA SP-7051 (2 vols.).

Petroleum Exploration and Production Thesaurus. (1985). Tulsa Petroleum Abstracts Advisory Committee, Tulsa, Okla.

Thesaurus of Engineering and Scientific Terms (TEST). (1967). U.S. Department of Defense, Washington, D.C., AD 672000.

U.S. Department of Energy. (1984). *Energy Data Base Subject Thesaurus.* Technical Information Center Office of Scientific and Technical Information, Oak Ridge, Tenn. DOE/TIC-7000-R6; DE84010568.

---

software available to help with development of a thesaurus for gems and gemology.

Proposals to Increase Subject Access

Particularly since users seem to be using online catalogs so heavily to search for subject information, there has been in the early 1980s an increase in attention to subject access. Mandel (1984) reviewed proposals for enriching the library cataloging record for subject access, and Lawrence (1984) reviewed system features for subject access in the online catalog. One proposal that has been revived is the addition of terms from tables of contents and indexes as subject access points in catalogs. This was first done in the Subject Access Project under a grant from the Council on Library Resources in 1976–77 (Atherton, 1978). The participants selected an average of 32.4 terms per book, taking an average of 10 minutes per book. They conducted searches (all in the humanities and social sciences) and found that the augmented record (called the BOOKS record) was superior on recall (2–3 times better) and computer time, but that both kinds of records were low on precision (less than 67%). BOOKS searches failed to retrieve 42 of the known relevant items, while MARC searches failed to retrieve 117. The "rules" developed by this project are summarized by Settel and Cochrane (1982).

Choice of a Classification Scheme

Classification is generally developed by taking a general topic and dividing it successively into its parts. A tree with many branches is a good image of a classification scheme, although faceted classification schemes do not really fit that image.

In U.S. libraries, classification is thought of mostly as a shelving device, a numbering scheme for arranging the books on the shelf. The function of the call number to help the user identify and locate a book on the shelf can be fulfilled by any method of numbering or marking so long as there is a correspondence between the number or mark on the document and that on the cataloging record. However, classification schemes provide the additional function of grouping like items together and allowing subject retrieval. For this reason, classification schemes are sometimes also used to organize bibliographic entries in catalogs and indexes.

In theory, materials could be grouped by author, physical form, size, or date of publication, but subject is the predominant characteristic for grouping.

All classification schemes use some kind of notation to represent the various classes and subclasses. For example, LC uses letters and numbers, and Dewey Decimal (DDC) uses numbers. For collections limited to a particular subject like science and technology, great detail has to be represented in the classification and LC has the advantage over Dewey because LC's notation does not become as lengthy. However, both have the advantage over less-used classifications, such as the Colon classification, that they are continually maintained and updated.

No classification scheme completely groups like items. Skolnik (1982) gives details of the way LC and Dewey scatter the four books in Table 7.2.

Books 1 and 2 are concerned with linear polymers of ethylene oxide, yet 1 is classified in "Applied Science" and 2 in "Pure Science", 1 in "Plastic Materials" under "Chemical Technology" and 2 under "Organic Chemistry" in the LC system, which places them very far apart in even a special library. These two books are separated similarly by the Dewey classification, with 1 classified in the "Plastics Industries, Resins, Gums" under "Applied Science" and 2 in "Polymerization" under "Organic Chemistry". Book 3, on the other hand, is shelved by LC and Dewey reasonably close to book 2, yet not so close that both could be spotted easily by browsing in the stacks. Book 4 is completely isolated from the totality of polymer science and technology books. (Skolnik, 1982, pp. 10–11)

Of course, a local library could choose to put these books all together in one number; the number selected by the Library of Congress for its own purposes need not be used blindly.

There are also a number of classification schemes developed specifically

TABLE 7.2.   Four Polymer Books with their LC and Dewey Classification Numbers: An Example of Scattering[a]

Book	LC number	Dewey number
1. Bailey, F. E., Jr., and Koleski, J. V. (1976). *Poly(ethylene Oxide)*. Academic, New York.	TP1180 P653 B34	668 .4'234
2. Vandenberg, E. J., ed. (1975). *Polyethers*. American Chemical Society, Washington, D.C. (ACS Symposium Series 6).	QD380 P63	547'.84
3. Kennedy, J. P. (1975). *Cationic Polymerization of Olefins: A Critical Inventory*. Wiley-Interscience, New York.	QC305 H7 K38	547'.8432'234
4. Schultz, J. (1974). *Polymer Materials Science*. Prentice Hall, New York.	TA455 P58 S36	620 .1'92

[a]Based on Skolnik (1979).

for sci/tech applications, including that of the Engineering Societies Library (which has one of the few classed catalogs currently existing).

The *COSATI Subject Category List* (U.S. Federal Council for Science and Technology Committee on Scientific and Technical Information, 1965) is one commonly used in sci/tech applications. NTIS has used it for grouping its announcements of technical reports in *Government Research Reports and Announcements* since 1964. It is occasionally used as a basis for arranging reports on the shelf. It was established in 1964 and was endorsed by the Committee on Scientific and Technical Information (COSATI) of the Federal Council for Science and Technology. The two-level list consists of 22 major subject fields (categories) and their 178 subdivisions (subcategories). In 1970 NTIS developed a new scheme which could be used with the COSATI classification (which did not cover the soft sciences adequately for NTIS's purposes). There are currently 39 major categories divided into 325 subcategories.

NTIS currently uses both schemes, assigning one COSATI code and up to four NTIS codes to each report. All of the subject categories consist of three-character codes; two numerics and one alpha character. The numeric codes represent entire categories such as "chemistry" and "biology." The alpha codes are used to designate subcategories within these broad categories. The COSATI subject categories range from 01 through 22. The NTIS subject classifications range from category number 43 through category 99. NTIS/SR-77/04 (1977) contains a description of the NTIS subject categories. Complete scope notes for the COSATI subject categories may be purchased from NTIS using the order number AD-612200.

Many of the producers of abstracts and indexes also have classification schemes which might be usable in some situations: for example, *Chemical Abstracts,* COMPENDEX, GeoRef, *Mathematical Reviews.* These classification schemes are more often used to group bibliographic entries than as shelf locations.

CHOICES TO BE MADE IN PRODUCING CATALOGS AND INDEXES

Whole books have been written on indexing (e.g., Borko and Bernier, 1978), and certainly anyone considering an indexing project should read one or more of them. No indexing task should be taken on lightly or without proper planning. Almost every librarian can cite an example of an indexing project for a small collection which later outgrew the indexing scheme.

The planning should start with the user of the index. It should also include decisions on the following aspects: (1) the content of the biblio-graphic description, (2) the subject description (which subject headings or index terms or classification scheme), (3) access points, and (4) what the index should look like, from its internal arrangement and filing order to its physical format. This includes the system interface if the index is online.

We have already discussed the content, the subject description, and some aspects of display formats and user–system interface. The important points include:

1. All important elements of the bibliographic description should be included. The temptation to use an abbreviated record is often regretted later.

2. It is useful to follow some already-established pattern for bibliographic description. If cataloging rules are not appropriate, it may be possible to follow an ANSI standard, the *UNISIST Reference Manual,* or the pattern of one of the abstracting and indexing services.

3. Controlled vocabulary for author, corporate bodies, and subjects is essential. Even in small files, the variation in the way even just a few people will enter and search for names and titles may create problems with finding entries if they are not controlled. Decisions should be recorded and cross references made from the not-chosen forms which people are likely to use.

Access Points

Title, personal author, corporate name, and subject headings are the common access points in both library catalogs and indexes. In card and book catalogs, any expansion of the number of access points is discouraged because of expense. Some hard-copy formats for special uses have had other access points, such as report number, project number, contract number (for technical reports), scale (for maps), or patent number.

With online catalogs, the choice of access points is much broader, and the user's access is dependent on how the system makes things available as well as on the choice of access points. Markey reviewed all the MARC fields which are subject-rich and how many current online public catalogs make them searchable (1984b, Appendix F, pp. 158–159). The use studies show that keyword access to titles and subjects is needed, and call-number access for browsing has a surprising amount of appeal. Contents notes can provide subject access. Other access points, such as date, publisher, and notes, are useful for limiting a search.

One type of access point which should be particularly useful in a sci/

tech library is keyword access to corporate names. Dickson (1984) described the difficulty of finding the *NASA Tech Briefs* in the NOTIS file because of the complexity of the cataloging, and suggested that making corporate names for serials searchable in the title file as well as the author file would be useful.

Online catalogs and KWIC (keyword in context) and KWOC (keyword out of context) indexes normally use stop words to keep certain common words (e.g., the, an, a) from producing large numbers of useless entries. Care should be taken to see that this does not make certain titles unfindable because they happen to be made up of all stopwords. Systems with large numbers of stopwords should have some way of compensating for this problem, perhaps by allowing the cataloger to tag titles at input when he or she recognizes that the title would be all stopwords.

A similar problem (unfindable titles) occurs because of past practice of not making a title access point if the title was the same as a subject heading (e.g., the title *Physics*), or if it began with a common word like Proceedings. Current practice is to make title an access point without exception.

As we have mentioned before, there is a problem with subject access when the access is precoordinated and not duplicated in reverse. For example, a book on water resources in Colorado would be useful to users looking under water resources and also to those looking under Colorado. When the cataloger is forced to choose between the two precoordinated headings (as he or she is with LCSH: Water resources—Colorado River, or Colorado River—Water resources) there can be an access problem. Access points for subjects are being widely discussed in the mid-1980s, as we have mentioned before (see Cochrane, 1984a,b,c,d,e,). Most experts are saying that in a machine-readable environment, it is both unnecessary and inefficient to precoordinate elements. The machine can postcoordinate in whatever manner the user wishes. (Hagler, 1982, p. 170.) The way this is handled in DOE's Energy Data Base is ideal. Each term is searchable separately, but the relationship is also recorded so that the terms can be searched as a "major descriptor pair" if precise retrieval is needed.

Files and Filing

. . . there are few absolutes in determining advantageous file structures in either the manual or the machine mode: there are only trade-offs in terms of both user service and economics. The issue of a structure for a bibliographic file is a very complex one because 1) these files tend to be very large, 2) the average record is consulted very infrequently yet is expected always to be instantly accessible, 3) each record contains an assortment of data elements whose interrelationships can be very complex, and 4) users have been trained to expect a high degree of organization in which works and subjects are intelligibly collocated, and persons and bodies easily identified. Cost-efficient file structures to meet all these demands simultaneously are probably non-existent. However, librarians have produced good (if expensive) results to date with manual files and are in process of perfecting computerized techniques for better service at lower cost. (Hagler, 1982, p. 157)

There are four basic choices for arrangement of entries in a file: alphabetical, classified, alphabetico-classed, and register-catalog. Other choices, such as chronological, are rarely used as the primary arrangement, but are sometimes used as a subarrangement (e.g., works on a subject may be subarranged in reverse chronological order to list the newest items first).

The usual arrangement of records in U.S. library card catalogs is an alphabetical arrangement. Very few classified catalogs exist—but the Engineering Societies Library has one. Classification is, however, commonly used as a way of arranging abstracts. NTIS, which arranges its abstracts in *Government Reports Announcements and Index* by COSATI category, is an example of a classified bibliographic file.

When entries are arranged by a limited set of subject terms rather than a notation, we have an alphabetico-classed catalog. A limited set of subject terms is chosen, and all terms assigned are subheadings of these main terms. *Engineering Index,* which has terms like "Mechanical properties—creep" and "Mechanical properties—strain," is an example of this kind of index.

Another arrangement has been used primarily in COM and online catalogs. The records in the main file are filed as they are entered into the file, and indexes provide the only access. This arrangement is sometimes called a register-catalog. It has the disadvantage in manual systems that the user always has to look in two places (the index and then the register). The advantage is that supplements can be added on to the original register catalog without redoing the whole thing—only the index needs to be updated. In online files, it is a fairly compact storage system. The whole entry is stored only once, but "pointers" from each of the indexes to the number of the record are created. The actual arrangement in this case is usually transparent to the user.

One of the reasons an alphabetical arrangement is often chosen for library catalogs is that alphabetical arrangement seems natural to users, who learn it when young. The disadvantage is that closely related topics, such as "earthquakes," "seismology," and "tectonics," are widely scattered alphabetically. A classification scheme like LC would group them together under QE 531–541. But classification schemes can also scatter material, as Skolnik's example in Table 7.2 illustrates. Any particular item might belong in several areas. For example, aircraft might be divided by kind of engine (e.g., jet), kind of user (e.g., military), or purpose of use (e.g., cargo). This problem is somewhat solved in the development of faceted indexes, where numbers for new subjects (like the space shuttle) can be synthesized out of the numbers for the old (e.g., transportation and space exploration). Classified arrangements require an alphabetical index to the classification, so most libraries choose an alphabetical subject heading arrangement as their primary catalog and rely on the arrangement on the shelf or in the shelf list to bring the materials together by classification.

Alphabetical arrangement, while familiar, is not always obvious. Libraries have developed extensive filing rules to deal with inherent problems. Yes, "a" goes before "b" but where does "$" go? Or "2"? Any filing code has to deal with questions like these:

1. Will filing be letter-by-letter or word-by-word? That is, will New York file before Newark (word-by-word) or after (letter-by-letter)? Most libraries file word-by-word, using the principle that "nothing comes before something," that is, that a blank precedes any other letter or character. This is also the ASCII sort sequence for computer filing, if the blank is not ignored. It either files before all other characters (in ASCII) or between punctuation marks and alphanumerics, producing word-by-word arrangement. If the blank is ignored in computer filing, the result is letter-by-letter arrangement. There are only three studies which tried to show user's preferences for word-by-word or letter-by-letter filing. Although letter-by-letter seems to be slightly preferred, Milstead says the results are inconclusive (Milstead, 1984, p. 49).

2. How will the following be filed: signs and symbols, numbers, abbreviations, initialisms, acronyms, diacritical marks, punctuation? Some schemes file abbreviations as separate words but acronyms as one word. Some spell out numbers; others file them at the beginning or at the end.

The increased use of computers has simplified the filing order by not allowing so many options. Machine filing has been unable to handle the intricacies of library filing rules (e.g., filing numbers as if they were spelled out), and therefore many of the so-called standards are not being used. Even before machine filing, there was not as much standardization about filing as there has been about descriptive cataloging.

CONTROL OF PERSONAL FILES

Many sci/tech libraries are being asked to help their users control their personal files. Most scientists have collections of reprints, and increasing the efficient use of those files should significantly increase a researcher's productivity. Much of what we have discussed in this chapter is applicable to personal files as well as library files, but the personal nature of the files and their uses makes it even harder to suggest "the perfect system." Books have been written on personal information systems (e.g., Jahoda, 1970; Stibic, 1980; Foskett, 1970) and considerable variety in approach is found because the system must be simple to maintain and work for a particular individual. Wanat summarized the principles as follows:

> Every system needs to plan for change; all researchers' interests will change over time, as indeed entire disciplines will change focus.
> Every system needs to plan for growth; even the most rigorously weeded collection will grow over time.
> There is an inverse relationship between the effort expended in placing a document into the file and the effort necessary to retrieve that document later.
> If a system requires too much effort to maintain, it will fail. Therefore, it is essential to develop the simplest system which still meets the researcher's needs. The payoffs must be worth the effort expended or the researcher will fail to maintain the file.
> (Wanat, 1985, pp. 255–256)

There are basically three categories of personal files: single-access systems, manual multiple-access systems, and computer-assisted systems.

The single-access system is the most common first approach. Papers are arranged in folders by subject or first author. The arrangement is simple, but retrieval can be difficult, since the filing heading may not be what is recalled.

Manual multiple-access systems supplement this basic arrangement with added indexes, often on index cards (e.g., if the files are in author order, the card index is by subject). Since 1983, Wanat has been teaching a seminar at the University of California at Berkeley on management strategies for personal files. She reports that interest in manual methods is still quite high in spite of the large number of computer-aided systems being developed.

The natural expansion of the manual multiple-access system is to use the computer to maintain the record. Database managers can be adopted for this purpose [e.g., see Palmer (1984) for directions for using dBase II or III for information retrieval], and a number of specialized programs are also appearing on the market. None of these yet meets all the needs of document management (e.g., some are good at retrieval but do not provide output, other provide good output but have a cumbersome input process, etc.) At the end of 1985, the available packages included Bibliofile, Bibliog, Bibliography Writer, Bibliotek, Bookends, CAIRS, Citation, FYI 3000, Library-Mate, IN-MAGIC, Microfile, Notebook II, Paperbase, Paperfinder, Professional Bibliographic System, PULSAR, Quick Search Librarian, Ref-11, Sapana Cardfile, SciMate, Searchlit, SIRE, Superfile, and ZyIndex. Seiden and Kibbey (1985), Wanat (1985), Lundeen and Tenopir (1985), Chiang *et al.* (1985), Burton (1985), Fetters (1985), and Hubbard (1986) all discuss more than one of these systems. This area is changing so quickly that scanning current journals is necessary to keep up with new developments. *Library Software Review, Library Hi Tech, Online,* and *Special Libraries* are good journals to watch for changes in these systems, but increasing numbers of reviews are also appearing in the journals directed toward micro users (e.g., *Byte, PC World*).

PHYSICAL CONTROL

Physical control covers all aspects of handling the physical materials, from deciding on a shelf arrangement, to shelving the materials, to circulation control.

The choice of a classification system was discussed earlier in this chapter. Although one of the standard classification schemes is often chosen for the books in a sci/tech library, it is frequently not used for collections of nonbook formats, unless the collection is small. Although the specific formats are considered separately in Volume 2, Part I, some general principles can be summarized here.

1. The principle of shelf arrangement is to arrange items the way people are likely to ask for them. Some other approach may be more important than subject, and it may be reasonable to keep materials together by form. For example, if well logs are usually asked for by the township/range location of the well, it makes sense to use township/range as the overall arrangement of the file and create indexes (perhaps a map, perhaps an alphabetical subject index including geographical terms) to handle the occasional question which has a different approach.

2. A new location or classification number should not be superimposed if one is already inherent in the material (e.g., Standard Technical Report Number for technical reports), unless there is some overriding reason to do so. One overriding reason might be if space were more pressing than staff time or the need for direct access. Superimposing on reports an accession number which makes new reports always shelve at the end means that shelf space can be used more efficiently than if empty space needs to be left all through the collection.

3. A simple accession number is fine for some types of material, if indexes are providing access from all the other possible approaches. The shelf arrangement need not be by subject.

4. The location number, whether an accession number, report number, or call number, should be *unique*. The number should be clearly and consistently marked on the document as well as in the indexes. And it should

be filed by this number. These principles may seem almost too obvious to be stated, but they are violated much too often. For example, filing systems are frequently devised which give the same general classification to a number of items, such as NP for all the material on National Parks. And the index from the Nuclear Regulatory Commission *(Monthly Title List of Materials Currently Received)* for a long time revealed that an item existed but did not give any unique filing location.

The components of circulation control include checking out materials (applying loan periods, which may vary by borrower status or type of material); checking in materials; shelving materials which have been returned (and usually also those which have been used in-house); accepting holds for particular items; sending notices (which may include overdues, recalls, notification that a book requested on hold is available, and bills for lost material); and searching for missing or misplaced items.

In general, circulation policies vary more between libraries which are in different settings (e.g., corporate versus academic) than between libraries which are in the sci/tech subject area and those that are not. However, the chapters covering specific subjects in Volume 2, Part II bring out a few of the special factors related to circulation for specific subjects. For example, photocopies of paleontology journals are often not satisfactory—the journals themselves need to circulate so that the scientists can compare the plates with specimens at the microscope. And mathematicians seem to prefer a noncirculating collection.

The matter of overdues is of great concern in academic settings, but not so much in corporate setting.

> The special library cannot force the return of materials which it owns but which users insist they still need, six months or two years later. It cannot win an argument over the question of what someone needs to do his or her job. However, the library can insist that since its copy is now being retained on what is effectively permanent loan, a second copy will have to be purchased and budgeted to the department in question. Not only do users find this fair, so do accountants. (Herbert White, 1984a, p. 51)

However, even special libraries should consider that books not on the shelf are not available to those browsing for them. Buckland's *Book Availability and the Library User* (1975) showed that loan policies strongly influence availability of library materials. He showed that decreased loan periods on the most-used items could greatly increase availability.

Circulation was one of the first functions to be automated, and there are a number of automated systems which function well for circulation. Boss and McQueen (1982) analyzed features needed in an automated circulation system and Matthews and Hegarty (1984) edited a conference proceedings volume covering available choices. Strauss *et al.* (1972) covered manual circulation systems, including the use of book cards and edge-punched cards, many of which are still in use.

MANAGEMENT ASPECTS

The aspects of collection control which we have discussed are usually split between a technical services organization (which does acquisitions, receiving, cataloging, and physical processing) and a circulation department which handles shelving, shelf maintenance, and circulation.

In libraries which are divided between public and technical services, circulation tends to be in the public services section, while acquisitions, receiving, and cataloging are in technical services. However, automation is changing this perception to some extent.

> The major task with regard to circulation is the file building associated with the marriage between the item record and the patron record. Since the catalog record is the basis for the item record, the creation and maintenance of these machine records within technical services argues for that functional unit to manage circulation. (Freedman, 1984, p. 1198)

The collection control function can also be split along "type of material" lines, rather than function. For example, monographs, serials, reports, government documents, etc. can be considered separately and have separate organizations to deal with all aspects of collection control. This "form versus function" argument has not been conclusively settled for any of the formats. As a general guideline, if a format is extremely important to the users of a collection (for example, serials and technical reports in sci/tech libraries), it may be more effective to organize along type of material lines. Otherwise, organizing by function (acquisitions, receiving, cataloging, circulation) will be more efficient.

"Acquisitions" includes verification of the item, choosing a source, ordering the item by creating a purchase order, and keeping track of funds. One of the prime functions is to interface with the purchasing department of the overall organization.

Verification for acquisitions includes price information, something not commonly included in a bibliographic record, so some of the sources are different from the verification sources discussed in the next chapter on document delivery. Common sources for verification include *Books in Print,* publishers' catalogs, and, increasingly, online databases. Sometimes a phone call is the most efficient form of verification, so acquisitions departments maintain directories of organizations. In a sci/tech library these must include not just trade publishers, but also scientific and technical professional societies and government agencies.

Several aspects of acquisitions were covered in Chapter 6. But the creation of the purchase order is the beginning of the bibliographic record for the item, so acquisitions is also part of collection control.

The order record needs to correspond both to cataloging practices (so it can be used as a base for cataloging) and to ordering standards. ANSI has standards for 3 × 5 forms and for multiple order forms (ANSI Z39.30-1982) (American National Standards Institute, 1984). The Book Industry Advisory Committee (BISAC) has been developing a format for transmitting order information from buyers to vendors. ALA's *Guidelines for Handling Library Orders for In-Print Monographic Publications* (American Library Association Bookdealer–Library Relations Committee, 1984) is a good summary of all the information that must be included in an order.

Acquisitions in sci/tech libraries differs from the same function in general libraries mainly because of dealing with special formats and special publishers (e.g., professional organizations) and because of the pressing need for timeliness.

The receiving function is similar to receiving in a purchasing organization. Receiving staff maintain a file of orders (in manual files by purchase order number, author, and/or title). They claim on orders which are not received after a certain time (either manually or automatically) and collect information to analyze vendor performance. They receive the materials (which involves verifying that they received what was ordered, and pulling and marking the order). They process payments.

The cataloging section produces the catalog record, conforming to whatever standards are chosen. Descriptive cataloging and subject cataloging can be separate administratively. Original cataloging can also be separate from "copy cataloging" (when "copy" is found in a source like OCLC, the *National Union Catalog,* etc.).

As Godden (1984), Striedleck (1984) and Neal (1984) point out, automation is bringing diverse changes in organization. There is more interrelatedness between parts of the control cycle (e.g., between acquisitions and cataloging) than there was with manual systems. Also, the availability of all bibliographic information throughout the library (instead of only at the one physical location of the card catalog) allows for the dispersal of technical services staff. Catalogers need not be centralized in one physical location.

Staffing for collection control will vary depending on whether the functions are centralized or decentralized, how large the volume is, whether a network is used, etc. Clericals and paraprofessionals have been utilized successfully for descriptive cataloging and copy cataloging. Professionals are generally required to do the subject analysis and classification and to solve descriptive problems.

Libraries should calculate the costs to acquire and catalog an item in the local system in order to compare other alternatives. A cost study requires data on the time it takes to order and catalog the item, the salary level of the staff involved, the costs of supplies and overhead, and the cost of producing the catalog. Costs will vary depending on the form of catalog, the proportion of cataloging which is original, what network or other services are used, the salary level of the staff, and the size of the file (authority work, as one example, is more complicated in large files). The factors are discussed in more detail by Getz and Phelps (1984), who found labor costs in three university libraries to range from $14.88 to $24.25 per monograph.

Statistics for the collection control function are necessary for management of the organization (workload figures) and for management of the resources. Some statistics can be collected continuously (e.g., number of books, etc. ordered; number of books, etc. cataloged; number of items circulated). Others can be gathered as needed in short-term studies (e.g., turnaround time for a particular vendor; turnaround time for cataloging).

Penniman and Dominick (1980), Arnovick and Gee (1978) and Cochrane (1981), among others, have discussed the need to include evaluation of the system by its users in guidelines for system design. We have discussed several kinds of studies which can help evaluate collection control. Catalog-use studies and failure-analysis studies were discussed earlier in this chapter. Document availability studies were discussed in Chapter 5. The questionnaire which was used in the CLR-sponsored catalog use studies is included in Matthews *et al.* (1983, pp. 183–189). It would be useful to have a sample of sci/tech libraries in various settings use the same questionnaire and compare the data to that for general libraries.

Journals which keep up with changes and issues in cataloging include *Cataloging and Classification Quarterly* and *Library Resources and Technical Services* (LRTS). Changes in LC's application of AACR2 are covered in the *Cataloging Services Bulletin*. Developments in automation are covered more broadly in the library literature (see Chapter 11 for further sources).

The primary professional organization focused on collection control is the Resources and Technical Services Division of the American Library Association. However, other groups focus on cataloging and access to the special formats that are so vital to sci/tech libraries. Special Libraries Association is a good forum for discussing control of patents, technical reports, standards, etc.

Chapter 8

Document Delivery

Stephen J. Rollins

Zimmerman Library
University of New Mexico
Albuquerque, New Mexico 87131

Document delivery is the last of the five main functions of a sci/tech library as outlined in this book. Document delivery is closely related to the functions of collection development, collection control, information retrieval, and current awareness, since document delivery routinely utilizes the end-products of each of these functions. Without a responsive and relevant collection development program, a document delivery service would be unable to satisfy requests from the holdings of the local library. Without adequate collection control, a document delivery service would be unable to determine what is available locally or what is the current status of the requested material. The processes of information retrieval and current awareness, if they are successful, assist the library user in determining what is relevant for a certain topic; and these processes will often generate document delivery requests. The remarkable growth of on-line databases (see Fig. 4.2) has made document delivery even more important to sci/tech libraries.

A precise definition of document delivery has not yet emerged in the library profession. James L. Wood points out that there is a "lack of published current information about document delivery in general" (Wood, 1982a, p. 1), but he attempts to define document delivery as "the entire process from generation of the request through shipment of the document or a copy including payment of fees and return of loaned documents to the lender." (Wood, 1982a, p. 2) Unfortunately, this definition also defines the concept of interlibrary loan (ILL), which is only one function of a document delivery service. Wood elaborates further by stating "that the term document delivery is used rather than ILL because organizations other than libraries engage in providing documents, both originals and copies to other organizations" (Wood, 1982a, p. 1). A Council on Library Resources (CLR) report defines document delivery as "the transfer of a document or a surrogate from a supplier, whether a library or a document service, to a requesting library." (Council on Library Resources, 1983, p. 1) Another distinction should be made, however, since a library document delivery *service* is designed to retrieve and to deliver a requested item from any source, including from the local library's holdings, whereas an interlibrary loan operation retrieves only materials not held by the local library. A library document delivery service involves much more than the mere transfer of a document or a copy from a supplier to a requesting library.

Perhaps the best way of defining a library document delivery service is to examine the components of the service. A document delivery service (1) accepts requests for materials, (2) locates a source or supplier of the material, (3) transmits the request if needed, (4) verifies the request if necessary, (5) receives the material, (6) delivers the material to the library user, (7) pays any or all fees associated with the transaction, and (8) secures the return of the material if a loan was provided. The term "document" includes any print or nonprint material, whether it is a photocopy, monograph, serial, technical report, government publication, cassette, sound recording, microform, or whatever. Figure 8.1 is a simplified flowchart of the document delivery process.

ACCEPTING REQUESTS FOR MATERIALS

Requests for materials can be generated as part of the information-retrieval function, the current-awareness function, or the collection development function. Requests for materials may be generated as part of an online search, an SDI program, or from a search of the local library's catalogs, indexes, or bibliographies. Requests may also be a result of less formal methods of information awareness, such as from discussions or referrals between colleagues. According to the King Research report (King Research, Inc., 1982), 75% of the requests generated by federal and special libraries were "job-related" and the requests in special libraries were primarily for serials. The King report indicates that 75% of the ILL requests transmitted by federal and special libraries were for serials, and these libraries generally request more recent publications than public and academic libraries. According to the King report, 65.2% of these requests were filled with photocopies.

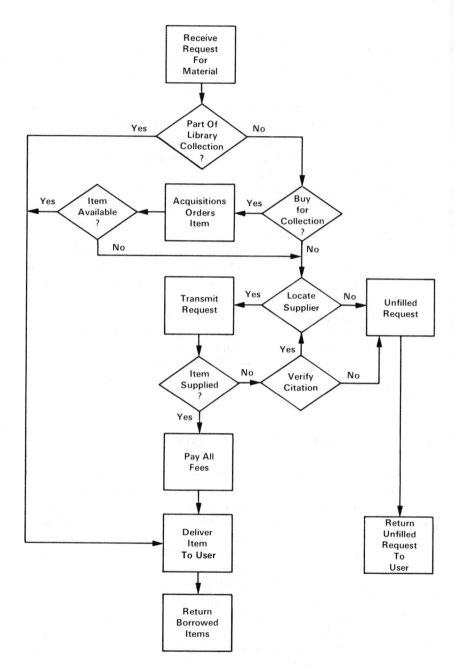

Fig. 8.1. Flow chart of the document delivery process.

Once a library user has established a need for a certain document, the request must be communicated to the document delivery service. A document delivery service has a variety of methods available for accepting requests, besides in person, by the phone, or by mail. A document delivery service can utilize a local courier service, a phone answering machine, or electronic transmission systems. There are two basic types of electronic transmission systems which may be used by document delivery services. One type is based on telecopiers or telefacsimile machines, and the other is the use of electronic mail systems. The major difference between these two types of electronic transmission systems is that telefacsimile machines are designed to convert by scanning printed materials to a format suitable for electronic transmission, while electronic mail generally does not utilize any scanning conversion method. Electronic mail is often generated by way of a terminal keyboard or by transferring data which already exist in machine-readable formats. Electronic mail is a direct user-to-user communication of typed information transmitted via a host computer.

While it is often more expensive than the U.S. postal service, electronic mail is faster and it can be completed whether or not an operator is available at the receiving end at the time of transmission. Electronic mail can be used in a local area network environment, or it can be supplied as part of a commercial service such as MCI's mail, the American Library Association's ALANET, GTE Telemail, OnTyme, BRS Mail, or the Source. Since requests usually do not require a substantial amount of text, electronic mail is an excellent method of accepting requests from the library's users. Electronic mail systems range from those which are restrictive in format and length of the message to those with flexible text-editing capabilities, word processing, dictionaries, and spelling-check programs. One electronic mail system, DIALCOM, offers a gateway connection where a searcher can automatically store output from a database in the electronic mailbox of a library user. If the procedures are established correctly, the recipient can then forward to the document delivery service any requests for specific documents (Association of Research Libraries, 1984, p. 1). While there are a variety of options available for electronic mail systems, the most desired features include:

Active, not passive system. Electronic mail users should be actively prompted that mail is waiting.

World-wide compatibility.

Hard-copy, paper delivery options such as offered by MCI's Mail.

A reply requested feature. The recipient is prompted to reply or must acknowledge receipt of the message.

Error checking to insure the accuracy of messages.

An outgoing mail log to review the status of the message.

Telecopiers or telefacsimile machines (FAX) can also be used to transmit a request. In order to utilize this technology in the fullest sense, it is important to purchase a machine which offers an auto-answer feature. Telefacsimile machines are easily installed at most sites where standard power and telephone outlets are available.

Besides printed materials, FAX machines can transmit sketches, charts, maps, or photographs, virtually error-free. According to Richard Boss's report (Boss and McQueen, 1983) in *Library Technology Reports,* there are three groups of telefacsimile machines on the market and a fourth group which is being tested.

Group I includes FAX machines which operate at 6 minutes per page with frequency analog modulation.

Group II FAX machines operate at 3 minutes per page with amplitude analog modulation. A 2-minute speed is also available. The minimum resolution for group II equipment is 96 × 96 lines per inch (Lpi).

Group III machines operate at 1 minute or less per page speeds and use digital, not analog, techniques. The minimum resolution for group III is 204 Lpi horizontally and 98 Lpi vertically.

Group IV machines are being tested, and they are expected to include high-speed digital transmission perhaps 10 times faster than Group III. The minimum resolution may be at least 240 × 240 Lpi. Group IV units may have the capacity to store and forward electronic mail and have the enhanced capabilities for interfacing with intelligent photocopiers and computers.

In terms of any future purchases of a FAX machine by a library document delivery service, only group III or perhaps group IV machines should be considered which offer compatibility with other manufacturers' group II and group III machines. The FAX machine should have the desirable features of sub-minute per page transmission, automatic feed, unattended answer, and very high resolution.

LOCATING A SOURCE OR SUPPLIER
OF THE MATERIAL

Once a request is received in whatever format, the first action taken by a document delivery service is to determine if the material is available in the local library. By using the products or systems maintained by the collection control function of the library, a document delivery service should be able to determine if the item is held by the local library. A document delivery service should be able to readily determine if the item is presently out on loan, at the bindery, on order, or missing from the collection. If the item is not held by the library, the document delivery service should work with the collection development unit to determine if the item will be purchased for the collection.

If the item is not available locally and it is unlikely that the item will be ordered in the near future, the next task is to identify an appropriate outside source for the material. This task is perhaps the most difficult for a successful document delivery service.

Until recently the task of locating an appropriate outside source was rooted in the time-consuming and labor-intensive practice of checking printed sources. The "traditional" tools used by interlibrary loan operations included such major printed sources as the *National Union Catalog* (NUC), the *Register of Additional Locations,* the *Union List of Serials* (ULS) and *New Serials Titles* (NST). These sources are based on the concept of a single or "main" entry for each bibliographic record. Attached to these bibliographic records are NUC holding symbols indicating a supplier for a particular title. There are also a wide variety of other specialized sources available which can be used to locate an outside source. These sources include such publications as *Chemical Abstracts Service Source Index* (CASSI), *INIS Atomindex,* local union lists, and printed bibliographies or indexes. Printed sources for locating outside suppliers continue to be a valuable resource. However, they are limited by their lack of multiple access indexing and by their timeliness. It is difficult to continually update accurate holdings information in a printed source.

There appears to be a shift occurring in the work flow of locating an outside supplier as more libraries gain access to online databases. Many sci/tech libraries have access to one or more of the major bibliographic utilities. OCLC, RLIN, UTLAS, and WLN provide access to large bibliographic databases and display holding information of other libraries. These major bibliographic utilities are a necessary tool for a document delivery service, since they can contain more holdings information than *NUC* and the *Register of Additional Locations* combined. In the case of OCLC, more than 13 million titles are stored online with well over 225 million holding symbols. With OCLC, one document delivery service found it could supply 75% of all science/technology materials that were not held locally (Rollins, 1981a,b, 1983a, 1984a).

Sci/tech libraries which do not have access to a major bibliographic utility may have access to other online systems. These systems may include BRS, SDC, INKA, Questel, ISI Search Network, Blaise, NASA/RECON, DOD/DROLS, CAS Online (STN), DOE/RECON, LEXIS, NEXIS, MEDLARS, NEWSNET, Westlaw, Vu/Text, Wilsonline, EasyNet, or other commercial or private databases. These online systems can often indicate a potential source for a document by adding availability information to the citation. An example of online availability information is DIALOG's file 77, Conference Papers Index, which includes specific information about how to obtain copies of the abstracts cited in the file (see Fig. 8.2).

While availability information from an online file can be useful, it can also be out-of-date. It is more important to determine if the database producers will supply the items they cite. More and more database producers are entering the business as document suppliers. In the case of DIALOG, a document cited in one of Chemical Abstract's files may be obtained from Chemical Abstracts (Rollins, 1984b, pp. 183–191), or the Engineering Societies Library can usually provide materials from files 8, 12, or 13. ERIC will supply ED documents cited in file 1, the American Geological Institute will supply from file 89 or perhaps 58, the American Society of Metals can supply many of the documents cited in file 32, ISI from file 34, NTIS from file 6, and perhaps 103 and 104, Data Courier's ABI/Inform Service from file 15, and University Microfilms International from file 35.

Besides online databases, there are other sources available which can assist a document delivery service in locating an outside source. USBE (formerly Universal Serials and Book Exchange) is a good source to consider for obtaining complete issues for requested periodicals. It is also useful to use an assortment of directories to find potential suppliers. Interdok's *Directory of Published Proceedings* lists availability information for conference papers and its indexes include access by location of the conference as well as by key word in the conference name. The *Ency-*

78047350   v6n6
Characterization of severity of valvular heart disease by
Doppler monitoring of brachial artery flow parameters.
Abrams, S.G.
Univ of ILL Med Cntr, Chicago IL
Association for the Advancement of Medical Instrumentation
Thirteen Annual Meeting & Exhibit 781   2032 Washington, D.C.
29 Mar - 1 Apr 78
Note
Availability
Information
Association for the Advancement of Medical Instrumentation
Abstracts (Eng) free at meeting; after meeting in "Proceedings
AAMI 13th Ann Mtg," 1 Mar 78, $12.00 members and $15.00
non-members: AAMI 1901 N. Ft. Meyer Dr. Suite 602, Arlington,
VA 22209
Descriptors:
Section Heading:

Fig. 8.2. Sample record from file 77 (Conference Papers Index) on DIALOG. The record includes availability information. [Courtesy of Cambridge Scientific Abstracts.]

clopedia of Associations provides phone numbers and addresses for association publications. The *National Faculty Directory,* the entire "Who's Who" series of publications, and the appropriate phone book can provide the phone number or address of a specific author.

It is also possible to dial-up the databases of other libraries by using a compatible terminal. The University of Illinois, Ohio State University, the CARL Network in Colorado, and the University of New Mexico, for example, all offer direct dial-up access to their library's machine-readable database. A survey of 115 ARL and Canadian university libraries found that 27.8% of this group offered dial-up access to their databases (Jamieson, 1985). Libraries are developing interface capabilities for linking different remote systems. These linking projects can reduce the confusion of searching on different databases by offering menu-driven or common search protocols. One example of such a system link is CLSI's recent announcement that their CLSI Datalink module is capable of establishing a communication link between the CLSI system at the Peabody Institute Library and the DataPhase ALIS II system at the University of Lowell. The University of Lowell's ALIS II system can connect via an auto-answer port on the CLSI LIBS 100 system at Peabody.

One of the more interesting trends in the area of document delivery is the emergence of many profit and nonprofit document suppliers who charge specific fees for providing specific documents. In contrast to the traditional interlibrary-loan practice of verifying known locations before requesting materials, these suppliers will accept and attempt to fill any request even if there is no prior knowledge on the part of the local library that the item is actually available, or, for that matter, that the citation is correct. Examples of this kind of document service include the for-profit companies such as Information on Demand, Information Store, Information/Documentation, Federal Document Retrieval, and Tracor Jitco Inc., to name a few. These companies establish a national, and, in some cases, an international network of "runners" who will retrieve documents directly from libraries or government agencies. These document suppliers will also purchase materials for their clients or they will acquire other publications by directly contacting the author, the association, or the professional society. Two good reference tools for identifying these document suppliers are the latest editions of *Document Retrieval, Sources and Services* published by the Information Store and the *Directory of Fee-Based Information Services.*

Besides the for-profit commercial document suppliers, there are many nonprofit suppliers who have established a document service for a fee. Some examples of nonprofit document suppliers include the BL Document Supply Centre (formerly the British Library Lending Division), the

Technische Informationsbibliothek, Universitatsbibliothek (Hanover, Germany), the Colorado Technical Reference Center (University of Colorado), the Royal Netherlands Academy of Arts and Sciences, and the University of New Mexico General Library.

Another distinction should be made in describing document suppliers. Some document suppliers attempt to fill any request, while others limit their retrieval to local resources only. The "generalists" are often the for-profit companies such as the Information Store, Information on Demand, or Information/Documentation, but this category also includes nonprofit institutions such as the Colorado Technical Reference Center, the University of New Mexico, Michigan Transfer Source (University of Michigan), and the Information Exchange Center at Georgia Institute of Technology. The other category of document suppliers will not attempt to fill a request unless it can be filled from within their own specific collections or holdings. This category also includes for-profit and nonprofit document suppliers. Examples of for-profit document suppliers who are part of this category are Chemical Abstracts, University Microfilms International, the Institute for Scientific Information (ISI), Data Courier Inc.'s ABI/INFORM Retrieval Service, and EIC Inc. Examples of nonprofit document suppliers include the BL Document Supply Centre, the Technishe Informationsbibliothek, Universitatsbibliothek, and NTIS.

While the nonprofit document suppliers are a major part of the total document business, it is interesting to note that the private for-profit document supplier is gaining an impressive share of the market. According to the CLR report (Council of Library Resources, 1983, p. 7) the "private sector back-copy delivery currently represents about 10% of the interlending, and it is growing at the rate of about 15% a year—a growth rate substantially greater than that experienced by libraries. The King study of ILL made a similar estimate that 10.5% of the nation's major libraries used a commercial document service at least once during 1980. Nearly 34% of the special libraries sent requests to a commercial document service, accounting for 95% of the $6 million in revenue of these suppliers" (Council of Library Resources, 1983, p. 7). The expanding role of commercial document suppliers has also been documented at the University of New Mexico. In 1980/1981, 9.8% of all document delivery requests filled by an outside source were filled by a commercial supplier. In 1984/1985, this percentage had increased to 19.8% (Rollins, 1985).

The CLR report also states that

> . . .among the 200 or more companies listed in the *Directory of Fee-Based Information Services,* 12 appear to be responsible for the bulk of the materials supplied to U.S. libraries by document services. Three of the six largest suppliers of materials—each filling more than 100,000 requests a year—are document services. Two of the three are commercial organizations. (Council of Library Resources, 1983, p. 8)

In locating a source or a supplier for the material requested, a document delivery service must pay particular attention to the type of material that has been requested. Patents, technical reports, and requests for foreign-language publications are examples of formats which may require special sources. These are covered in Volume 2, Part I.

TRANSMITTING THE REQUEST

After a potential outside source or supplier is located for an item which is not available at the local library, a request must be transmitted to the supplier. A request can be sent by mail, Federal Express Zapmail, UPS, or by other delivery services. The request can also be transmitted by telephone, TWX, TELEX, electronic mail, or telefacsimile machine. The major bibliographic utilities offer interlibrary loan systems which can be used at the same time that a supplier is located. OCLC's and RLIN's interli-brary-loan subsystems are examples of excellent systems where a request can be searched, suppliers listed, and the request transmitted all in the same session. BRS's mail can send requests to hundreds of libraries or to other suppliers, as can ALA's ALANET. DIALOG's DIALORDER can transmit requests to more than 60 document suppliers.

The prices of electronic transmission can be very competitive with other mail methods. By using file 1 in DIALOG, for example, a request can be sent to the BL Document Supply Centre in 1985 for an average cost of 70 cents. OCLC provides an electronic routing procedure for polling up to five libraries for one request for a cost of approximately $1.50. Western Union's Net Express claims that 15 seconds of transmission of one page would cost $2.00. In today's electronic world it is possible, with the appropriate system connections, to transmit almost all science/technology requests electronically. At the University of New Mexico, 99% of all science/technology document delivery requests during fiscal year 1984/1985 were sent to outside suppliers via electronic methods. (Rollins, 1985) At MIT during 1980, 90% of the outgoing requests were verified and located online (Ferriero, 1981, p. 2)

In transmitting the request by any method, another factor to consider is the method of payment of the document. The method of payment can make a difference in how the request is transmitted, to whom, and in the delivery time of the material. The best payment methods are either by deposit accounts (which can offer substantial discounts) or by billing after receipt of the material. The worst payment method is by coupons or by prepayment. Both coupon and prepayment methods require that the request be sent by the U.S. postal service or by another mail delivery system, since few document delivery services can electronically transfer coupons or prepayments to the suppliers.

Prepayment also adds additional processing time for drafting money orders or checks. While the prepayment method can be avoided in most cases by selecting a different supplier for the document, it is unavoidable when dealing with certain suppliers, such as the United States Geological Survey or the American National Standards Institute.

In transmitting a request, it is also necessary to consider the issues of copyright if the request is for a copy of an original which may be copyrighted. Under what authority is this particular request performed? Does this copying request violate the principles of the copyright laws? Is the copying serving as a substitute for purchasing the document? These are important questions which must be resolved by the library before an active document delivery service is established.

While this chapter will not discuss all the issues of copyright, there are a number of important considerations for a document delivery service. Section 108 of the Copyright Statute of 1976 is the critical section for a document delivery service. This section permits a library to create and to distribute a copy of a copyrighted work from its collection if the copying is performed "without any purpose of direct or indirect commercial advantage" and if the library's collections are "open to the public" or "to persons doing research in a specialized field," and if the "notice of copyright" is included on the copy.

According to section 108, a library can also produce a copy for a library user or may, on the behalf of the library user, request a copy from another library. This copying should be for "no more than one article or other contribution to a copyrighted collection or periodical issue or a small part of any other copyrighted work." The copy should not be retained by the library, but it should become the property of the library user. Section 108 also directs that the library will have "no notice that the copy or phonorecord would be used for any other purpose than private study, scholarship, or research" and that the library has posted a "warning of copyright where orders are accepted" and this "warning" is included on the library's order forms.

A library may provide a copy of an entire copyrighted work "if the library or archives has first determined, on the basis of a reasonable investigation, that a copy or phonorecord of the copyrighted work can not be obtained at a fair price."

Section 108 also restricts library copying to "isolated and unrelated reproduction of a single copy or phonorecord of the same material on separate occasions" and this right to copy

. . .does not extend to cases where the library or its employee 1) is aware or has substantial reason to believe that it is engaging in related or concerted reproduction

or distribution or multiple copies or phonorecords of the same material, whether made
on one occasion or over a period of time and whether intended for aggregate use by
one or more individuals or for separate use by the individual members of a group or
2) where the library or its employee engages in the systematic reproduction or distribution of single or multiple copies or phonorecords. . . .

While these statements appear to prohibit the copying activities of a
document delivery service, the section continues by specifically stating
that

nothing in this clause prevents a library or archive from participating in interlibrary
arrangements that do not have, as their purpose or effect, that the library or archives
receiving such copies or phonorecords for distributions does so in such aggregate
quantities as to substitute for a subscription to or purchase of such work.

The wording of section 108 must be interpreted in conjunction with the
Conference Report of September 29, 1976 (H.R. 94-1733), which further
defines library copying. While the Association of American Publishers
believes that copying in libraries owned by for-profit organizations is very
strictly limited, the Conference Report did include the statement that

. . .isolated, spontaneous making of single photocopies by a library or archives in a
for-profit organization without any commercial motivation or participation by such a
library or archives in interlibrary arrangements, would come within the scope of section
108.

The Conference Report also incorporated the Guidelines developed by
the National Commission on New Technological Uses for Copyright Works
(CONTU) (1976). The CONTU Guidelines outline specific regulations for
library copying. These include:

1. A record-keeping system based on a calendar year.

2. Unfilled requests for copying are to be counted, but they do not
apply to the limitations on the number of copies received each year.

3. Limitations on copying from periodicals apply only to those issues
published in the last 5 years. The guidelines do not limit copying from
issues published more than 5 years ago. This distinction of publication
year does not apply to monographs.

4. A statement should be included on the copying request indicating
compliance with section 108 and with other copyright sections.

5. A library may receive no more than five photocopies per year of
articles from the same journal published within the last 5 years, whether
these five copies are for the same article or for five different articles.

6. The library is expected to maintain its records of copying for 3 years.

7. The supplying library must examine each incoming copying request
to determine if it carries the statement of copyright compliance. The request should be rejected if it does not have this statement.

Copyright should not be seen as a barrier for a library's document delivery service, and it is possible to obtain copies of all requested materials
if valid procedures are followed. Repeated requests for copies from a specific journal, monograph, or conference proceedings are a clear sign that
this title should be considered for purchase by the collection development
unit. If numerous copies for one particular issue are requested, it is possible
to secure permission from the publisher, pay the necessary copyright roy-
alties, or attempt to borrow the entire issue. USBE may be able to provide
the entire issue for a small fee. Most commercial document suppliers have
established procedures for paying the necessary copying fees, and they
can act as the library's agent in satisfying the publisher's concerns about
copyright violations.

VERIFYING THE REQUEST IF NECESSARY

In an attempt to secure a document, it is possible that the request itself
is in error or is lacking important bibliographic information. Even in those
cases where the information has been retrieved from an online source,
errors will appear in the citation. Verification work can be very time-
consuming, especially using the traditional tools of printed indexes and
bibliographies. Unfortunately, many indexes provide only subject access
and, as such, cannot readily assist in correcting a citation. Once again it
is the online databases offered by DIALOG, BRS, ISI, etc. that can be
extremely valuable for correcting a citation. This point can be illustrated
by the following example:

> A request was received as: Tectonic Analysis of Megafractures by I. G. Gol'Braykh,
> V. V. Zabuluyev, and G. R. Mirkin (1968).

Initially this request was interpreted to be a monographic citation lacking
the publisher information. When this title could not be located in a variety
of sources, a search of file 89 (GeoRef) in DIALOG was performed in the
following manner:

B89: ss au = gol'braykh, I? and megafractures/ti

From this search, it was determined that the request was actually a
chapter in a monograph entitled *Morfostrukturnyye Metody Izucheniya
Tektoniki Zakrytykh Platformennykh* and it appeared on pages 87–136.
This kind of verification work could not have been performed easily without the use of an online system which offered both author and title keyword
search capability.

Another example will serve to illustrate the powerful retrieval options which are available in using online databases:

A request was received as: Strong, J. P. Basic Image Processing Algorithms on the Massively Parallel Processor. Academic Press, 1982.

Since this title could not be located in a variety of sources, a search was performed in file 470 (Books in Print) in DIALOG by using the publisher and the publication date.

B470; ss PY = 1982 and PU = Acad Press

This search produced 616 hits, and these citations were combined with the term Image?, which produced a final set of four citations. By examining the four citations, it appeared likely that the publication by J. P. Strong would appear in the following publication:

Multicomputers and Image Processing: algorithms and programs, based on a Symposium held in Madison, Wis. May 26–29, 1981.

Fortunately in this case, the local library had this symposium in its collection. By checking the table of contents, it was discovered that Strong's publication was a chapter in this symposium.

Sometimes it is useful to combine several online files with a printed bibliography. For example, this sequence of steps occurred in processing an actual document delivery request (during 1983):

1. The request was received as E. Rose *et al.* (paper) in Proc. IFAC 1975 World Congress, Boston, Mass. August 1975.

2. Since nothing was found in *NUC* or on OCLC, this publication was searched by location and year in Interdok's *Directory of Published Proceedings.*

3. In the *Directory* a listing was found which indicated that the proceedings were published in four volumes and each volume had its own unique title and ISBN. The question to be resolved was, in which volume did the Rose paper appear?

4. By searching in file 77 (Conference Papers Index) in DIALOG by author and conference location, the title of the paper was determined but the specific volume number was not listed.

5. By searching in file 13 (Inspec) in DIALOG by the paper title and by the author, it was discovered that this paper was listed as number 39.5 and it was in a volume of 766 pages, but file 13 did not list the specific volume number.

6. By checking in Interdok's *Directory* again, it was determined that volume 2 had 766 pages and the other three volumes had other paginations.

7. By using the ISBN for volume 2 as listed in the *Directory*, the proceedings were located on the OCLC database, and a request for the specific paper was transmitted to another library.

Cross-file searching of relevant databases is also an efficient method of verifying and correcting citations. If, for example, a request is received for a title in the area of chemistry and it is lacking its publication date, a quick search in DIALOG's DIALINDEX could verify the article.

A request for Seigo Kishino's article "X-ray topographic study of lattice defects. . ." listed the journal as *Proc. Conf. Solid State Devices*, pages 303–307, but no volume or publication years was given.

By using file 411 (DIALINDEX) on DIALOG, the citation could be verified by the method in Fig. 8.3.

RECEIVING THE DOCUMENT

Most items ordered by a library document delivery service arrive by the mail. According to the CLR reports, 95% of all requests filled are delivered by the U.S. Postal Service.

Fewer than 20% ever use an overnight service such as Emery, Federal Express or U.S. Air. Almost all of the use was by Special Libraries and most was for materials sent between two libraries of the same company or for materials requested from a commercial document service. Virtually all photocopies are sent by first class, but books are sent by a wider variety of document delivery means including USPS priority mail and library rate mail. (Council on Library Resources, 1983, p. 33)

Special Libraries rely heavily on first class mail because they request photocopies of journal articles more than three-fourths of the time and because they place a premium on reducing the satisfaction time. . . . The Special Libraries also account for almost all of the use of overnight services according to the commercial document services. (Council on Library Resources, 1983, p. 36)

The commercial document services report that they sent only 10% of their materials by surface courier such as UPS or Purolator and another 7 to 12% by an overnight service such as Emery or Federal Express. One large commercial firm reports while it receives 60% of all its requests electronically, it sends out 78% of its materials by first class mail. . . . Chemical Abstracts reported only 1.4% of requestors stipulate that USPS Express Mail or an overnight courier service is to be used to transport the materials. (Council on Library Resources, 1983, p. 34)

It is also, of course, possible to utilize telefacsimile machines to receive materials, but these machines, to date, have not been widely used by libraries.

Only 2% of the libraries and 10% of the document services have ever used telefacsimile for document delivery and only on an experimental basis. Two commercial document services with telefacsimile capability reported that while they receive a substantial

```
Welcome to DIALOG

Dialog version 2, level 5.7.15
LOGON File001 05aug85 15:46:34
** FILE 13 IS NOT WORKING **
 ** FILE 207 IS NOT WORKING **
------------------------------------------------------------
Phase 5 DIALNET?B411; SF CHEM; SS AU=KISHINO, S? AND
LATTICE()DEFECTS()RELATED/TI
        05aug85 15:47:13 User008283
        $0.25     0.010 Hrs File1
        $0.10     Telenet
        $0.35     Estimated cost this file
        $0.35     Estimated total session cost     0.010 Hrs.

File 411:DIALINDEX(tm)
(Copr. DIALOG Inf.Ser.Inc.)
File 308:  CA Search - 1967-1971
File 309:  CA Search - 1972-1976
FIle 320:  CA Search -- 1977-1979
File 310:  CA Search - 1980-1981
File 311:  CA SEARCH 1982-85 UD=10304


File     Items     Description

308:  CA Search - 1967-1971
         22          AU=KISHINO, S?
       3314          LATTICE/TI
       1819          DEFECTS/TI
       6492          RELATED/TI
          0          LATTICE/TI(W)DEFECTS/TI(W)RELATED/TI
          0          AU=KISHINO, S? AND
                     LATTICE()DEFECTS()RELATED/TI

309:  CA Search - 1972-1976
         88          AU=KISHINO, S?
       4207          LATTICE/TI
       2623          DEFECTS/TI
       8158          RELATED/TI
          1          LATTICE/TI(W)DEFECTS/TI(W)RELATED/TI
          1          AU=KISHINO, S? AND
                     LATTICE()DEFECTS()RELATED/TI

320:  CA Search - 1977-1979
         29          AU=KISHINO, S?
       3029          LATTICE/TI
       2168          DEFECTS/TI
       6139          RELATED/TI
          0          LATTICE/TI(W)DEFECTS/TI(W)RELATED/TI
          0          AU=KISHINO, S? AND
                     LATTICE()DEFECTS()RELATED/TI
```

Fig. 8.3. Sample search on DIALINDEX (continued on pp. 220–221).

```
310:   CA Search - 1980-1981
               13     AU=KISHINO, S?
             2056     LATTICE/TI
             1605     DEFECTS/TI
             4819     RELATED/TI
                0     LATTICE/TI(W)DEFECTS/TI(W)RELATED/TI
                0     AU=KISHINO, S? AND
        LATTICE()DEFECTS()RELATED/TI

311:   CA SEARCH 1982-85 UD=10304
                8     AU=KISHINO, S?
             3978     LATTICE/TI
             3188     DEFECTS/TI
             8435     RELATED/TI
                0     LATTICE/TI(W)DEFECTS/TI(W)RELATED/TI
                0     AU=KISHINO, S? AND
                      LATTICE()DEFECTS()RELATED/TI
?SAVE TEMP

Temp Search-save "TA003" stored
?B309; EX TA003
        05aug85 15:48:49 User008283
    $0.95     0.027 Hrs File411
    $0.27     Telenet
    $1.21     Estimated cost this file
    $1.56     Estimated total session cost    0.037 Hrs.


File 309:CA Search - 1972-1976
(Copr. 1984 by the Amer. Chem. Soc.)

    Set      Items              Description
               88              AU=KISHINO, S?
             4207              LATTICE/TI
             2623              DEFECTS/TI
             8158              RELATED/TI
                1              LATTICE/TI(W)DEFECTS/TI(W)RELAT
                              ED/TI
                1              AU=KISHINO, S? AND
                              LATTICE()DEFECTS()RELATED/TI
    S1          1              Serial:  TA003
?T 1/3/1


    1/3/1
    85200730     CA:  85(26)200730e     JOURNAL
    X-Ray topographic study of lattice defects related with
degradation of gallium arsenide-gallium aluminum arsenide
(Ga1-xAlxAs) double-heterostructure lasers
    AUTHOR(S):  Kishino, Seigo; Nakashima, Hisao; Ito, Ryoichi;
Nakada, Osamu; Maki, Michiyoshi
    LOCATION:  Cent. Res. Lab., Hitachi Ltd., Tokyo, Japan
    JOURNAL:  Proc. Conf. Solid State Devices  DATE:  1975
VOLUME:  7,  PAGES:  303-7 CODEN: PCSDD   LANGUAGE:  English
?LOGOFF
```

Fig. 8.3. (Continued.)

```
        05aug85  15:49:42  User008283
   $1.14       0.015 Hrs File309
              $0.018  1 Types in format 3
   $0.18      $1 Types
   $0.15      $Telenet
   $1.47      Estimated cost this file
   $3.03      Estimated total session cost     0.052 Hrs.
 Logoff:  level 5.7.15  15:49:42

 415 20D DISCONNECTED   00 40
```

Fig. 8.3. (Continued.)

number of requests by telefacsimile—40% of all requests in the case of one major service—materials are rarely transmitted by FAX. (Council on Library Resources, 1983, p. 34)

Files where the entire article or publication is stored online offer an exciting source for document delivery. These databases provide full text searching of the complete article and retrieval of the entire article from a remote terminal. Examples of this type of database include BRS's American Chemical Society Journals Online, Drug Information FullText, Harvard Business Review/Online, IRCS Medical Science Database, and HAZARDLINE. DIALOG also has or will have a stable of such files, including IAC's ASAP files (647 and 648) and 13 McGraw Hill periodicals such as *Electronics Week, Byte, Chemical Week, Aviation Week and Space Technology,* and *Nucleonics Week.* While this method of accessing entire articles online can be expensive, it is certainly very fast in retrieving requested materials. This type of document delivery is still a new phenomenon in the library world, as indicated by CLR's report which stated that "only three libraries and six document services have had experience receiving the full text of a document online from a remote database" (Council on Library Resources, 1983, p. 34). A major drawback for these files at the present time is that the production of graphics, illustrations, and colors is limited.

The actual delivery time or turnaround time for a requested document varies substantially depending on a number of variables, including the method of transmission, the type of material requested, the method of payment, the year of publication, and the critical factor of selecting a supplier. There are several studies which attempt to identify the average turnaround time for interlibrary loan requests. These studies are summarized in Table 8.1.

Two different studies were conducted at the University of New Mexico which examined turnaround times for OCLC and for DIALORDER (Rollins, 1981b). These studies are summarized in Fig. 8.4 and 8.5 and show the percentage of requests that are filled within a particular time frame.

TABLE 8.1.   **Average Turnaround Time According to Various Studies**

Study	Average time
Council on Library Resources (1983)	10–16 days
	8.5 days for instate requests
	18 days for interstate requests
	6.3 days average transmit time
Dodson *et al.* (1982)	11.6 days
Selenk (1976)	10.5 days
	8.5 days for instate requests
	18 days for interstate requests
Tallman (1980)	11 days for monograph requests on OCLC
Baker (1981)	12 days for OCLC
	17 days for TWX
	24 days for U.S. mail
Rollins (1984b)	6–8 days for Chemical Abstracts via DIALORDER
	7–11 days for Chemical Abstracts via U.S. mail
Popovich and Miller (1981)	10.5 days for DIALORDER vendors

It is not surprising that the DIALORDER vendors in the University of New Mexico study would have a slower turnaround time than OCLC, since the requests that were sent to the DIALORDER vendors were for materials not located on OCLC and could be considered as more difficult to fill than the OCLC requests.

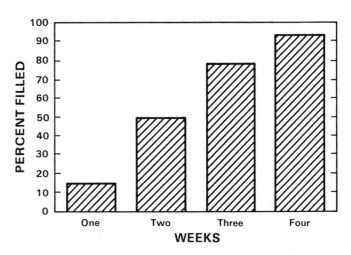

Fig. 8.4. Average turnaround time for OCLC, based on studies performed at the University of New Mexico; 93.51% of all requests were filled within 4 weeks.

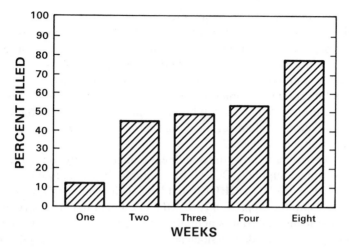

Fig. 8.5. Average turnaround time for DIALORDER vendors (from studies performed at the University of New Mexico); 77.55% of all requests were filled within 8 weeks.

Besides efficient turnaround time, a successful document delivery service must also establish a high rate in completing requests. The CLR report states that "a large majority of requests (89.5% for serials and 86.5% of books) were filled according to the King study. Data gathered by several others have confirmed that the rate is generally over 80%" (Council on Library Resources, 1983, p. 13). This rate of under 90%, however, should not be considered satisfactory for a document delivery service involved with filling requests for science/technology materials, and it should be possible to fill at least 95% of all requests received for science/technology materials if the document delivery service has a strong local collection and the appropriate connections to the relevant document suppliers (Rollins, 1981, 1983a, 1984a,b, 1985). It is important that the document delivery service is skilled in bibliographic verification, as outlined in the previous section.

DELIVERING THE DOCUMENT TO THE REQUESTER

After receiving the document, a document delivery service arranges for the delivery of the document to the requester. This task can be performed in a simple manner by informing the requester that the item is available at the library. This approach is the most passive of all the methods available. A more service-oriented document delivery service arranges for the delivery of the material to the requester by using the U.S. mail or a local courier service, or by transmitting the material via a telefacsimile machine or by electronic mail. The major problem of electronic transmission is preparing the material for transmission. If the document consists of photocopied papers, it is not difficult to transmit these papers via a telefacsimile machine. However, this method is not necessarily suitable if the document is a monograph or if its format is not compatible (e.g., videocassettes, microfiche, or microfilm). Entering the entire document via a keyboard for electronic mail can be very time consuming. At the present time, local courier systems or local mail services serve the needs of most document delivery services.

It is also possible to have the document mailed directly to the requester by the document supplier. OCLC, for example, has a separate "ship to" address field, as do DIALORDER and BRS mail. The one drawback with this direct "ship to" approach is that the document delivery service may never know if the material is actually received by the requester without checking with the individual periodically. In the case of OCLC, a document delivery service must update the ILL record as received, so it is imperative to establish a "receive date" of the document. If the request involves handling charges billed by the supplier, it is also important to determine the receipt of the material when the invoice is paid. In other words, while it may appear more efficient to have the document shipped directly to the requester, this approach actually can require more follow-up by the document delivery service in determining the receipt of the material.

One other consideration in delivering the document to the requester is the document's format. While the format should be of little concern for a document delivery service in obtaining the item (unless it is subject to telefacsimile transmission), the format could be of great concern to the requester. For example, a technical report on microfiche is of little value to an individual who does not have a microfiche reader.

PAYING FEES ASSOCIATED WITH THE TRANSACTION

Staff salaries should be the largest cost in providing document delivery service, since document delivery is labor-intensive in retrieving in-house materials, in locating a document supplier, in transmitting a request, and in verifying the request when necessary. Equipment costs can vary depending on what type of digital equipment is purchased, and whether telefacsimile machines are used. Besides the staffing and equipment costs of a document delivery service, there are many direct costs associated with each transaction. These direct costs include:

1. Photocopying or reproduction costs.
2. Costs of postage or delivery of materials by courier services.

3. Direct charges billed by the outside document supplier.
4. Telecommunication costs for transmitting or receiving documents or requests.
5. Royalty charges for photocopying copyrighted materials.

Costs of photocopying materials in-house and for postage or courier services generally have a fixed rate which can be easily determined. It is not difficult to calculate that it costs less than 1 cent per page to copy an article in-house or to determine the U.S. mail service's rates for a package. The other charges associated with a document delivery service can vary significantly, however. Direct charges billed by an outside supplier can range from as little as $1.25 (academic libraries) to $25.00 minimum (commercial suppliers). As of July 1984, a sampling of major document suppliers gave the costs shown in Table 8.2.

While this sampling indicates the wide range of charges and options, the situation becomes even more confusing with periodic increases in prices and with the introduction of new vendors. Since July 1984, Chemical Abstracts raised their minimum charge from $12.00 to $13.00 and the American Geological Institute (GeoRef) has entered the market place with a $14.00 minimum charge. The United States foreign exchange rate will affect the pricing schedules of overseas vendors, except for the British Library, which recently established a fixed rate in U.S. currency. Inter-library-loan charges of academic, public, and special libraries in the United States are varied and constantly changing. These libraries can charge a variety of fees for photocopying, and more and more libraries are charging

TABLE 8.2. **Costs of Some Major Document Suppliers, July 1984[a]**

Supplier	Costs
BL Document Supply Centre	$7.50 per item
	$5.50 for first 10 pages with account
	(effective 6/1/86)
ABI/INFORM	$7.95 per item
	$6.50 with desposit account
Engineering Societies Library	$6.00 per item plus 40¢ per page
	$5.00 with deposit account
Information on Demand	$14.00 per item up to 20 pages, discount for volume
	users being at $3.00 per item.
ISI	$7.50 per item
Linda Hall Library	$2.00 per item plus 25¢ per page
USBE	$10.00 per item, $7.00 per item with subscriber
	account, $4.25 with "member account"
University Microfilms International	With account, $4.00 and up

[a]From Data Courier (1984), p.2.

for lending monographs. While this trend of rising interlibrary-loan charges is perhaps understandable, it is disturbing in that it signals the rapid decline of "free" document delivery and interlibrary loans among United States libraries. While an interlibrary loan transaction was never literally "free," the increase in interlibrary-loan charges will be borne by the library patron or by a library's document delivery service if the service is funded appropriately. At best, the situation is confusing in that different rates can be charged even for the same document, and a substantial amount of staff time can be devoted to "shopping around" for the best buy. Perhaps the best solution for a local library is to establish as many no-charge reciprocal agreements as it can with other libraries. Both the AMIGOS (OCLC) and the BCR (OCLC) interlibrary loan codes are excellent examples of such agreements established on a regional basis. No-charge agreements not only reduce the direct billed charges for a transaction, but they eliminate the cost of paying and billing for interlibrary loan, which reduces accounting overhead. It is amazing that a library will charge as little as $2.50 for an article and actually pay more than this charge in billing the requesting library. It is not unusual for a library to pay more in billing and in collecting its interlibrary-loan fees than what it actually collects from these bills.

Telecommunication costs for transmitting requests and documents also vary depending on what system is utilized. Local electronic mail systems may not cost the library directly if the system is maintained by the library's parent organization. However, most electronic mail systems have a direct cost associated with each transaction. In 1981/1982, OCLC charged $1.20 for each interlibrary-loan request produced on their system, and this rate can be higher depending on the local OCLC network surcharges. A study performed at American Cyanamid determined that the average real cost for 100 documents for using OCLC's interlibrary-loan system was $3.82, as compared to $6.09 for manual interlibrary-loan procedures (Willet and Finnigan, undated). King Research, Inc., found in the Pennsylvania study that the average processing costs for completed requests ranged from $6.25 for monographs and $6.93 for serials (King Research, Inc., 1985). However, "a review of 20 interlibrary-loan studies by Ferriero at MIT determined that many of the studies which place ILL costs lower than $10.00 for borrowing and $6.00 for lending have not included all direct costs" (Council on Library Resources, 1983, p. 14).

DIALOG's DIALORDER has a rate established by the specific file selected, the method used for ordering, and the communication node accessed. UNINET, TYMNET, TELENET, and DIALNET all have connection costs between $6.00 and $10.00. A request sent from file 1 (ERIC) can cost as little as $.70 in 1985, but a request sent from a Chemical Abstracts file can cost as much as $2.85 depending on the method used to locate and to transmit the citation to a document supplier (Rollins, 1984b).

The method of isolating and transmitting the citation to a document supplier can determine the cost. If a citation is to be sent to Chemical Abstracts via DIALORDER and if the document delivery service has the CA number available, it is inexpensive to order the material by using the KEEP command and order sequence. This method is illustrated in Fig. 8.6 using UNINET.

If the CA number is not available and if the citation must be isolated by searching by author and title, then the cost will be greater.

The cost of utilizing telefacsimile machines depends on the machine's speed, the number of pages transmitted, and the local phone rates. In 1983 a CRL report stated that telefacsimile costs "are such that only high volumes of thousands of pages a month can bring the cost down to $.50 or so per page. At $.50 per page, the average journal article, which contains fewer than 10 pages, would cost $5.00 to reproduce and send; a cost which might be attractive to a large number of users. However, a volume of only a few hundred pages a month would force costs up to $1.00 to $3.00 per page, a figure few libraries or their patrons would accept" (CLR, 1983, p. 54). A report in the University of Texas (Austin) *Library Bulletin* supports this premise that high volume FAX operations produce low unit costs. It was found that during the 1984/1985 fiscal year, 1,175 requests were filled and "17,840 pages were FAXed at an average cost of $.22 per page in telecommunication costs" (University of Texas at Austin *Library Bulletin*, 1986, p. 3).

While it can be assumed that most libraries abide by the CONTU Rule of Five and pay royalty fees if necessary, there is little information available regarding the amounts of money spent by document delivery services for royalty fees. James L. Wood notes that "during September 1980–June 1981, CAS paid the Copyright Clearance Center, Inc., an average of $2.18 per copyrighted, registered document copied" (Wood, 1982a, p. 22). Information on Demand claims that they pay an average Copyright Clearance Center charge of $2 to $3 per request (Saldinger, 1985). If a document delivery service follows the copyright guidelines, royalty charges may be avoided. However, there will always be those cases when the royalty charges should be paid. If commercial suppliers are used in these cases, then the royalty fees should be paid as part of their bill for the document.

RETURN OF THE MATERIAL IF A LOAN IS PROVIDED

While in the sci/tech area many of the document delivery requests will be for photocopied materials from periodicals, technical reports, or conference proceedings, requests for monographs continue to be part of the

```
@C 415 48D

415 48D NOT REACHABLE 05 ED

@
uninet pad 06a8 port 04
service : DLG;2


*u001 000 connected to 41500007

DIALOG INFORMATION SERVICES
PLEASE LOGON:
?XXXXXXXXX

Welcome to DIALOG

Dialog version 2, level 5.7.15
LOGON File001 07aug85 08?
                              B320; K 87113898;   .ORDER CASDDS
           07aug85 08:36:45 User008283
           $0.15           0.006 Hrs. File1
           $0.06  Uninet
           $0.21  Estimated cost this file
           $0.21  Estimated total session cost      0.006 Hrs.

File 320:CA Search 0 1977-1979
(Copr. 1985 by the Amer. Chem. Soc.)

           Set    Items    Description
           S0       1      87113898
1 item Ordered
Order# 46436
Item 1
           87113898       CA: 87(15)113898p       PATENT
           Catalyst for biochemical reaction and method of preparing
it
           INVENTOR(AUTHOR):  Wood, Louis Leonard; Hartdegen, Frank
J.; Hahn,

?LOGOFF
           07aug85 08:37:21  Users008283
           $0.76           0.010 Hrs. File320
           $0.10  Uninet
           $0.86  Estimated cost this file
           $1.07  Estimated total session cost      0.016 Hrs.
Logoff:  level 5.7.15  08:37:22
```

Fig. 8.6. Use of file 1 on DIALOG to order an item from Chemical Abstracts Service inexpensively.

service. If the monograph is retrieved from the local library and delivered to the requester, the return of this item to the local library is part of the collection control function. If the monograph has been borrowed from another institution by the document delivery service, it is the document delivery service that is responsible for securing the return of the material to the borrowing library. It is important to maintain a good reputation with all potential lenders, so all items loaned to the document delivery service should be returned in a timely manner. If loans are not returned in a timely manner, it is very likely that the document delivery service will eventually lose a valuable source of materials. In those unfortunate cases when a loan is lost in transit, even if it is not yet received by the document delivery service from the lender, it is important that the document delivery service pay any replacement costs for the material as quickly as possible.

MANAGING DOCUMENT DELIVERY

A document delivery service can be placed within a library organization as a separate unit or as part of the collection control unit or the information retrieval unit. There does not appear to be any major advantage in any one of these options, as long as the document delivery service has a good working relationship with the collection control, information retrieval, and collection development units.

Particularly important are the skills of the document delivery staff. The document delivery unit must be familiar with the catalogs and files maintained by the collection control unit and the rules of filing cataloging entries for the library's collection. The document delivery staff should be skilled in bibliographic verification procedures and in online information retrieval. The document delivery staff should be able to utilize the telecommunications equipment available to the library. Above all, the staff of the document delivery service should be creative, efficient, and accurate in processing requests.

The factors in evaluating a document delivery service are cost, fill rate, and turnaround time. One document delivery service reported that it could fill all sci/tech requests at an average rate of $6.85 in 1984. This figure does not include indirect overhead costs, but it does include staff salaries, photocopying costs, in-house retrieval, ordering and receiving materials from outside sources, paying all invoices from outside suppliers, and all telecommunications costs (Rollins, 1984). This average is acceptable for any document delivery service. An average cost in the $10.00 to $12.00 range for a library can be considered too high, since this range is within the rates charged by many commercial services. If a library document delivery service is averaging a $10.00 or higher cost for filling a request, it may be more cost-effective to contract out the document delivery service to a commercial vendor.

A document delivery service which has the necessary bibliographic skills and access to a range of outside suppliers should be able to fill 95% to 99% of all requests received (Rollins, 1981a,b, 1983a, 1984a, 1985). A lower rate indicates that the local collection does not support the mission of the organization or that the document delivery service is unaware of or unable to access the appropriate outside sources.

A document which does not arrive within an appropriate time frame can be considered virtually useless to the library user. A document delivery service's success is often measured by its ability to deliver the material quickly or within a specific deadline. In-house materials should be available in less than 24 hours, even if photocopying is required. If an outside supplier is necessary, it is difficult to provide next-day service unless a courier service such as Federal Express or a telefacsimile machine is used. More likely than not, if the U.S. mail is utilized, it will require between 1 and 3 weeks to receive most items from outside suppliers.

Another factor in the success of the document delivery service is staff training. Since the information industry seems to be changing daily, the document delivery staff has to keep current with the emergence of new outside suppliers, with the establishment of new databases, and with the utilization of new telecommunication technology. The document delivery staff should take advantage of training sessions offered by online database vendors or by computer companies. Library conferences are an excellent place to learn of new services provided by information suppliers, and a wealth of information can be found both in articles and in the advertisements of appropriate professional journals such as those listed in the bibliography.

PLANNING FOR FUTURE TRENDS

During the next 3 to 5 years, a number of factors will have a substantial impact on the future direction of library document delivery services. These factors include full-text databases, increased use of high-density storage devices, more international communications, increased use of electronic transmission, increased cost of obtaining documents, an increase in the number of commercial document suppliers, increased use of FAX machines, improved telecommunications, the impact of the marketplace, attempts to standardize search commands and system protocols, the introduction of gateway systems, and the demands of the library user.

Full-Text Databases

The number of databases which have the capacity to store the entire text of an article or of a publication online will increase. BRS and DIALOG have set the pace in this area by offering a variety of full-text files. Two European operations attempted to establish similar retrieval systems. ARTEMIS (Automatic Retrieval of Text from Europe's Multi-National Information Service) and ADONIS (Article Delivery Over Network Information Systems) were projects which were designed to supply "printing on demand" of articles published by a select group of major scientific publishers. Neither system has become operational at this time, but the concept is valid.

Increased Use of High-Density Storage Devices

The use of high-density storage devices will increase as they become more cost efficient and more standardized. New high-density storage mediums have recently arrived in the market place, as indicated by GEAC's Gigadisc (GC1001), which supposedly will hold 400,000 pages of print on one disk, and by Sony's Compact Disc (read-only memory), which may hold 540 megabytes of information on a single 4.72-inch optical disk. This amount of memory is equivalent to 500,000 MARC records. The British Library reports that it is studying the use of a CD-ROM which holds up to 600MB on a 12cm disk (British Library Newsletter, p. 2). In recent projects at the Library of Congress both optical and video disks have been used. LC reports that a 15-square-foot "juke box" stores in digital form the equivalent of 1 million pages of text and that this storage capacity should quadruple in the near future. A "juke box" can hold 100 disks, with each disk storing the text of approximately 300 books (American Libraries, 1984, p. 768). The National Archives is also testing optical disk storage, and they estimate that 80,000 cubic feet of documents can be compressed to 300 square feet. University Microfilms International (UMI) has announced that they are investigating optical disk technology for a method of storing and for producing distribution copies (Hendricks, 1985, p. 2).

These storage media are suitable for print as well as for pictures, photographs, or graphics. Besides the advantage of a better price per byte, these high-density storage mediums offer the opportunity to access and develop local databases. Such local databases could replace or supplement the current use of accessing remote databases presently being offered by commercial vendors such as DIALOG or BRS.

Information Access Corporation recently announced such a local system, InfoTrac, a laser disc system which permits up to four users to simultaneously search the database stored on videodisc. Each 12-inch videodisc can index nearly 500,000 articles published in more than 1000 publications. The system includes four terminals, one laser disc player, one disc interface, one printer, and all maintenance for a price of $16,000 (*Information Today*, January 1985, p. 1). Each videodisc may contain more than 530,000 printed pages of information, and subscribers are provided with a new videodisc each month to update the database.

> After the user locates a citation or a list of citations of interest, pressing a button produces a printed copy. An accession code is included if the full text of the article is available in one of IAC's companion products, Magazine Collection or Business Collection. Thus if the library subscribes to that service it is possible for a searcher to obtain full text in a matter of minutes—full document retrieval. (Melin, 1985, p. 26)

Besides InfoTrac, International Standard Information Systems (ISIS) offers a system based on a 4.75-inch compact optical disk. ISIS's new SilverPlatter Service contains the databases of ERIC and PsycLit. Digital Equipment Corporation's CDROM Database Publications include portions, subsets, and entire files of some major technical databases. One 4.7-inch disk has the capacity to store 600MB or the equivalent of 1600 floppy disks. The CDROM collection at this time includes 10 databases produced by Engineering Information Inc. (Compendex), NTIS, Chemical Abstracts Service, the Royal Society of Chemistry, and the publisher Fraser Williams. Brodart's Le Pac CD ROM system and Bibliofile are two other examples of optical disk technology.

More International Communications

Document delivery services in libraries will be faced with the task of establishing more international sources as the use of international databases increases by the information retrieval unit of the library. The need for more direct and more open electronic mail systems to Japan, China, and the Soviet Union is already apparent to many U.S. sci/tech libraries.

Increased Use of Electronic Transmission

More requests will be sent and received electronically on a national and international level. According to one study, more than 99% of all sci/tech document delivery requests can be transmitted electronically to an appropriate outside source (Rollins, 1985). As the cost of first class postal rates continues to increase, this method of transmitting requests will be used by more libraries more frequently.

Since not everyone has access all of the time to a keyboard and to a terminal, it will be interesting to watch for the emergence of "voice mail." Voice mail is "a computerized system that digitizes the human voice and stores a spoken message in a file that can be stored for several days or longer, retrieved, or forwarded" (Black, 1985, p. 21). Since the only equipment that is needed for voice mail is an ordinary Touch-Tone phone, this system can offer a convenient method of submitting requests to a document delivery service or to a document supplier.

> Voice mail can be a service provided by an outside service bureau, a standalone turnkey system, or an integrated module of a PBX (private branch exchange). It can cost anywhere from less than $20 per mailbox per month from a service bureau to hundreds of thousands of dollars for a large PBX-integrated turnkey system. (Blackwell, 1985, p. 27)

International communications will become more routine as indicated by GTE's announcement that its Telemail will support the X.400 standard. The X.400 standard has been adopted by the Consultative Committee on International Telephone and Telegraph (CCITT), and this standard should permit world-wide compatibility for TELEX machines. GTE Telemail's new interface allows "two way communications between TELEX machines currently in use in the U.S. and abroad. Telemail users are now able to receive TELEX messages on their own terminals and do not need to have access to a TELEX machine" (GTE Telemail, 1985).

Increased Cost of Obtaining Documents

A document delivery service can expect to pay more for items supplied from outside sources. This price increase will include academic, public, and special libraries, as well as nonprofit and for-profit commercial document suppliers. According to one study, the average cost for an item supplied by an outside source for science/technology materials in 1982/1983 was $2.58, while in 1984/1985 this average billed amount increased to $5.88 (Rollins, 1985).

Increase in the Number of Commercial Document Suppliers

The number of commercial document suppliers will increase. These commercial document suppliers will include those who produce databases which cite the materials they provide, as well as those suppliers who will handle any requests. As more public and academic libraries charge for interlibrary-loan services, these commercial document suppliers could become more attractive by offering competitive pricing. The commercial sector, however, will only offer what is profitable in the marketplace, and libraries will continue to play a major role in the document delivery area if they collectively decide to support such a mission.

Increased Use of FAX Machines

The use of telefacsimile machines should increase, especially for high-volume and rush-service operations. Their use, however, will not be dramatic as long as the U.S. mail service or private courier services remain cost effective. The feasibility of FAX lies in the area of journal articles. Today's telefacsimile machines have improved substantially over the older models, and they are more compact, easier to use, more reliable, and more compatible across different product lines. The price for certain group III models has decreased in the last several years. For example, the RAP-ICOM machines had the following price drop (Reynolds, 1984, p. 1):

Machine	Late 1982	May 1984
RAPICOM 3100	$ 4,850	$3,795
RAPICOM 3300	$ 5,250	$5,000
RAPICOM 6100	$10,000	$6,259

This type of price reduction should continue as newer and faster models are introduced.

The major disadvantage with present telefacsimile machines is the need to photocopy an article from a bound journal before the article can be transmitted via FAX. This disadvantage may disappear with the introduction of the "communicating copier machine" (High-speed communicating copier demonstrated, 1980), which is not just a FAX but an intelligent copier which can double as a FAX (Arnett, 1981). Until the cost efficiency of a FAX or of the high-speed communicating copiers is improved, the FAX machine will be an appropriate tool for those document delivery services which handle a volume of 500 to 1000 pages per month or which decide that it is worth paying a premium price for rapid delivery (Council on Library Resources, 1983, pp. 45–46). Unfortunately, systems which have been developed for transmitting directly from bound journals are still too expensive for most library document delivery services (*Library Systems Newsletter,* 1983).

Improved Telecommunications

Technology will offer more capabilities with improved quality at a reduced unit cost for document retrieval. As more and more information is created and stored in machine-readable or digitized form, this technology can be exploited. The next few years will see the increased use of optical scanners (for converting print to digital form) and improved output with high-quality graphics, better quality, and less expensive printers, and perhaps even holographic image memory for unique material storage.

Digital and electronic communications will expand by utilizing fiber optics, cable TV, and satellite transmissions. The overall result will be lower-cost, faster transmission methods. The British Library Lending Division, for example, has recently announced a pilot project with Western Europe as part of the APOLLO program (Article Procurement with On-Line Local Ordering) which uses the EUTELSTAT satellites. British Telecom is developing an Integrated Services Digital Network which can support group IV facsimile transmission, which is at least 10 times faster than group III (The Dishy Side of Interlending, 1985). Another example is the use of a commercial satellite for exchanging cataloging information at the University of California. Edwin Brownrigg of the University's Division of Library Automation is quoted as saying,

> . . .this is a first. . .the first time satellite communication will be used in conjunction with an online union catalog (MELVYL), the first time that the UC library system has used satellite technology for two-way data communication, and the first time in the UC system that satellite communication has been connected to data communication using packet switching. (Backyard Satellite Dishes as Checkout Counters, 1985.)

Combining existing technologies of computer graphics and telecommunications could create efficient computer-assisted retrieval systems. Current systems offered by Access, Infodetics, Ragen and Teknekron retrieve microforms by computer assisted methods, and these systems are capable of handling 18 to 30 million pages per system. By interfacing such systems with telecommunications packages, a display could be sent via satellite link, telephone lines, or cable transmission. The major problem is again the cost factor, but such computer retrieval systems could be cost-effective to a group of libraries (Council on Library Resources, 1983, pp. 49–50).

Impact of the Marketplace

The impact of the marketplace on library document delivery services should be of particular interest to librarians. There are three trends to monitor in the coming years. These areas include the role of the for-profit document supplier, the rising cost of telecommunication systems, and the privatization of public information.

For-profit document suppliers will continue to enter the market place in those areas where it is predicted to be profitable. UMI's dissertation copy service is an interesting case. Some time in recent years, academic libraries began to restrict the loaning of their institution's dissertations to other libraries and referred requesting libraries to UMI's service. For a fee, UMI provides a copy of the dissertation, and over the years this cycle has escalated to the point where more and more libraries have stopped loaning their dissertations. Will this trend continue as libraries refuse to supply specific articles because of increasing ILL costs, since these articles are available from ISI or UMI or CA? Librarians should be careful about surrendering their role as document suppliers.

While the unit cost may drop in terms of the cost per minute with the introduction of faster transmission methods, the total cost for such systems will be a financial burden for many libraries. Walter G. Bolter of Bethesda Research Institute predicts that

> . . .deregulation will bring unstable prices to libraries. Library use of telecommunications, he said, accounts for only .3 percent of income but the impact on libraries will be all out of proportion to their tiny contribution to the market. (ARL at Colorado Springs, 1985, p. 16)

Librarians must continue to lobby to convince the Federal Communications Commission (FCC) that lower library rates are necessary and desirable.

The move toward privatization of public and government information should be another major concern. The recent interest of the private sector to index and repackage government information can have a negative impact on already-overburdened library budgets. "There is a lot of money to be made by constricting the role of libraries who hurt the market by giving away free the use of databases" (ARL at Colorado Springs, 1985, p. 16).

Attempts to Standardize Search Commands and System Protocol

The need to standardize search commands and system protocol is already apparent, as a new Tower of Babel has emerged in the areas of telecommunications and online bibliographic searching (Rollins, 1983b, p. 233). There are no standards for electronic mail formats, and many electronic mail systems cannot communicate with each other as telephone systems do. There are more than 20 electronic systems, which support different formats. The new optical disks are appearing in a variety of sizes from 4.32 inch to 5.25 inch. Online bibliographic searching can be very confusing, since even the act of signing on to a system requires different procedures. Each online system and even each file within a particular online system can require different search commands for searching the same fields. As more online systems are made available, the Tower of Babel continues to grow.

> In the next twenty years we should not expect much standardization in telecommunications. Hopes of standardizations have to be tempered by the reality of the technical problems and fierce competition among a multiplicity of systems. Network development in the future will be bottom up rather than top down; lots of small networks will appear that may eventually coalesce but these will not be a single massive network designed by a small group of people. (Bellardo, 1985b, p. 52)

Introduction of Gateway Systems

While it is perhaps naive to think that all or most online systems or electronic mail systems will have similar search commands and protocols, it is not difficult to predict the emergence of gateway systems which will act as "front-end translators" for the end user. Gateway systems promise a variety of services, including common search protocols for different systems, vocabulary translation, and simultaneous searching of several databases. Few gateway systems actually deliver what they promise at this time. However, one of the more interesting is Easynet, a gateway project of Telbase, Inc., and the National Federation of Abstracting and Indexing Services. Easynet offers a menu-driven interface to NewsNet, Orbit, Questel, Pergamon/Infoline, Vu/Text, DIALOG, and BRS. Easynet does not require a microcomputer (most terminals with a modem will suffice) and no previous online searching experience is needed. The user is simply directed through the search after entering a credit card or Easynet account number. Easynet also offers the option of selecting a particular database for the search, and it is expected that the system will provide copies of the full documents cited for a fee. Such a gateway system would allow any individual with a terminal and a modem to bypass the local library and order documents directly.

Demands of the Library User

The role and expectations of the library user will also be significant factors in the future development of library document delivery services. The user who develops a higher level of expectation for rapid delivery of documents will direct the library's document delivery service to use faster and more reliable delivery systems. If the user remains satisfied with 1- to 3-week turnaround times, then the document delivery staff will have little incentive to develop or to utilize new or more expensive approaches or to lobby for more library funding to support new technologies. Technical feasibility does not in itself guarantee acceptance or usage of a certain system without a stated or perceived need for this technology. If the price of faster transmission methods is higher than the benefits as perceived by the users or the staff of the document delivery service, then these methods will not be incorporated into the sci/tech library.

Part III

Secondary Functions

Whether the five primary functions are accomplished well or poorly is often dependent on the secondary functions to be discussed in the next four chapters. Secondary functions include management, space planning, automation, and the purchase and maintenance of equipment for storage, duplication, and access to information. They are secondary functions only because they exist to support the primary functions, not because they are unimportant.

Chapter 9

Management

Librarians often end up managers when they have little academic training or experience in the field of management. And yet, accomplishing the library's goals and providing good service can be more dependent on management skills than on technical skills. A similar problem is that faced by technical organizations who promote scientists and engineers based on their technical skills: the education of a good technical person to be a manager. Because the problem is similar, some of the solutions can be applied to librarians as well: courses and workshops in management skills (such as communication, delegation, motivation, time management, assertiveness, planning, interviewing, and evaluating staff) are valuable to both. Often the librarian and the engineer are encouraged to get MBA degrees in addition to their technical qualifications. Although many courses in an MBA program are heavily slanted toward profit-making settings, the techniques and skills which are taught are valuable in all settings. And more and more courses and texts for management in the nonprofit sector [for example, see Kotler (1982)] are being added to business-school curricula. Management training usually includes experiential learning such as practice in group dynamics. As De Gennaro says, "management . . . cannot be taught in the classroom any more than swimming can be learned by reading about it." (De Gennaro, 1983b, p. 1320)

Many aspects of the management of particular functions have been discussed in the earlier chapters. For example, we have discussed the organization, staffing, budgeting, and evaluation of each of the major functions. Also, the chapter on automation is largely concerned with management. One particular aspect of management which is of universal concern to libraries, space planning, has a chapter of its own. In this chapter we will discuss management from a library-wide vantage point, rather than from the perspective of a particular function. The following activities are commonly considered a part of management: planning and budgeting, organizing, communicating, and personnel management (which includes hiring, training, motivating, and evaluating). In addition to each of these, we will also consider collection and analysis of statistical data, leadership, and marketing.

Management is a very broad topic, covered by voluminous literature geared toward both professionals and the general reader. This chapter can only provide an introduction for those new to management of sci/tech libraries and suggestions of further sources. There are three library-school texts on library management (Evans, 1983; Rizzo, 1980; Stueart and Eastlick, 1981) which are worth consulting for further background. Person (1983) also collected a number of useful articles and reviewed the literature of management as it applies to libraries. Although there are a large number of useful books on management, McGregor's *The Professional Manager* (1967) and the writings of Peter Drucker (1974) are notable.

American management is coming out of an era of management science which was removed from the actual production of goods and services and preoccupied with finance and marketing (Abernathy and Hayes, 1980). For libraries, this means that management techniques such as management by objectives, participatory management, zero-based budgeting, program planning and budgeting, and decision theory still have something to offer but are only good if they produce results—for the users.

> Just as American business is going back to basics, rediscovering the importance of the factory floor and production, so we librarians need to go back to basics and rediscover that our main function is serving users, not building collections. It is not our main function to devise and implement new cataloging codes, or online catalogs, or national networks. Like collection building, these are all means of serving users and not ends in themselves. (De Gennaro, 1983b, p. 1321)

It may be useful for librarians to think of management skills split into two parts: skills useful for managing the library itself, and skills useful for operating in the larger environment, whether that is the next layer of management within the library, a corporation, a large library system, or a government agency. Skills within the library are those that help get the major functions accomplished. These include hiring, training, motivating, evaluating, and communicating with staff, organizing the library so the work can be done effectively, planning, setting goals, and communicating those goals. Skills within the larger organization are usually used to get the resources which are needed to accomplish the library's functions. The leadership is focused upward and outward, rather than within the library. Skills required include budgeting, planning, politics, and communicating effectively with upper management. Communicating effectively includes being able to explain the library's needs in terms of the goals of the organization as a whole. Politics, though often ignored by librarians (Herbert White, 1984a,b), can be an essential determinant in the resource allocation process.

PLANNING

Planning is essential in managing the library, and it is also one of the most effective ways of communicating with the larger organization. Budgeting is closely tied to planning, too, as is personnel management. Planning as a topic covers such relatively well-defined tasks as forecasting what periodicals will cost next year and planning to order a microcomputer for online searching, as well as much less well-defined and broader tasks such as what is the goal of the library?; what changes are coming in the company (or the university) which will affect the library?; what will new technology (e.g., optical and video disks) mean for us 5 years from now? Strategic planning is the type of planning which deals with the broad questions. For example, Reneker (1983) summarized a number of factors in the funding environment that are affecting university sci/tech libraries. These include inadequate levels of funding (for both libraries and the universities themselves), decreased holdings of periodicals, uncertainty about the level of federal funding of scientific research, increases in enrollments in the sciences and engineering, and the need to strengthen relationships with the corporate sector. She also summarized changes in the production and transmission of scholarly information (Reneker, 1983).

Dealing with these broader questions is difficult but also has great potential to improve library services. The experts recommend that strategic planning (the broad approach) be structured, that it have input from users, staff, and management, and that it be written. The actual arrangement of the strategic plan must depend on the local situation. It might be structured by the major organizational divisions in the library, or perhaps by the major functions and secondary functions we have discussed.

For university libraries, the Association of Research Libraries (ARL) Office of Management Studies has sponsored several self-study programs which provide a methodology for planning. The Management and Resources Analysis Project (MRAP), the Collection Analysis Project (CAP), and the more recent Public Services in Research Libraries project are all examples. One self-study which was specific to a sci/tech library was conducted by the University of Arizona Science–Engineering Library (1983) with the assistance of ARL OMS staff. The library staff analyzed existing information about the library and its environment, but also gathered additional data about user needs, use patterns, staff requirements, workflow, space needs, and faculty utilization of the library. Riggs (1984) and A. Wilson (1979) both provide further information on planning for libraries.

The four steps of the planning process are: (1) where are we now?; (2) where do we want to be?; (3) how do we get there? and (4) how are we doing at reaching our objectives? The first two aspects are commonly part of strategic planning, and the second two of operational planning.

Where Are We Now?

Although what will be considered in any of these steps will be dependent on the particular setting of the library, one way to begin is to consider each of the major functional areas and assess what the library is doing in each one. For example, is there ready-reference service? Literature searching? What is collected? Is everything cataloged, and what access do the users have to it? What is provided through document delivery? How do users keep current? Is there a space problem? Do users know about library services? This examination will certainly suggest some areas for improvement. It will also point up some things that are being done well which need to be preserved. Statistical data are helpful in stating "where we are now."

Where Do We Go from Here?

The second step, determining where to go, should start with an overall statement of the library's mission and goals. The mission statement is normally a broad statement of purpose, such as, "The Library's mission is to provide in a timely manner the information needed by the staff of Organization X to perform their jobs."

Goals are a little narrower and may focus on one function, such as document delivery. "To maximize the exposure of our user population to published information of relevance" is a goal suggested by Lancaster (1977) which ought to be useful in any kind of library.

Once the goals are stated, the more concrete objectives can be developed. Objectives should be more concrete and measurable, and might include statements like "to develop a collection which will satisfy 75% of requests for materials"; "to respond to information requests with an average turnaround time of 24 hours"; and "to satisfy 85% of the information queries within 2 days."

De Gennaro suggested:

> I believe the right goal for a research library in the next decade is to plan and implement a comprehensive program for using computer and communications technologies to add

a powerful new electronic dimension to supplement and enhance its traditional collections and services. At the same time, it must also continue to strengthen its traditional collections and services and provide the necessary physical facilities to house them. . . . The first objective is to implement an integrated system with an online catalog and appropriate internal and external network interfaces. The second is to convert the library's card catalog records to machine-readable form and add them to the online catalog. The third is to continue to strengthen the library's own book and journal collections while also developing its capacity to provide access to scholarly resources elsewhere both in traditional and electronic forms. The fourth objective is to provide for the library's growing and changing space needs during this time of transition. (De Gennaro, 1984, p. 1206)

Developing these goals and objectives should involve input from users, staff and management. The input can be formal (i.e., a formal needs assessment) or informal. Past statistics of user interactions can be useful at this point for forecasting trends. It is also important to consider changes that are occurring in the overall organization (e.g., numbers of users, new programs, moves of users to remote locations). Trends in the way science is funded and the way information is transferred should also be considered.

How Do We Get There?

The third step, figuring out "how do we get there," requires considering alternatives and assessing resources. Resources include staff, funds and space. How many staff are there? What are their strengths and weaknesses? What motivates them? If they are lacking necessary skills, are there training courses available?

Tactics of how to achieve the plan will be greatly dependent on the larger environment. The actual system of budgeting is almost always defined by the corporation, the university, or the government agency of which the sci/tech library is a part. Personnel positions and salaries may also be defined. It is thus very important for the library manager to understand the budgeting process and the hiring and evaluation process within the organization. These two processes provide the resources with which the library's functions can be accomplished: people and money.

For further information about libraries in particular settings consult Bond (1984) for planning, budgeting and personnel management in the federal setting, Ferguson and Mobley (1984) and Herbert White (1984a) for the special library setting, and Rogers and Weber (1971) and Mount (1985) for the academic setting.

In some environments, the budget process is the appropriate place to propose a new service and ask for the funds and staff to initiate it. But the budget process is not the only way to get resources, and if it fails, sometimes resources can be found outside the normal channels. For example, another organization which is doing part of a records management job might contribute staff or funds if the library could do the whole thing better; alumni might set up a special fund for expanding the collection; a department might be willing to fund computer searches for students as an educational endeavor; or a laboratory might have an extra terminal they didn't need. The ARL SPEC Kit 94, fund raising in libraries (Association of Research Libraries, 1983), reported that three-fourths of the responding libraries conducted efforts independent of the university to obtain outside moneys. The requests were for funds for automation, building renovation, preservation, and equipment purchases, as well as for collections. If outside resources are not available, it is possible to reorganize within the library to get additional resources to do something new which is important. Instituting automation rarely saves staff, but can provide much greater service for the same amount of staff. A staff member who is motivated can often reorganize his or her own work in order to take on a special project or new service which will satisfy him or her personally. Perhaps some function that the library is performing could be dropped, combined with something else, or accomplished more efficiently. For example, if a new acquisitions list is being typed, perhaps it could be produced as a spin-off of an automated cataloging system.

How Well Are We Proceeding toward Our Objectives?

Follow-up, the fourth step in planning, is the management process which monitors progress toward the objectives. Follow-up can be simple or complex. There are project management processes like PERT, CPM, and Gantt charts (see Dougherty and Heinritz, 1982) which help keep complex projects on track. But simple tactics like checking with the person responsible for each part of the plan to check progress can be effective, too. In budgeting, it is important to check at various times of the year to see how the projections are doing. This is not always a simple matter, because funds are rarely expended evenly throughout the year (for example, a large proportion of periodical subscription funds is commonly expended at the end of each year). But if patterns at least stay the same from one year to the next, last year's can be compared with this year's. Performance measures like those developed for public libraries can be of use in measuring progress (Zweizig and Rodger, 1982; Schrader, 1980/1981; Oklahoma Department of Libraries, 1982; Palmour *et al.*, 1980; Manthey and Brown, 1985). Five of these measures are particularly applicable to the corporate environment. These include the community's awareness of library services (established by a survey); the number of users in the community population; the number of reference transactions per capita; the reference fill rate; and the timeliness of information delivery (McClure and Reifsnyder, 1984).

Part of the follow-up process includes revising the objectives when they become unsuitable or unattainable.

BUDGETING

Although we have already discussed budgeting as part of the planning process, there are two aspects left to discuss: methods of charging back costs, and guidelines for the appropriate proportions between various items in the budget.

University libraries are recharging costs of online searching, although Koenig (1984) and Atkinson (1984b) both say this is primarily a consequence of the difficulty of budgeting for online services in a for-free library environment. Special libraries are recharging many activities, as the management theory that unallocated cost should be minimized finds its way into practice in the corporate environment. Koenig (1984) and Herbert White (1984a) both recommend charging back an estimated amount rather than forcing users to pay by the transaction.

> The reluctance to charge for services at the time they are rendered, even if that charging consists only of getting a supervisor's signature on an authorization form, is very well taken. Such "charging" can have a chilling effect on use of the library or information center. Further, it is almost certainly not only not in the librarians' interest, but also not in the interest of the organization as well. . . . "Charging back" for services implies that they are in some fashion charged against the department, unit, or project supported, but not necessarily at the time of service. This charging can be before the fact or after, rather than as the service is rendered. Politically, it is often most advantageous to charge back, as it were, before the fact. That is, one negotiates budgetary support for the library from its using departments for the upcoming budget period, based upon an anticipated level of use. . . . The procedure has a number of political advantages. Most importantly, there is no chilling effect as the individual researcher interfaces with the library. Furthermore, there is something of a "we've paid for it, we might as well use it" initiative built into the procedure which rebounds to the library's benefit. . . .negotiating for budget support at the department level ultimately results in greater security, because it demands an attention to the needs and to the satisfaction of those department heads. Forcing continuous attention to those needs and to their perceived satisfaction is very much to the librarians' advantage in the final analysis. (Koenig, 1984, p. 96)

The second factor to consider is what guidelines are available in the budgeting process for how much to spend on various categories. For example, how much should be spent on periodicals? Is there a "right" ratio of staff costs to collection costs? Of online costs to document delivery costs? Of professional staff to clerical staff? Of library staff to users?

In fact, there are no standards and little to guide the sci/tech librarian in this area. Mount (1984) found that of the 16 university and college libraries he visited, the average breakdown of collection funds in sci/tech libraries was 24% to books, 67% to periodicals, and 9% to binding. (The schools were Reed, Swarthmore, Georgia Institute of Technology, UC Berkeley, UCLA, University of Georgia, University of Washington, California Institute of Technology, Columbia University, Harvard University, Johns Hopkins University, Massachusetts Institute of Technology, Northwestern University, Stanford University, Tulane University, and University of British Columbia.) Others have found that the periodicals budget in sci/tech libraries is in excess of 90% (Koenig, 1984, p. 90). Mount also found that sci/tech libraries were getting 17% of the total salaries for academic library staffs, and 28% of the collection funds.

ORGANIZING

Organization skills seem to be more common among librarians than planning skills. Organizational choices include whether to organize around form, function, or geography, whether to centralize or decentralize, and the size of the span of control. Drucker favors a goal-oriented structure which would give an academic library two units, one for research support and one for instructional support, but this structure has not been used much in libraries (L. Martin, 1984, p. 203). Organization by user group is also a possibility. For example, most sci/tech libraries will face the decision whether to make technical reports a separate section, which includes all aspects of handling reports, including acquisitions, cataloging, information retrieval, and current awareness, or to have the cataloging section catalog technical reports as well as all other formats. We have discussed the pros and cons of these organizational decisions under the separate functions, and all these possible organizations are discussed in detail by L. Martin (1984). Although there are no concrete rules for any organizational issue, some generally accepted management considerations related to decisions of structural organization include:

The classical span of control (i.e., the number of people a person directly supervises) is from four to eight people. However, the rate of change in the environment should be considered in determining a workable span of control. When things are changing quickly, a smaller span is necessary; when things are relatively stable, a larger span is possible, and gives people more autonomy (Evans, 1983, pp. 100–102).

Each person who is responsible for a particular area should have the authority, as well as the responsibility, to do the job.

The organization should allow decisions to be made at the lowest level possible.

People's strengths and weaknesses should be matched with the requirements of the job. A person who is quick and likes to get around rules should not be given a job which requires a methodical approach (like control of secret documents). On the other hand, someone who is very strict about rules should probably not be in a public service position.

Another aspect of organizing has to do with organizing over time. If a new project is planned (e.g., starting up an SDI service), the implementation must be organized. In this case the organization is a planning through time rather than structurally, and most planning techniques, such as Gantt charts and setting target dates for various pieces (especially if different people are to do them), are useful.

COMMUNICATING

Managers must be able to communicate well both orally (one-on-one and with groups) and in written form. In communicating upward, it is useful to figure out how one's manager prefers to communicate, and then use that mode, especially in critical interactions. Most people have a preferred mode: either oral, written, or "show me." There are many mechanisms for communicating: meetings, memos, reports, conversations, and formal presentations. The Library Committee can be used to aid communication (Katayama, 1983), and so can the annual report (Rathbun, 1974).

The annual report is a common means of formally communicating what has been accomplished during a year and pointing out trends which will influence future performance. It can also be used to point out problems to management and users and to prepare the ground for budget requests. Graphically presented statistics are a great help in annual reports and in presentations to higher management in general. Annual reports are usually structured around achievements and problems for the year in question.

Meetings are often misused, but when well run they are effective communication mechanisms and help provide two-way communication (Doyle and Straus, 1976). Running good meetings is a skill that can be developed. If one considers the salaries of all the participants for the time they spend in the meeting, one realizes that meetings are expensive and deserve attention. Meetings experts (such as Margo Trumpeter of Signetics, Inc.) suggest attention by the person running the meeting before the meeting, during the meeting, and after the meeting.

Before the meeting consider the type of meeting, participants, place, time, required materials, and results expected. Ask questions like these: What should I raise at this meeting? What unfinished business should I raise at this meeting? What pitfalls should I be aware of? What do I need to find out? When should the meeting be held in view of workloads? When should I give advance notice? When should a particular topic be brought up? Who might be upset by the topic? Who might be most helpful? Who might be able to give me information I need for this meeting? Why would they be interested in this topic? Why is it important? Why is policy what it is? Why might they be reluctant to discuss this topic? How should I raise the topic? How should I handle unforeseeable adverse reactions? How can I get the subject across quickly but clearly? How can I get information I want?

During the meeting, the person in charge of the meeting is responsible for getting off to a good start, starting the discussion, keeping the discussion on the subject, helping an individual express meaning, stimulating discussion, controlling the discussion, keeping the meeting rolling, and summarizing. Some phrases that come in handy are: All of us have had some experience with this. Give me an example of something that has happened to you. What particular information would you like to get from this meeting? Now that I have described it to you briefly, what do you think? Are we a little off track? Let us get back to the question Bill raised a few minutes ago. Would you mind holding that question for a few minutes? Shall we first list all the answers to that question? Then probably we will be better prepared to go on to the next question. Have we finished discussing that subject? If so, let us move on to the next one. Did I understand you correctly? Did you mean this (summarize)? Does everyone understand what Susan said? Does anyone agree with that? What makes you think that? Do you all agree with Tom? Will you restate that in two or three sentences? Let us take these questions one at a time. We will start with JoAnn. Are we ready to move on to the next subject? We have here a number of questions. Do you feel that is enough? Have we reached our objectives? Has everyone had a say on this particular topic? Is there any comment or question before we move on to the next one?

After the meeting, the manager should ask her or himself questions like these: What subjects aroused most interest? What important questions were raised? What should I report up the line? When did things go particularly well? When were times particularly rough? When can I get answers for them? When should I reply individually? Who was most interested, upset, or otherwise strongly affected by the discussion? Who seems to have something on his/her mind I should try to find out more about? Who contributed most? Who contributed less? Who was helpful? Who was troublesome? Why did it go well or badly? Why was the group's attitude good or poor? Why were outside pressures apparent? How did I handle the topic successfully? How did I handle the topic unsuccessfully? How did they react to the subject? How did they react to each other?

A few other basic rules include starting the meeting on time and stopping on schedule. If it looks like the meeting will run overtime, summarize the progress of the meeting to fix a starting point, schedule another meeting to continue the discussion, and ask yourself whether you attempted to cover too much, permitted the meeting to wander, permitted someone to monopolize the meeting, or spent too much time on minor points.

Indeed, the manager needs communication from the staff as much as, or more than, the staff needs the communication from the manager. Managers depend on communication from their staff to let them know problems which need to be dealt with, since they are usually at a level which is removed from direct contact with users.

MANAGING PERSONNEL

Personnel management contains the areas of hiring, training, motivating, and evaluating staff.

Hiring

Hiring is often structured by the policies and procedures of the overall organization. But the library will have to decide what level position it needs, write the job description, and often recommend to the personnel department where to advertise and which applicants to interview and hire. If no structure is set, ALA's (1985) "Library Education and Personnel Utilization" could be useful.

Ivantcho's 1983 book *Position Descriptions in Special Libraries: A Collection of Examples* is a good starting place for job descriptions. The results of the King Research New Directions in Library and Information Science Education project (which set out to determine the competencies required by information professionals to perform different types of information handling activities) will also be helpful.

In the chapter on information retrieval, we discussed the advantage of reference staff having science background or experience. There is no question that a subject specialist has an advantage in understanding what scientists and engineers need, but there is a shortage of people trained in both science and library and information science. Since scientists in general get higher salaries than library and information workers, it is difficult to attract people trained in science into information work. So, libraries must often compromise on this issue, and in fact, Mount found that only 32% of practicing science librarians in his 1983 study had a degree in a sci/tech subject (although another 51% claimed some collegiate training, and 75% claimed some sort of job experience involving sci/tech subjects) (Mount, 1985, p. 50). In any case, job descriptions for reference personnel should at least list science background as a desirable qualification.

The following journals carry advertisements for open positions for professional librarians: *Library Journal, American Libraries, Chronicle of Higher Education, Specialist.* The Online Chronicle, an online file on DIALOG, also lists positions, and there are a number of joblines, both national and regional, sponsored by library organizations. These latter two sources have the advantage of speed. A comprehensive list of library placement sources, including these joblines, appeared in the *Bowker Annual* (Myers, 1984). The local newspaper can also be a good source, particularly in large cities. Most of the professional associations have placement centers at their annual meetings, and this is a particularly good place to recruit librarians who are fresh out of school. Most library schools also post job openings which are sent to them, too. All ALA-accredited library schools are listed in the *Bowker Annual.* Experienced people are often found through professional contacts and word-of-mouth, as well as through advertisements.

If the salary schedules set by the organization's central agency are too low, the salary surveys performed by periodically ALA and SLA can be used to help justify raising them (e.g., see Learmont and Van Houten, 1984).

There are several other issues related to hiring that managers should know. There are certain questions that cannot be asked of candidates (e.g., are you planning to become pregnant?) and certain qualities that cannot be taken into account in hiring (e.g., age, sex, religion, and race). The legal aspects of personnel management are discussed in *Personnel Administration in Libraries* (Creth and Duda, 1981). The *Librarian's Affirmative Action Handbook* (Harvey and Dickinson, 1983) is also helpful.

The issues of women in management should also be of concern to any manager of a library. Librarianship has been a women's profession, and this is one reason for salaries which are low compared to the educational requirements. Librarianship is thus a candidate for campaigns on equal pay for equal work and comparable worth, and library managers should be informed about the issues involved. Gasaway's pamphlet *Equal Pay for Equal Work* (1981) summarizes these issues. In companies, a woman library manager occasionally finds herself the only woman at her level, and all the issues of women in the workplace are very relevant (see, for example, Schuman's 1984 article "Women, power and libraries"). Other women librarians find themselves passed over for promotion. For example, women still comprise about two-thirds of the professional population in ARL libraries, but fill only 44.5% of executive positions (Irvine, 1985).

Training

A good orientation to the larger organization is an essential part of any new employee's training. If the larger organization does not provide one, the manager should arrange it. This means tours of the facilities and talks with people about what their various groups do. Although this may seem time-consuming, it will pay off in the long run. More information about training is included in the chapters on the five major functions.

Motivating

There are various theories of motivation (theory X, theory Y, theory Z, Maslow's needs hierarchy), and it is clear that different people in different settings react differently, partly because they have different values. Some are motivated by achievement (and those will respond to increased responsibility), some by social responsibility (they want their jobs to have some impact on society), some by security (they may be the ones who are working primarily for the paycheck). Library-school students have a personality profile that includes being more intelligent, critical, imaginative, and anxious than other equivalent students (Rothstein, 1985). Also, professionals as a group often wish to stay in positions which use their background and professional preparation, and this may make them seem inflexible (Drake, 1977). Most people want to know what is expected of them and try to live up to other people's expectations of them, and the manager who makes his or her expectations clear can capitalize on that tendency. Most people also prefer to have a voice in changes that affect them. Numerous studies show that people who feel they have participated in decisions will be more supportive of the changes after they are implemented (e.g., Coch and French, 1968). So managers who use some of the principles of participatory management can gain cooperation from these people. But not all decisions require full participation or consensus, and innovation can be stifled in an environment where all decisions are group decisions. As White says, "Innovation does not come from groups, it comes from iconoclastic individuals who are well ahead of the group and frequently unpopular" (Herbert White, 1985b, p. 62).

Dowlin summarized the issues of motivation like this:

> Among all of the various management styles reported on in management literature (hierarchical or pyramid, decentralized or flat, theory x, theory y, theory z), there is

only one that has the flexibility to thrive in an age of change—the one that has the ability to use the appropriate management at the appropriate time. An organization must have a pyramid structure, otherwise there will be no accountability and little direction. Yet, an organization must be able to decentralize functions that require decision making at the local or functional level, otherwise the time for the organization to adjust to purely local situations will be too long for effectiveness. *An organization must view some employees as motivated, willing contributors to the organization, yet at the same time be prepared for the employee who views the job as only a place for a paycheck in return for putting in a specified amount of time.* (Dowlin, 1984, p. 43)

Further background on motivation can be found in the book by Steers and Porter (1983).

Evaluating

Performance evaluation can be one of the tools of motivation. It is also a tool of planning, as employees can be reinforced for behavior which fulfilled plans and can be discouraged from continuing unproductive behavior.

The best performance evaluations are those focused on goals and objectives that were set at the beginning of the evaluation period (which should be no longer than a year). Unfortunately, some organizations force their managers to rank people against each other. This teaches people that winning is competing against their peers rather than cooperating. However, this kind of system is popular (particularly in for-profit corporations) because it is seen as a way of forcing managers to discriminate and rank their employees (instead of saying everyone is "very good"). Unfortunately, as was pointed out in *In Search of Excellence* (Peters and Waterman, 1982), very few people think they are below average at *anything,* and a ranking system which forces managers to rank people that way can be demoralizing.

Whatever the system for evaluation, the evaluation should be honest. People universally want to know how they are doing, and it is the manager's responsibility to tell them—honestly and tactfully. The evaluation should be focused on the job performance, not on personal traits unless those are interfering with the job. Any failure which is not communicated to the person cannot (in all fairness) be considered in the evaluation. For example, if an employee is chronically late, the manager should speak with him or her about it and provide a chance for the employee to change the behavior.

The evaluation process is closely tied to communication. One of the most important factors in people's performance is whether or not they understand clearly what is expected of them. Critical behavior problems should be documented in writing and signed by the employee. This is one way to be sure that the employee hears and takes seriously what is said; it also shows that the communication occurred, and helps document the case should it come to the need for dismissal.

COLLECTING AND ANALYZING STATISTICAL DATA AND OTHER MANAGEMENT INFORMATION

A 1983 ANSI standard covers statistics which should be collected for national reporting on libraries (ANSI Z39.7-1983) (American National Standards Institute, 1983). The *Library Data Collection Handbook* (Lynch and Eckard, 1981) also gives guidelines on what and how to count. Nevertheless, libraries do not have a consistent method of collecting statistics. Use statistics and collection statistics are normally collected in any library. National reporting also requires reporting of staffing statistics and budget facts. ANSI recommends counting full-time equivalents (FTE) by the following categories: administrative services, collection development services, technical service, user services, and total.

Use statistics are the best statistics to use in evaluation, as they are normally measures of output. Those recommended in the ANSI standard for national collecting purposes include (1) circulation (books, journals, reports, maps, etc.); (2) requests resulting in "database reference transactions"; and (3) number of interlibrary loans (both loaned and borrowed). Also, the ANSI standard suggests sampling for a week each year to determine: hours, users, circulation, uncharged use of materials, information service (reference and directional, counting each contact), shelf availability of library materials, use of seating, processing time for ILL, and turnaround time for ILL. Many libraries will keep at least the reference statistics constantly, rather than sampling.

Collection statistics are also often kept in detail. The ANSI standard recommends keeping statistics for:

Type of material by both title and physical piece. (Methods for estimating are included.)
Expenditures by type of material.
Periodical titles currently received.

Many of these statistics which are collected for national reporting purposes are useful for management information. In fact, one of the simplest uses of statistics is to simply graph past-use statistics and use the trends to forecast the future. However, these statistics become even more useful when they are combined with other information. Strain (1982) used statistics to develop ratios of number of staff per user in the company, number of staff per information request, information requests per user, etc. Calculations like these can be much more useful than the raw data, and they would be even more useful if they were widely used and published so they could be used for comparison.

Just what information will be needed by management will depend on the local library. For example, Georgia Institute of Technology Library found it necessary to know what proportion of library costs went to research and what proportion to teaching (Citron and Dodd, 1984).

Automation of library functions often produces as a by-product statistics which can be used for management information (Lancaster, 1982b). It also sometimes produces an overload of information (Olsgaard, 1982), so it's important to select statistics which are important to the local situation and present them in such a way that they are meaningful.

Bell Labs has perhaps been the most innovative at producing special management information reports from its automated systems. Kennedy's 1982 article "Computer-derived management information in a special library" discusses statistics which are gathered from the numerous automation modules of the Bell Labs Library Network. Collection levels are listed for detailed Dewey classifications, and a performance report shows, for each subject, the number of items purchased and the dollars spent by each library. A system which monitors items in process produces a list of items which have not moved in a certain time, monitors funds, and provides information on vendor performance (average order costs, discount percentages, delivery times, and number of claims and cancellations). The serials system produces lists of current serial subscriptions and their prices, and a price-ordered list of all serials. Alerting bulletins are kept on-target by systems used to process the article request traffic and to record copyright royalty payment obligations. These give information about numbers of articles announced, total requests, announced/request ratios, and an analysis of the demand for journal issues by the date of publication. The circulation system identifies weekly all book titles for which there is a total of five or more people waiting in the network, and annually all items which have no recorded use. It also produces a loan history report which is a subject-classed listing of the materials borrowed by a specified technical department. All of these offshoots from automation are useful, but most libraries purchasing commercially available systems will not be able to get this range of special reports.

LEADERSHIP

One of the primary characteristics of a good manager is leadership, although this is rarely listed as job requirement.

Peter Drucker says that the most important qualities a manager must possess are integrity of character, courage and vision. These are not qualities a manager can acquire.
Managers bring them to the job, and if they don't bring them, it will not take long for
their people to discover it and they will not forgive the manager for it.

Drucker says somewhere else that it is better to do the right thing than to do things
right. He also says that Management by Objective works if you know what the objectives
are, but 90 percent of the time you don't. To identify the right objectives, to know
what the right thing to do is, and then to have the courage to decide to do it in the
face of uncertainty, and have the trust and confidence of your people so they will
follow you—that is what management and leadership are about. They depend on character, courage, and commitment, not on techniques and tools. To be a leader is to
have a vision, a goal, and a determination to reach it. (De Gennaro, 1983b, p. 1320)

The authors of *In Search of Excellence* emphasized that the excellent
companies always had a goal that everyone in the company recognized
as the reason the company existed. And the goal was never just to make
lots of money. The advantage of the goal was that people who had inter-
nalized the goal knew what behavior would accomplish that goal and could
have a sense of autonomy about their own jobs. That goal was always
clearly important to the company's leaders—the leaders did not say one
thing and act as if something else was more important.

MARKETING AND BIBLIOGRAPHIC INSTRUCTION

Libraries offer a valuable service, but unless the users know about the
service, it will not be used. Special libraries and academic libraries have
different approaches to letting the user know about the library. Special
libraries generally think of their approach as marketing. Academic libraries
think of their role as *teaching,* and thus focus on bibliographic instruction.
Librarians often hesitate to sell their services, however. For-profit businesses can turn increases in business into increases in staff much more
easily than libraries can. To illustrate the difference, W. Miller (1984)
modeled an ad on a recent one for a pizza parlor which appeared in his
local paper. It read:

Metropolitan University
Metropolitan, Ohio

As a result of its successful public service program, Metropolitan University Libraries
has created the following new positions: 5 general reference librarians; 4 bibliographic
instruction librarians; 4 database searchers; 2 circulation librarians; 2 inter-library loan
librarians; 2 system librarians. Appropriate clerical and student staff are also being
hired. For full job descriptions and more information about our highly successful and
heavily-used library, please contact our Personnel Librarian.

The chances of seeing an ad like this are so small that reading it is
humorous. But the result of a successful for-profit organization is expan-

sion, and it is no wonder that in the absence of expansion as an option, nonprofit organizations hesitate to "sell."

Special Libraries

There are a number of methods which are commonly used in special libraries for marketing library services.

Brochures are often designed which describe library services. These can be given out within the library or mailed to target audiences. A common logo and color scheme on all library publications help people identify the brochure as being from the library.

Orientation and tours can be a marketing tool. Libraries can schedule a certain time of the week or year during which tours are given (and advertise this), or they can give them on demand. A few libraries have put together cassette tape presentations for self-guided tours.

Detailed presentations about library services can also be effective in marketing services. The librarian can ask various departments if they would like a presentation about library services. If there is a lot of demand, a videotape can be prepared. A more individual approach is to ask to talk about what the users *do,* and conclude the listening session with a description of library services which sound like they would be appropriate. Online search demonstrations are often a part of these presentations. The presentations should be tailored to target audiences: new employees, management, technical staff.

In addition to these direct techniques, library staff should be reminded that every response to an information query is part of marketing the library.

Most of these suggestions would be considered "selling." Modern marketing has moved beyond selling to a focus on the user and actually begins in the planning stage with a focus on the user's needs, wants, and perceptions and studies of the various market segments in the potential audience for the library's offerings. More segmentation, the use of new techniques such as focus-group studies and panel research, and the use of a sales approach to training library staffs are all newer techniques which Andreason (1980) suggests could be profitably used in libraries.

Academic Libraries

Brochures, tours, and other presentations are all used in academic libraries, but "bibliographic instruction" is usually added to fulfill the academic library's teaching role. Lubans (1974) edited the classic introduction to bibliographic instruction. Bibliographic instruction in sci/tech libraries is normally focused on the graduate student rather than the undergraduate, however. In contrast to humanities and social science students, science students rarely need library instruction at the undergraduate level, since they rarely do research or write papers at that level.

A bibliographic instruction program for a sci/tech library should include both overall orientation sessions and sessions oriented toward specific audiences (e.g., organic chemistry graduate students). The general library orientation should cover the major indexes and abstracts, encyclopedias, and other special tools. It should also include the use of the library catalog and the potential of online literature searches.

Most literature shows that effectiveness of teaching is determined by its relevancy to the student. Sessions that are focused narrowly on the subspecialty of particular classes (e.g., geochemistry, high-energy physics, organic chemistry) can be effective, particularly if class assignments require the use of the library. Rather than waiting to be asked, sci/tech librarians should offer these sessions. If large numbers of students from a particular class suddenly appear doing papers, the teacher for the class should be contacted and the offer of instruction made.

The faculty is often a difficult audience to reach with bibliographic instruction. They have established invisible colleges and feel they know all about libraries, and they are usually very busy. However, they may not be using available services effectively, because libraries have changed so much. The introduction of an online catalog or online literature searching sometimes allows an approach to the faculty. Seminars on keeping current or on organizing personal files (particularly on microcomputers) have the potential for attracting faculty attendance (Wanat, 1985). Also, if the graduate students use the service effectively, it will get back to the faculty.

Conclusion

Managing a sci/tech library is not terribly different from managing any other kind of library. Good planning which is focused on the needs of the users works in all settings and the resources developed for general libraries can be usefully applied to the sci/tech setting.

Chapter 10

Space Planning

Sci/tech librarians may be faced with any or all of three types of space planning. One is the rearrangement of current space. This could be to shift or move collections or to reorganize to accommodate automation, improve efficiency, or make space for a new service. The second is planning—that is, planning for growth for the collections, space for new services, etc. The third is the planning of completely new space, either for a new building or new space within a building.

All managers of libraries will have to deal with the first two, rearrangement and planning. These are normally within the control of the library, and growing collections and the increased needs for equipment will force their consideration. However, planning completely new space is an opportunity only some sci/tech library managers will have. But as Ferguson and Mobley say, "The canny library manager keeps a desired floor plan in a top desk drawer to be pulled out when opportunity knocks" (Ferguson and Mobley, 1984, p. 108).

REARRANGING CURRENT SPACE

A floor plan for the library is essential for considering rearrangement or remodeling. A common scale for blueprints (which should be available for the building) is ¼ inch = 1 foot. Although a ⅛ inch = 1 foot scale is better for planning concepts (like drawing circles for the location of reference, circulation, user space, etc.), the ¼-inch scale is good for considering the details of layout.

This is also a commonly available scale for graph paper, which can be used to cut out "paper dolls"—pieces which represent the furniture that needs to be moved around. Also, sheets of common office furniture (e.g., desks, tables, chairs, filing cabinets) at this scale can be purchased commercially. By using masking tape to hold the pieces down, they can easily be removed and rearranged (it is much easier than moving the furniture if it does not fit!). Although preparing the pieces seems like a lot of trouble, they can be reused when the space (inevitably) needs to be changed again.

Layouts for office space need to take into account some principles about people: for example, most people do not like to face the wall or to have their backs to a door. So an arrangement of staff space or carrels which has the side of the desk or carrel to the wall will be more comfortable. The Cohen and Cohen book *Designing and Space Planning for Libraries* (1979) is particularly helpful in pointing out this sort of behavioral factor.

Blueprints at this 1 foot = ¼ inch scale are also useful for planning a collection shift (a common occurrence in sci/tech libraries). The first step in a collection shift is to count the shelves available for the collection, then to measure the linear feet of the collection that must be shelved there. Blueprints normally show the stacks, and the amount of shelving can be counted more easily from the plan than by walking around and counting shelves. This is only true, however, if the shelf spacing is even (e.g., six shelves per stack, or seven per stack). If it varies for whatever reason, it should either be adjusted for the shift, or the variations should be recorded on the planning document. Measuring the linear feet of the collection sounds like a monumental job, but it is normally not as bad as it sounds. Two people can measure a collection of about 50,000 volumes in less than 2 hours, one measuring and the other recording the beginning call number for each shelf and the number of inches of material on each shelf. With these figures, a total amount of linear inches of material from any call number to any other call number can be figured out, as well as the total linear feet of the collection. By dividing the total linear feet of the collection by the total feet of shelves available to house it, one gets the percent of empty space available to distribute. By distributing this amount evenly on each shelf, the collection can be shifted accurately.

For example, let's say there are 204 sections with six shelves each available for the collection and each shelf is 36 inches. The collection measures 32,666 inches. Dividing the collection by the space available to house it gives us 32,666 divided by 44,064 or 74%. If we want to distribute the space evenly on each shelf, we should fill 74% of the shelf (or 27 inches), leaving 9 inches empty.

The original figures measuring the collection can be used to figure out where the collection will break at any point, so the materials can be double-checked at various points to be sure the shift is occurring properly. For example, we can find the halfway point in both the shelving and the collection.

There are certain kinds of collections (e.g., periodicals) in which the space should not be distributed evenly. For example, in a continuous run of journals, there is no point in leaving 6 inches on each shelf—all the space is needed at the end of the run.

Planning for shifts of collections is adequately covered in the literature. The standard reference is Metcalf's *Planning Academic and Research Library Buildings* (1965), but Fraley and Anderson's *Library Space Planning* (1985) is a shorter, practical guide.

PLANNING FOR THE FUTURE

Planning for space needs is one essential part of managing a sci/tech library. It is the first step in requesting new space, and since getting new space is usually a lengthy process, space planning should focus at least 5 years ahead, and 10 or 15 may be more appropriate, especially if a new building is the proposed solution. Evans reported that a 1971 study of 350 new library buildings found that the average time from first request to the occupation of the building was 8.35 years. He suggests that 5 years is a planning minimum (Evans, 1983, p. 149).

A useful way of planning is to consider the following areas: user space, staff space, collection space, and systems and special equipment space.

For each of these areas, the first step is to conduct an inventory of the current status. After the inventory, the next step is to consider whether what exists is adequate. For this, standards are useful. Finally, we use the inventory and standards information to forecast space needs for 5, 10, or 15 years.

User Space

How much user seating is there now? In which areas of the collection? Of what kinds and types? Large tables to spread out materials, or just chairs to sit in and browse journals? How many square feet are available for user space? User space to use special equipment (e.g., microform readers, patron access terminals for the online catalog) is usually considered under "system and special equipment space."

Guidelines for user space have been to allow about 30 feet per user in academic libraries and 50 net square feet per user station in research libraries. For user stations in which equipment will be used, the guideline may have to be increased to 75 net square feet. There are also guidelines for user seating. Libraries have been expected to provide seating for from 5 to 25% of the user population. The high figure is commonly used for academic libraries with heavy usage and the low figure for corporate libraries where most materials are used in nearby offices.

Staff Space

How many staff are there and what do they do? What is the total amount of square footage assigned to staff space?

Guidelines for staff space have suggested 100–120 square feet per staff member. As a result of automation and the increased space taken by terminals, printers, and modems, Beckman (1982a) suggests this should be increased to 175.

Collection Space

The stack space currently available for the book, periodical, and hardcopy report collections has to be inventoried, and the actual linear feet of these collections has to be measured. It is also possible to estimate space based on the number of volumes in the library (a statistic most libraries collect) by using formulas for the number of volumes per shelf (see Roberts, 1984) but actually measuring the collection is more accurate for immediate purposes like stack shifts. If stack space needs to be converted to square feet, the total linear feet can be used to estimate square footage needed, based on formulas (see Metcalf, 1965; Daehn, 1982).

Collections of materials which are not housed on stack shelving also need to be considered—in fact, any collection which is kept separate in the library should be considered separately in this inventory. Microfiche shelved in cabinets or lectrievers, maps shelved in map cases, laboratory notebooks kept in locked cabinets—all need to be considered.

Guidelines for collections have suggested from one-third to one-fourth empty space, to allow for growth and easy shelving.

Space for Systems and Special Equipment

How many microform readers are there? Which kinds (fiche, film, cartridge)? How many public terminals (or card catalog cases or read-only memory readers for the catalog)? What kind of tables are they on? Photocopy machines, the library's computer (if there is one), and any other special equipment should be considered here, and the amount of square footage they occupy should be recorded.

Special equipment requires more space per user and involves other considerations. For example, the use of cathode-ray terminals (CRTs) brings with it a large group of new considerations: glare, appropriate furniture, wire management, increased air conditioning and acoustical control. Map and microform cabinets can cause floor loading concerns. Equipment for terminals is discussed in Chapter 12. Cohen and Cohen (1981) also give guidelines on the determination of adequacy for all these types of space.

The Space Forecast

If the library is inadequate in some area (e.g., user space), the inventory should be adjusted to be adequate before beginning to forecast for 5, 10, or 15 years. For example, if there are only 10 user seats right now and you should have 20, the projections should start from 20.

In addition to the inventory information, to forecast space needs we need to know growth rates for the different collections. What is the current size of the collections (e.g., number of volumes, number of reports, number of microfiche)? How many of each are received each year? The number received each year divided by the number currently in the collection is the growth rate. For example, if we have 60,000 volumes and we buy 5000 a year and weed out 3000 a year, the growth rate is 2000 divided by 60,000 or 3.3% per year. The growth rate allows us to forecast how many volumes we will have in 5 years or 10 years, so we can predict the amount of space we will need. For example, at 3.3% per year we will have about 70,600 volumes after 5 years.

Consider user space and staff space the same way. Will the library's user population be growing or shrinking? Are there other factors which require increased user seating (e.g., more people located at long distance who will be visiting rather than using materials in their nearby offices)? Will the library be adding new services and staff?

In corporate settings, there may also be guidelines (sometimes remarkably strict) about how much space certain types of people are allowed. If these are not adequate, sometimes adequate space can be acquired by separating out processing areas (which are physically part of people's workspace) and allocating space for them separate from the calculation of workspace per staff member.

Systems and special equipment is a little harder to forecast because the technology is changing so quickly. Even a library which maintains a no-growth collection by weeding will probably have increased space needs for systems and special equipment, for optical and video disks, and for increased automation. And the whole arrangement of space will probably change. As Atkinson said,

> The idea that centralized library services should be emphasized because they may, in some theoretical, mass sense, be able to answer more reference questions, catalog

more books, or stay open more hours downtown or at the main campus library is flying in the face of new societal demands. Electronics provides for independence from distance. The catalog, order file, and periodical check-in files are no longer available only at one physical location. . . . This destroys one of the great imperatives of present organization—the need to group around files. (Atkinson, 1984a, p. 556)

PLANNING NEW SPACE AND NEW BUILDINGS

Planning new space is a great opportunity, and librarians who have clearly inadequate space should not be afraid to draw up a layout of the ideal library and work to get the project approved, using the planning figures collected to justify it. In a corporate setting this normally requires estimating costs and getting the funds approved in the budget. In other settings it may not require cost estimates by the librarian, but just a request to get it into the list of worthy projects—which come out of some other department's budget for construction or remodeling.

The process of planning new space varies depending on the setting of the library, but almost always involves a number of people besides the librarian (such as an architect, one or more contractors, and higher administrators). Thus, the librarian is only one source of input to the plan, and the challenge is more one of communication of the library's needs than actual laying out of the space. Often the librarian will have a chance to write a "program" which describes the library's requirements. But normally an architect will actually design the building. Unless the architect is knowledgeable about libraries, the librarian will have a need to educate the architect. The importance of this cannot be overestimated. Sometimes what is appealing aesthetically is a disaster for the library. For example, Albuquerque Public Library, which is very attractive from the outside, has its door in an unexpected place, and users have sometimes walked all the way around the library without finding the door. Rovelstad (1983) suggests that Metcalf (1965), Mount (1972), and other library-planning literature should be required reading for architects.

The figures gathered in the planning process will be essential in planning the new space. But combining all the factors involved in making a library function is not a simple matter. Accessibility of services, flexibility, the relationships between functions, between staff and the collections, between users and services and users and the collection—all these must be balanced. And all the physical factors must be suited to the planned and potential uses of the space (e.g., floor loading, lighting, acoustics, temperature, etc.).

Factors particularly relevant to planning for sci/tech libraries include the need for larger tables for materials, the need for electricity for calculators etc., and the importance of the journals and the current-awareness function. Display areas for current journals should be accessible and comfortable.

Rovelstad (1983) describes the planning process for a new facility and suggests various ways librarians can get the background for planning a new facility, including reading, visiting library buildings, and attending workshops. Metcalf (1965) and Mount (1972) both include checklists which should be helpful in being sure that nothing is overlooked.

Because of the size of the task, planning new space is an area where a consultant is useful. He or she will be knowledgeable of many factors with which librarians have little experience (for example, lighting, floor loading, acoustics, temperature) and also will know about library aspects with which the architect may have no experience. Consultants can also help in the communication process, particularly in cases where the library's needs are not being taken seriously enough. Often an outsider's opinion is more influential than an insider's.

Suggestions for selection of a consultant can be obtained from some issues of *American Libraries* and its publication *Library Building Consultant List*.

Chapter 11

Library Automation

Sharon Kurtz

Sandia National Laboratories
Albuquerque, New Mexico 87185

Automation is clearly one of the most important secondary functions of sci/tech libraries. A good automated system can make a library truly a service organization; a bad one can be much worse than none at all.

Automation has made an impact in all five main functions. Collection development has probably been affected the least of the five. Online catalogs offer the potential for statistics on the use of various areas and rates of acquisitions. Online acquisitions systems offer the potential for making the acquisitions process more efficient. But the selection process itself has not been amenable to automation. In information retrieval, online database searching has greatly improved productivity of literature searching and opened access to the world's literature to anyone with a terminal and a telephone (and money) (see Chapter 4). The current awareness function has been greatly altered by automation, too. The availability of computerized SDI service offers an efficient way of tailoring output to individual needs. Even library acquisitions bulletins are being produced in new ways, often as spin-offs from the automated cataloging systems (see Chapter 5). It is in the area of collection control that automation has made its biggest impact, since the catalog and the circulation system have often been the first functions to be automated.

The purpose of this chapter is to look at automation of library functions as a topic in itself. We will consider the major activities pertaining to automating a library—deciding whether or not to automate, systems analysis, evaluating alternatives, and acquiring a system [the request for quotation (RFQ), evaluation, selection, and implementation]. After reviewing these basics, we will discuss system evaluation, management aspects of automation and recent trends. The library's interface with office automation is considered under "trends." The chapter is geared to someone who knows about libraries, but has not gone through the process of automating one.

WHY MIGHT A LIBRARY CONSIDER AUTOMATION?

Any number of factors can precipitate the automation decision. A library's first step toward automation can stem from a desire for increased productivity or cost savings/containment. Automation may be considered to address needs to perform tasks less expensively, more accurately and more quickly; to handle tasks which can no longer be accommodated by the manual system due to higher volumes or costs; to perform tasks which cannot be done by the manual system; to prevent duplication of effort; to improve control over operations; or to provide better management information.

Indications that existing systems need improvement include the occurrence of operational delays, inadequate inventory control, a high volume of work being processed (especially if the work is repetitive), inadequate or unreliable managerial reports, and inaccurate and untimely information or data.

Automation can be done as a pilot project to gain familiarity with computers and software, or because of the availability of equipment or because of pressure from management.

A library already having some automated functions may wish to upgrade its system for various reasons; to accommodate increased work or data loads, to expand a single automated function into an integrated system, to network with others to share resources, or because a point has been reached when the cost or frequency of repair makes it more advantageous to replace all or parts of the system.

Libraries traditionally have complex requirements, such as (1) large data files, (2) variable and lengthy records, (3) the need for many index points, (4) the need for output sorted many ways, and (5) a large amount of alphabetic, textual, or abstract data. These requirements are exactly the things for which computers are best suited. Handling large volumes of data (the census information) was a primary reason computers came into existence in the first place. Automating bibliographic records allows libraries to retrieve data via a multitude of avenues and can provide immediate access to data.

As an example, 15 years ago, the first library in which the author worked had four manual card files (author, title, subject, and shelf list), and the titles were retrievable only by the first significant word of the title. Not many access points were available, it took time for new entries to be filed, and it required a lot of effort to keep the files in sync. By contrast, the most recent library (with an automated system) in which I worked could access its bibliographic information in all of the following ways (and more), within moments after the data was entered into the system: author, subject, any significant title or abstract word, publisher, ISBN or ISSN, classification and/or report number, vendor, and patron names. It could apply full Boolean search criteria, such as searching for a particular author and a certain publisher but limiting the search to only those books published between 1982 and 1984 and including only those items which were not currently circulating. All operations were performed using a combination of screen menus and commands.

Automation, especially because computers keep getting more powerful and cheaper, is within the reach of any library and provides numerous benefits to the library staff and patrons.

SYSTEMS ANALYSIS

Deciding to consider automation, for whatever reasons, is just the first step. The next step is to perform a systems analysis. While systems analysis is frequently associated with automation, even libraries with no automation ought to perform a systems analysis periodically. Systems analysis can provide a thorough understanding of the present system, reveal ways to improve the existing system immediately, serve as a foundation for defining and designing a new system, and form the basis for a cost justification of proposed changes. An analysis, therefore, should be carried out with no prejudices for or against automation.

These tasks are extremely important, especially if automation is the direction in which the library is heading. Table 11.1 shows the extent to which certain information gathered during the systems analysis affects hardware and software decisions.

Many systems analysis methodologies and tools exist, including some automated ones. Systems analysis and design methodology has become a business of itself for which numerous books, articles, and classes are available. No attempt will be made to list, endorse, or elaborate on specific methodologies. The next section will, however, identify the major aspects of systems analysis.

TABLE 11.1.   Effects of Systems Analysis Information on Hardware and Software
Decisions

Information	Size and speed of computer	Number and type of terminals	Amount of storage (disks and tapes)	Type of software
Variety of work (different functions being done)	Yes	Yes		Yes
Quantity of data	Yes	Yes	Yes	Yes
How fast data grows	Yes		Yes	
Quantity of work (total and peak)	Yes	Yes		Yes
Amount of concurrent work (people doing the same work at the same time)	Yes	Yes		Possibly

Planning the Systems Analysis Project

The systems analysis phase can take months or even years of work for very large systems and may require the full-time efforts of a person or team and additional tools and materials. A small library may find that minimal work is required for some of the tasks. Whether a large or small system is being analyzed, the project planning should not take longer to complete than the analysis. The systems analysis phase is an area in which use of a consultant may be helpful. (Consultants will be discussed more fully under management aspects.)

The planning phase should include management approval and funding for the project; communication with all staff regarding the purpose of the project and findings; a project plan with objectives and constraints; a schedule of activities to be completed; and a timetable for completion and assignment of activities. Corbin (1981) includes a chapter on project planning and management and an appendix on the phases and activities of a project.

Analyze the Present Operations

This task should result in a complete description of the current system. It consists of collecting, organizing and analyzing data about the existing system's objectives, inputs and outputs, processing operations and resources. The results are information on the quantity and type of work being done, staff productivity (how much time is being spent on doing the work), whether the work is being performed in logical order, whether every operation is necessary, what reports and paper products are being generated and whether they are necessary, what reports ought to be but are not being produced, preliminary cost estimates, etc.

The Markuson *et al.* (1972) book, although published nearly 15 years ago, provides general guidelines for systems analysis, as well as guidelines for specific tasks (i.e., file analysis) which are still pertinent. Matthews (1980) offers methods of gathering information and library performance measures.

A variety of tools and techniques including flowcharts, data and operation flow diagrams, and interviews can be used in gathering information. Major activities are:

1. Identify the scope and purpose of each function. Use written descriptions, procedure manuals, and former feasibility studies.

2. Identify the system inputs/outputs and work flows. Required information includes number and sizes of files, number and format of data items, items which are used for searching and sorting, input editing requirements, transaction rates (number of transactions performed during a period of time), and input and output forms. Inventory the personnel, equipment, and supplies necessary to operate the system.

Graphic and pictorial representations showing input, processes performed on the input, sequence of processing steps, and outputs can be helpful.

Identify the System Requirements

Use the information gathered from the previous steps to identify the mandatory, highly desirable and desirable features required of a system.

Tenopir (1980) lists questions to be considered in assessing needs.

It is important not to simply restate the requirements as they apply in the current system and to be realistic about requirements; do not develop a mandatory requirement for an infrequently encountered situation. Projecting requirements 5 years into the future may prove worthwhile for planning and budgeting. It is a good idea to rank the requirements. Which mandatory requirement is most important, and why?

The requirements definition is the heart of any systems analysis and/ or automation effort. Many problems encountered later can be avoided if the library has a clear idea of its needs and the importance of those needs. Absurd as it seems, systems have been purchased before considering what the system will be doing.

Analyze the Collected Information

Review the information collected to tell how well the current system measures up. Is data being recorded or stored that is never used? Are there illogically sequenced processing steps? Is output still being generated, even though the reason it was originally produced no longer exists?

After deciding what (if anything) is wrong with the current system, the library has three options for making a better system and may wish to do a cost–benefit analysis of each option before deciding upon one (King and Schrems, 1978). Matthews (1980) provides formulas for cost/benefit components and several articles about costing are included in the Matthews (1983b) book. The three options are:

1. Stay with the existing system. No change will occur. The current system is meeting all requirements and performing well.

2. Enhance the existing system to solve whatever problems were identified. Improvements could be hiring more staff, a reordering of the processing steps to make procedures more efficient, or purchasing a new microfiche reader to replace an inadequate one. The library may opt for this choice simply as a stopgap measure because lack of funding or other reasons prevent the library from replacing the existing system as its first choice.

3. Replace all or part of the existing system with an automated one.

Write the Project Report

A project report should always be written showing findings and recommendations.

EVALUATING AUTOMATION ALTERNATIVES

If automation appears to be the most viable option, some immediate questions are, what should be automated, are software and hardware available to do the job, and are there other factors which affect the choices?

Ideally, the software choice should be made first, and that decision will usually dictate or limit the type of hardware required. However, several decisive factors can affect the goal of acquiring an automated system or can prevent the ideal situation from occurring. Mostly these factors have to do with the availability of existing resources and the political climate. They include:

Already having hardware. Existing hardware will affect the choice of software, since software products are developed to run on particular types of hardware.

Already having software. Existing systems software will affect or limit the choice of additional software, since packaged software is written to run with specific systems software on certain machines. In addition, existing software may affect the purchase of additional software if an integrated system is the goal. (Systems software is defined later in the section on software alternatives.)

Already having systems staff. In-house or available management information systems (MIS) or computing staff provides a library with more flexibility in making a software choice. Such staff can modify purchased software, write custom programs, and maintain complex systems.

Receptivity. No matter how cost effective or desirable an automation decision may be, if the timing is wrong from a political standpoint no change may occur.

Three primary alternatives exist for automating traditional library functions (acquisitions, cataloging, searching, circulation, serials): automate one function at a time, automate with an integrated system, or automate one step at a time. If one or more of the factors mentioned above apply, the choice of which functions to automate may be predetermined. For example, if the library must use the university computer and no integrated library software package has been written for that type of computer, integrated systems automation cannot be accomplished with one purchased package.

Automate One Function at a Time

One function at a time or several functions separately can be automated, perhaps using different hardware.

The advantages of single-function automation are that (1) automation can begin with a small, easily managed system; (2) fast delivery and implementation of the system can occur; (3) automation may be less expensive, particularly if a small computer is used; and (4) the library may have more flexibility to change direction as technology changes.

Some disadvantages are that (1) difficulties may be encountered later in integrating separate functions and (2) the choice today of a single-function automation may limit a library's future choices.

[Matthews (1983a) states that libraries choosing an automated circulation system 1 to 5 years ago are now finding that the choice of an acquisitions system or an online catalog has already been determined by the previous decision.]

Single-function automation has been the most common method of library automation, because more single-application library systems were available and circulation systems are the most prevalent single function. Examples are CL Systems (CLSI), Dataphase Systems, and Geac.

Much has appeared in the literature in the past several years on microcomputers and their applications for libraries. Several articles in the bibliography list microcomputer hardware/software vendors and describe library applications using microcomputers. Nolan (1983a) identifies 175 programs specifically for library applications. The majority of articles describe single function automation using microcomputers. More discussion on microcomputers appears later under trends.

Automate with an Integrated System

A totally integrated system has all principal library functions automated, perhaps in a network or distributed system. This can even mean using hardware from different vendors or separate similar machines, as long as the functions can "talk" or interrelate to one another.

According to Matthews (1982b), an integrated system should provide (1) acquisitions and acquisitions accounting, (2) circulation control, (3) a public access catalog and authority control, (4) an interface to a bibliography utility, and perhaps (5) serials control. It may also interface with other functions such as word processing. Epstein (1984) provides a long (and probably unrealistic) list of features desired in an integrated system.

Advantages are (1) all functions are automated at once, and a large amount of manual record-keeping can be eliminated; (2) functions interface with one another; (3) data is not replicated with some integrated systems (data duplication requires a concentrated effort in order to keep records in different files in sync and uses more storage than nonduplication); and (4) many access points to data are usually provided.

Disadvantages are (1) a higher initial cost may be incurred than with other systems [however, Matthews (1984) states that the costs for each function declines since the expensive fixed hardware costs are used for several functions]; (2) MIS or computing staffing is usually required (integrated packages are only beginning to be available as turnkey systems); (3) conversion costs may be high since many functions are involved; and (4) the library staff will be adjusting to much change in a shorter period of time.

Some integrated systems advertised on the market include DOBIS/ LEUVEN from IBM, Techlib from Battelle, NOTIS developed by the Northwestern University Library, the NLM Integrated Library System available from NTIS, and LS2000, a version of the ILS system, available from OCLC. However, De Gennaro (1985) states that "there are no furnished integrated online library systems available on the market today for any size library." His point is very valid. For example, a package which offers cataloging, circulation, and authority but not acquisitions should probably not be regarded as an integrated system. Nor should a system which offers those functions but replicates data across functions. De Gennaro (1981) has also previously said, "libraries cannot take full advantage of automation until they can implement integrated systems," adding that they cannot have integrated systems while a substantial portion of their records are in manual format.

Automate One Step at a Time

Stepwise automation differs from single-function automation in that each function is automated separately but in a sequence planned to evolve into a completely integrated system. Stepwise automation thus eventually provides the same advantages as an integrated system but at a lower initial cost and with less trauma than may occur when automating all functions at once.

The integrated DOBIS/LEUVEN software is an example of a system that may be purchased in modules. The cataloging/searching module may be purchased separately from the acquisition/circulation module.

SOFTWARE ALTERNATIVES

Once a decision has been made regarding what functions are to be automated, the next step is to find out what software is available.

Software is the programs that make the computers do their intended jobs. D. Shaw (1981) gives easy-to-understand, often amusing analogies to explain software. The two types of software are applications software and systems software.

Applications software consists of step-by-step instructions to perform a particular job (order entry, circulation discharge, display of bibliographic information on a terminal screen). Programming languages commonly used include COBOL, PL/I, BASIC, and FORTRAN. All of the packaged software discussed later falls under the category of applications software.

Applications software always needs some systems software in order to run. In fact, several separate systems products may be required, depending on the library's plans.

Systems software can be quite expensive, so it is important from the beginning to find out the extent of systems software required and to budget and plan accordingly.

Systems software performs the translation, loading, supervision, maintenance, control, and running of computers and application programs. The more important systems programming products are operating systems, language processors, utility systems, and file-management systems. These products do tasks such as interpreting the instructions fed into the computer; telling the computer what programs and data files exist, where they are located, and which ones are available for which application to use; keeping track of the number and type of input and output devices; and controlling the number of system users and their authorizations. Assembly language has been the traditional language for systems programs. Some specific examples are:

Compilers, processors, interpreters. All programming language (source code) needs to be translated into machine language in order to run. These translations are done by compiler or interpreter software, and each language has its own compiler. Applications programs may need to be recompiled or rewritten to be used on different hardware or operating systems. Compiler software is necessary if the library intends to write its own programs or to modify purchased software.

Utilities. Programs which store functions (i.e., sort or copy utilities) so commonly performed that to reprogram them every time they are needed would waste programmer time and effort.

Specialized. One specific type is protocol conversion software, which can make terminals of one manufacturer "look like" or emulate terminals of another manufacturer so that the installation does not have to get rid of useful equipment just because a new system is implemented.

Sources of Information about Software (and Hardware)

The sources include:

Services like Datapro Research Corporation and Auerbach Publishers, whose purpose is to identify and evaluate software and hardware.

Literature. A number of citations in the bibliography list software and hardware vendors. Several journals (*Library Technology Reports, Library Software Review, PC Magazine, PC Week, PC World*) report on specific products.

Other libraries. Call libraries having automated systems and ask questions.

Vendors. Call prospective vendors, discuss your requirements, and ask whether the vendor has products which would meet those requirements; visit vendors' exhibits and booths at professional conferences; ask to see demonstrations of products.

The responses to a request for information (RFI). An RFI is a brief document used early in the acquisition process to identify vendors who can meet the minimum requirements, to obtain preliminary cost estimates, and to reveal questions and terminology misunderstood by vendors which can need clarification when the formal bid document is issued. It can provide some idea of what is available in the marketplace.

How to Obtain Software

The functional requirements specified during the systems analysis define what the software must do, the functions and tasks it must perform. Regardless of what functions are to be automated, only two ways of getting the software exist—the library must either purchase/lease it or write its own. With the great number of software products available on the market today, a decision for a library to write its own custom software is probably only justified if the library has unique requirements and has found no acceptable software to purchase. Generally, packaged software is a better and more economical choice even if outside MIS assistance is required to operate or adapt the software.

Purchase/Lease Software

A wide variety of software packages suitable for library applications are currently on the market. Prices can vary widely, although you will probably find the "you get what you pay for" adage true. The more functions provided, the more expensive the software. Also, software which runs on larger machines will probably be more expensive. The ease with which the software is installed and used can also vary greatly.

Software acquisition is a big decision which can have impact for years to come. Try not to make a hasty decision. Take time to look at the software in operation, preferably in a library similar to yours. Ask lots of questions. Librarians are not hesitant about saying exactly what they think (both the good and the bad) when asked about particular software. It behooves you to get as much information as possible before purchasing software.

One consideration which will have much impact, especially in the staffing area, is the type of control the library will assume regarding the software. With standalone software, the buyer assumes control over the installation, maintenance, and running of the software. Assistance from the vendor can vary greatly. Vendors may provide extensive help (for which the library may pay extra) or may take the "you bought it, it's yours" attitude.

Library software can be placed into five categories: integrated (multifunction software), turnkey (single- or limited-function software), software from another library, software not specifically written for libraries but which has successfully been used for library applications, and service bureau software. Examples of each, with their benefits and potential problems, are as follows.

Integrated-Function Packages. Integrated packages are obviously ones that supply integrated systems. See the section on p. 274 for a discussion of integrated systems and examples of integrated packages.

Turnkey (Single- or Limited-Function) Packages. Turnkey systems have been designed, programmed, and tested by a vendor who sells or leases the product ready to be installed and operated. Sometimes the vendor continues to operate the system for the library. In other cases, the library assumes operational control.

Widely known turnkey products have come from CL Systems Inc., Dataphase Systems, Gaylord, Geac, and Universal Library Systems (ULISYS). All began with circulation systems and now have or are developing acquisitions or cataloging functions. One turnkey system, Dobis/Libis SSX, of the integrated function type was recently announced by IBM in Europe and may be available shortly in the United States.

Advantages of turnkey systems include that (1) the initial purchase price may be lower than the costs of developing custom software, because all vendor costs (development, testing, marketing) are being distributed among many users, or the price may be lower than the price of an integrated system package because it has fewer functional capabilities available; (2) no maintenance costs may be incurred (depending on the vendor and type of purchase); (3) the system can be installed and operational quickly; (4) systems staff are not required; and (5) it is likely that changes to the software will be made in a coherent and documented manner.

Disadvantages include that (1) the needs of an individual library may not be exactly met; (2) operational procedures that do not fit the system may need to change since the software cannot be modified; (3) the library is dependent on the vendor, which can present problems if the vendor goes out of business, changes methods of operation, or does not fulfill obligations; and (4) the library may be paying for capabilities not needed or used.

Another Library's Software. Software developed and sold by another library may be adapted for the purchasing library's needs. Any adaptations require the software programs, so acquiring the program source code must be made a mandatory requirement.

Examples of software written for specific libraries which are being marketed are NOTIS from Northwestern University Library, Maggie's Place from Pikes Peak Regional Library, Virginia Tech Automated System, and NLM Integrated Library System from NTIS.

Costs for changing the software can vary greatly, depending on whether the required adaptations are minimal or extensive. Advantages of adapting another library's software are (1) the initial purchase price might be lower than that for other types of packages, since the seller frequently is not seeking large profits and (2) the software may meet many of the buyer's requirements, especially if the libraries are similar.

Disadvantages are (1) acquisition or access to hardware and systems software is necessary; (2) maintenance and training may or may not be provided; (3) the software may not be easily adaptable to the purchasing library's needs; and (4) adaptions require systems staffing.

Non-Library-Specific Software. Database management software (i.e., BASIS from Battelle) has been successfully used for library applications. Common database management features, such as audit trails, journal backup, and recovery and performance monitoring, can be a bonus. Adapting this type of software has the same advantages/disadvantages as adapting another library's software.

Service Bureaus. Service bureaus offer data-processing services for a fee. The library purchases the use of the software (and possibly some hardware such as terminals), not the software itself. Examples are individual library files stored on DIALOG, or the common database and cataloging services available from OCLC, Inc.

The advantages are that (1) the only costs incurred are those actually used (such as the computer time and storage); (2) automation benefits (i.e., online searching) can be obtained while avoiding the responsibility of owning and operating a system; (3) this can be a very cost-effective way of automating infrequently updated files; (4) interim automation results can be realized while the library is waiting for its own system; and (5) the library can have access to a larger set of data than the library owns.

The disadvantages are that (1) the bureau retains complete control and maintenance of the software, so no software modifications are possible; (2) the library is dependent on the bureau; (3) integrated capabilities are not prevalent; and (4) the library may be paying a premium price for services which might cost less on its own system.

Write Custom Software

The alternative to purchasing software is writing custom software. The library has complete control over the design, modification, maintenance, and upgrading of the software. A developed system can exactly meet the library's needs, and an integrated system can be achieved. Some disadvantages are that (1) developing software is liable to take a long time (especially if an integrated system is developed), such that the needs may have changed by the time the system is operational; (2) the library must acquire and maintain or have access to hardware and systems software; (3) systems staff expertise is required; (4) costs for development and continued maintenance are likely to equal or be higher than for any other method; (5) customized software may present problems for networking; and (6) the resulting product may be a poorer quality system than what could be purchased.

The library can either write the software in-house or contract with a software house to write the software. An advantage of the second method is that in-house systems staff are not needed (therefore, costs may be less), but the library must clearly communicate its needs. The library must be sure that the software house understands library functions in general, as well as this library's specific requirements.

HARDWARE ALTERNATIVES

Hardware is a general term covering the entire range of computing machinery. The main components of hardware are the CPU, storage and peripherals.

Central processing unit. The central processing unit (CPU) consists of the control unit, arithmetic/logic unit, and main memory. The control unit interprets the instructions for processing data and coordinates the interconnections between the various parts of the computer. The arithmetic/logic unit adds, subtracts, multiplies, and compares data according to program instructions. Main memory, also called core or primary storage, is the high-speed, rather costly, "working" storage. Main memory houses the operating system software, resident data and programs, and other programs or data temporarily moved into and out (swapped) of auxiliary storage.

Storage. Storage is a slower, less costly form of auxiliary or secondary storage than memory. Typically, these are disks or tapes. Storage devices may need additional pieces of hardware called controllers in order to interface with the CPU.

Peripherals. Peripherals are input and output devices that transport data into and out of the computer, such as terminals and printers. Peripherals can account for a surprisingly high portion of the total system cost. A high-quality printer, for example, can cost in the tens of thousands of dollars.

Figure 11.1 shows hardware components. For further information on hardware, consult Shaw (1981).

Information collected during the analysis phase (i.e., transaction volumes and file sizes) is used to determine the amount of memory and storage required. This type of statistical information goes into the bid request document (the RFP, discussed later with acquiring the new system) to give the vendor the data needed to make the best proposal. For example, each automated record requires about 2000 characters or bytes of disk storage. (A full MARC record averages approximately 800 bytes; associated indices and pointers account for the remaining bytes.) If the library has 40,000 books, a minimum amount of proposed disk storage must be 80 million bytes (80 megabytes).

It is practical to purchase the minimum amount of memory required to get the job done but acquire hardware which can be upgraded or increased in size later, if that proves necessary. Backup needs are an important factor in the amount and type of storage purchased. If continuous uptime is very important, the library may wish to have backup disk copies of its data. In that way, if the main disk(s) go down, the system can be switched quickly to the backup disks, rather than restoring from tape backups, which would take longer.

The two main means of acquiring computing resources are to buy your own dedicated computer or to share resources on someone else's computer.

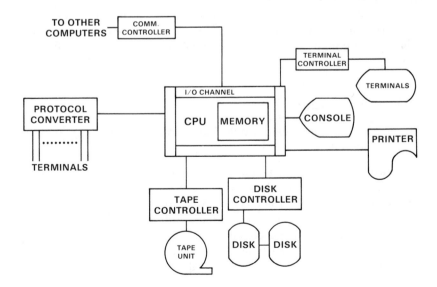

Fig. 11.1. Hardware components in a typical computer installation.

With a *dedicated computer*, the library owns and operates the computer and is the only or principal user. The benefits are that (1) the library is freed from dependency on an external computing center; (2) the library does not compete with other computer users who may have higher use priority; (3) the library relies on its own computing staff, whose first work priority and training is the library, not another user; and (4) excess computer resources (if a library is not using the entire system) may be sold to others and some costs recovered.

The disadvantages are that (1) most dedicated computers represent a substantial investment; (2) computer operations staff, systems, and applications programmers are required; and (3) having dedicated hardware often gives a library an appetite for more sophisticated applications, and it may soon exhaust the resources of the machine, whatever its size.

With a *shared computer,* the library uses a portion of the computing resources owned by another organization. Advantages are that (1) only those computing resources actually used are charged for, so the cost is less than having a dedicated computer; (2) the library may have use of a larger, better system than it could afford on its own; and (3) the assistance of very capable, knowledgeable computing personnel may be available.

The disadvantages are that (1) the library is dependent on the computer owner, and if the computing center decides to change hardware or operating systems, the library must accommodate to those changes; (2) other users will be competing with the library for computing time and resources; (3) the library may have to rely on computing staff with no prior knowledge of libraries and whose sole job responsibility is not the library; and (4) expansion of computing capabilities or the size of the library's files may be more difficult.

Additional Cost Factors

Buying a new system can be quite expensive. The following costs need to be considered:

1. Hardware cost. This can be a substantial cost, dependent on the size of the machine and amount of storage acquired. The library can purchase, lease, or rent hardware. A library may wish to investigate purchasing used equipment, such as government surplus. Substantial dollar savings can be realized by purchasing used equipment, and the items may perform as well as new equipment, particularly if warranted as new. Any hardware is a capital asset and may be depreciated.

2. Software cost. Software will account for approximately 50–80% of the total system cost. The library may have to purchase systems software in order to run the desired applications software, especially if the library is buying a dedicated computer or plans to modify purchased software.

3. Personnel cost. This will be a factor if the library needs to hire or contract for additional staff to install or run the system. The College Placement Council reported that the 1984 average beginning salaries for data processing personnel were $2000 to $2500/month. Personnel costs also include the cost of staff who are diverted from their regular job assignments.

4. Hardware maintenance. This is an ongoing cost, estimated to range from 0.75% to 5% of the initial purchase price for each year the hardware is operational.

5. Software maintenance. This is an ongoing cost, estimated to range from 20 to 35% of the initial purchase price. Maintenance is an especially important factor if the library chooses to modify purchased software or to develop custom software.

6. Site preparation. This is a one-time cost, although it may be a substantial one depending on the size of machine acquired.

7. Conversion of records. Converting records to a new system can be a large expense, especially if existing records are not in machine-readable form. Conversion costs depend on the quantity of data. Lawrence *et al.* (1983) cite a conversion cost of $0.03 per record, if the records are in machine-readable form, while Matthews (1980) estimates the cost as high as $1.65 per entry.

8. Supplies. This will be an ongoing cost for computer paper, printer ribbons, bar code labels, magnetic tapes, etc.

9. User fees. This may be a factor if the library joins any software users organization or attends user conferences.

10. Upgrading the system. This is an expense which the library will have to address if it wishes to obtain additional storage, memory, peripherals, etc., in the future.

11. Training. Library staff may need to take classes and the library may need to purchase user manuals. Some systems include training and documentation in the initial purchase price.

12. Telecommunications. This is a cost which can frequently be postponed or delayed but which must be addressed if the library needs to connect remote sites such as branches in another building or city. Included are considerations such as communication avenues (phone lines), signal processing equipment (modems, communications processors), and sometimes additional software (protocol convertors).

13. Service fees. If a service bureau is employed or computer time is shared, fees will be assessed for computer time, storage used, etc. It may be necessary to purchase or lease terminals.

For more information on costs, consult the article by Drabenstott (1986).

ACQUIRING THE NEW SYSTEM

The RFP

After it has some idea of what software and hardware are desired, the library is ready for the next phase: to go about acquiring the system. The first step is writing a request for proposal (RFP)—a formal document asking vendors to propose a system which meets the library's needs, as specified in the document. If the library is planning to acquire a turnkey system (in which the software and hardware are bundled as one system), a single RFP will suffice. Otherwise, separate RFPs should be issued for the hardware and software. As previously mentioned, ideally, software should be selected first.

The types of RFPs are open/competitive, closed/sole source, or partially competitive. A competitive RFP is open to anyone and may elicit totally unsuitable replies in addition to reasonable bids, and this can mean extra evaluation work for the library. A partially competitive RFP requires more work beforehand to narrow the responding vendors to only those having workable solutions. A sole-source RFP further limits the respondents to only one vendor. Issuance of an RFP or the type of RFP may be mandated by law or the library's governing body. If the RFP is required, the library should find out what rules apply before writing it.

A library may be tempted to bypass the RFP. Even if an RFP is not required, it is a good idea to write one. An RFP provides vendors with a written list of needs and requirements (information they need in order to propose the best system) and facilitates comparing vendor proposals (vendors can be expected to respond in the same manner; that is, they answer the same questions and answer them in the order requested).

The RFP does not always need to be a massive document. Match the solution to the size of the problem. A 40-page RFP is clearly inappropriate for a small word-processing system but may be needed for a $200,000 system. Too detailed an RFP may scare off vendors, while a too simple one may not help the vendors determine what the library needs.

The literature indicates that some libraries have merely copied RFPs from other libraries. Another library's RFP should only be used for reference, to ensure that no major points are neglected. Each library has individual requirements and its own priorities. Also, an RFP should be written around the library's requirements, regardless of what is available on the market.

Several avenues exist for knowing to which vendors to send an RFP. They are the same sources listed in the section on where to find out about software (p. 276). Prior to actually issuing the RFP, the library may narrow its field of prospective vendors by doing some vendor evaluation. Some pertinent questions are: What is the vendor reputation for quality products, reliability and responsiveness? How long has the vendor been in business? Has the vendor previously done business with libraries? What support (installation, training, problem-fixing assistance) does the vendor provide? Does the vendor have a local or nearby support office? How many and who are some other customers for any proposed products?

One further point should be made before discussing the components of an RFP. A request for quotation (RFQ) differs from an RFP in that an RFQ is strictly a price quote. A customer wishes to buy a specific product (i.e., a certain model of disk drive) or a service and wants vendors to say how much this would cost. Strictly speaking, an RFP does not include the cost. However, the terms are frequently used interchangeably, and many organizations issue an RFP and an RFQ as sections of one request document.

RFP Components

The following authors have discussed RFP content and/or presented sample RFPs or samples of RFP tables of contents: Matthews (1980), Corbin (1981), D. Shaw (1981), Merilees (1983), and Epstein (1983b). None of them structured the RFP in exactly the same manner, but all recommend including the following information in the RFP:

1. Overview or introduction: a brief description of the library, background on the existing system, objectives in writing the RFP.
2. Instructions and conditions. These include such points as:

a. Pertinent dates (expected system installation date, contract award date, last date allowed for questions, proposal due date, etc.). Allow the vendor plenty of time to respond: 4–6 weeks is not unreasonable.
b. Instructions on how to prepare and submit the RFP. One such instruction might be that the vendors may cite reference manuals to answer questions, but only those answers written in the response document will be evaluated. (This prevents vendors from citing a huge manual which the library must sift through to find the answer.)
c. Whether a performance or benchmark test will be required.

d. The name and address of the library contact who will answer questions regarding the RFP. No matter how thoroughly the RFP has been prepared, questions will always occur.

e. Evaluation criteria. This tells the vendors how they will be evaluated.

3. Functional requirements. This is the heart of the RFP, for it is where much of the systems analysis effort will be put to use. Desirable, highly desirable, and mandatory requirements are listed and identified as such. Telling a vendor the mandatories lets them know what features they absolutely must propose, since failure to do so will disqualify them or place them at a serious disadvantage during the evaluation. Numbering the requirements is a good idea. Include also performance requirements such as file sizes, transaction rates, processing rules, critical timing considerations, archival record needs, or any other pertinent data which could help inform vendors of the library's needs.

4. Technical requirements. These state whether the proposed software must run on existing hardware or be compatible with certain other software. Request descriptions, attributes, and specifics of proposed hardware, applications and systems software, and vendor support; ask vendors to supply financial references and names and phone numbers of customers using the proposed products.

Evaluating RFP Responses—System Selection

Once the vendor responses have been received, the library must evaluate the responses. The product evaluation is difficult. No easy method or pat guidelines are available to tell the buyer how to evaluate a system. Matthews (1980) and Corbin (1981) introduce evaluation techniques, but the literature is weak in this area. Evaluate each product against how well it meets the requirements before comparing products with one another. Do remember to distinguish between features actually working and those planned to be operational in the future. Here are some of the techniques to consider.

Mandatories Met. If a vendor does not meet all mandatory requirements, it is disqualified immediately. If several vendors meet all the mandatories, use one or more of the other evaluation methods to evaluate desirables. Note that cost can be a mandatory: i.e., the library absolutely will not spend more than $75,000 for the product.

Points Scoring. The library preassigns points to the requirements and selects the system earning the most points. For example, a mandatory requirement is assigned 10 points, a highly desirable requirement has 5 points, and a desirable requirement gets 2 points. If a vendor provides a mandatory, that vendor will get 10 points added to their total score; if not, they get zero points.

Awarding of points need not be all-or-nothing. If the vendor partially meets a requirement, give partial points. If one vendor meets a requirement in a more efficient or preferred manner, give higher points. Do have documented and justifiable reasons for awarding points.

Cost Only. If the system cannot meet the library's requirements, its price is irrelevant, so selecting a system simply on the basis of the lowest bid is a mistake. Selecting the lowest-cost system which meets the most requirements is the goal. Therefore, cost should be examined after another evaluation technique has been employed.

Cost in Combination with Another Method. The library's governing body may require this and insist that cost be no less than a certain weight (i.e., no less than a 20% factor in evaluation). Some cost methods are:

1. Cost-effectiveness ratio. This is used with the points scoring method. The library divides each vendor's cost by the total of the points scored for the desirable requirements.

2. Least total cost. This method considers all costs to the library to acquire and maintain the system projected for a period of 5 to 7 years.

Benchmarking or performance testing can be made part of the evaluation; that is, a selection will not be made until the benchmark results are evaluated. A benchmark is a demonstration simulating the purchaser's actual workload and processing to determine whether the proposed system will perform as claimed. However, it is possible for vendors to skew benchmarking to their favor. For example, if the vendor is allowed to benchmark at its own site or is given complete control over the library system, it can optimize the benchmarking environment in a number of ways (allocate a large amount of memory, set a high run priority, scatter data over several disks to use several I/O paths) to increase the speed and efficiency of performance. The library should maintain control over the environment and use the same site for all of the benchmarking to make comparisons easier. Even so, be careful of using benchmarking as a determinant.

The library should document the evaluation selection. Such documentation can serve to justify the selection if a rejected vendor protests.

If the library finds no adequate responses after evaluating the responses, it has several options. The library could reevaluate its requirements and reissue a new RFP. Maybe the requirements were unrealistic. The library could reevaluate its original options and decide to stay with or enhance the original system. The library could bide its time and reissue the RFP at a later date when, perhaps, more products are available to meet its needs. The library could elect to design its own custom system. If the library does decide to write its own software, the first step is a systems design. A discussion of systems design will not be included in this chapter. Corbin (1981) includes a chapter which is a starting point on the subject.

Contract Negotiation

After the RFP responses have been evaluated and a system has been selected, the library needs to negotiate a contract with the successful bidder. Actually, several contracts may be negotiated: one for the software purchase, one for software maintenance, etc.

Contracts should contain incentives for the vendors to make good their proposal claims and detail penalties and remedies should either party default on their contract obligations. Important points to remember are:

1. Do not automatically sign the standard vendor contract. Vendors negotiate contracts as part of their regular business and could write the contract to favor themselves. The library may change sections as appropriate.

2. Do not believe everything the vendor says. Vendors are in business to sell products.

Auer and Harris (1981) discuss the important area of contract negotiation. Some of the topics are popular vendor ploys, negotiating skills, used-equipment acquisition, contract provisions, and a contract checklist (including a checklist for turnkey systems). Matthews (1980) also offers contract guidelines.

Implementing the New System

After the software and/or hardware have been selected, a whole new set of decisions crop up, such as preparation of the computer site, testing the new system, system conversion, and data conversion. Each area will be briefly covered.

Site Preparation

Depending on the type of system, site preparation can be an extensive job requiring a lot of lead time or a minimal effort. Personal computers (PCs), microcomputers, and terminals require little preparation—a table and some work space, perhaps a phone and a modem for dial-up access to a central computer. Configuring a central computer site is a bigger task for which many vendors will provide assistance. Preparations include:

1. The site. The site must be large enough to accommodate all equipment (which may include storage space for such items as tapes and service manuals) and allow room for opening cabinet doors for servicing equipment. A raised floor may be necessary to cover cables. The location should be convenient for moving equipment into and out of the room (i.e., no stairs, wide enough entrance). A phone in the computer room is handy for the operator.

2. The environment. All computers, except microcomputers, require temperature, humidity, and dust controls and a constant electrical power source.

3. Security and safety. The computer room should be under restricted access. An offsite storage location is recommended for backup copies of the system and the data files. If the system will contain private or other restricted data, extra access precautions must be considered.

Testing

An acceptance test might require the system to operate for a minimum of 30 days with 2% or less downtime, performing all the functions described in the RFP. It differs from the benchmark because it takes place after the system has been installed at the library's site and the library has begun training and/or using it. If the library has machine-readable records, loading some portions of those records would provide a more realistic test environment.

System Conversion

Several methods of moving to a new system include:

Total. As of a designated date, the library entirely switches operations to the system and ceases using the old system.

Phased. The library gradually moves to the new system. For example, circulation is brought up first, followed later by acquisitions. Another method is that books are checked out on the new system while discharging existing charges continues on the old system until the old system gradually fades away.

Parallel. Concurrent operations are performed on the old and the new systems for a period of time. This means duplicate work for the staff.

Pilot. A branch or department is selected to bring up a "mini" version of the system before it is implemented library-wide.

Corbin (1981) talks more about these conversion methods.

Data Conversion

Usually, the library's records must be converted to run on the new system. The conversion method chosen will depend on the purpose of conversion, the available people and time, and the accuracy and condition of the existing records. Conversion techniques include manual keying of records, writing a program to convert records, and capturing records from an automated bibliographic database after the library has matched its records against those in the database.

Epstein (1983a), Lisowski and Sessions (1984), and Butler *et al.* (1978) discuss data conversion. Literature sources suggest that it is important to retain full MARC format even if the immediate application does not require it, because the data elements eliminated today may be desired tomorrow. There is less pressure for abbreviated records today than in the past, since computer storage is cheaper.

SYSTEM EVALUATION

After the system has been operational for several months, it is worthwhile to evaluate how well the system is meeting the library's needs or to assess its productivity. The system evaluation, or postimplementation review, is similar to the systems analysis in several respects and should include the following:

1. Hardware utilization assessment. A number of software products have measurement facilities to analyze information such as the average terminal response and wait times, the number of times programs and data files are accessed, and the average number of concurrent users.

2. Attitudinal assessment. What are the users (staff and patrons) perceptions about the project's success?

3. Needs assessment. Is the system meeting the library's needs and objectives?

The system evaluation should be done annually. The average lifetime of a computer system is 5–10 years. Markuson *et al.* (1972) offer the following guidelines for improving the life of a system: use standard codes and terms (i.e., CODENs), use standard formats (MARC), include hooks into other systems (ISBN, LC card numbers), make accurate projections for file and index growth, and pay attention to human factors.

MANAGEMENT ASPECTS

The management aspects of planning, decision-making, and evaluation have already been discussed as part of the automation process. Staffing and budgeting will be discussed further. All of these aspects vary with the setting and size of the library, but in many cases the difficulty in managing automation projects varies more with the complexity of the task than with the size of the library. For example, periodicals is a more complex automation task than circulation, because periodicals has a large number of variables and exception situations.

Two guidelines for management of automation projects of any size are:

1. Keep the library staff informed during each step, even if only partial information is available or a decision has yet to be made. Information sharing will promote feelings of ownership and participation so that when the new system is implemented, the staff will have less of a tendency to subvert the system or blame it for problems it has not caused.

2. Allow plenty of time for everything. Do not try to rush or push people into acceptance. The principles of change management, discussed by Dowlin (1984) and Malinconico (1983a,b,d,e, 1984b,c,d), are also relevant to automation projects.

Staffing

Most automation projects will be organized around a project team and a project manager. The project manager should be at a high enough level to view the entire system picture, not simply one small aspect, and have authority to make decisions. The library director should not be the project manager, since the project can be a full time assignment.

It is wise to use staff with experience in computing, whether by (1) using librarians with computer training, (2) having in-house MIS staff, or (3) hiring a consultant. (Types of MIS staff are discussed later.)

Librarians with Computer Training

Repeatedly, the point is being made in the literature that microcomputers are making it possible for libraries to develop and maintain systems without relying on MIS staff to program and enter their data. However, many library problems are quite complex, and solving them may require more complex skills than the average librarian wishes to acquire. It can be virtually guaranteed that something will go wrong when using computers, regardless of the size of the machine or the brevity or simplicity of the software, and the primary responsibility of librarians is not data processing. A specialist in any field who masters data-processing skills has the best background to ensure that any automation of that field is implemented correctly.

Libraries using a centralized computing staff should have a library staff person who can serve as a liaison between the library and the computing organization. Communication skills, a knowledge of the library's operations, and familiarity with various aspects of data processing are important. The liaison may interact with the central computing manager, as well as with the data-processing staff. Therefore, the liaison must address managerial issues (such as articulating the library's requirements and constraints, understanding technical and economic factors of the dp project, negotiating the libraries needs and time tables with those of other users), as well as confer with the data-processing staff about more detailed issues such as file structures and keep the library personnel informed about the new system and its changes.

In-House MIS Staff

A number of benefits result from a library having staff with MIS expertise dedicated entirely to the library's computing system. Having in-house staff tends to erode the "us versus them" attitude common when the user and central computing organizations are separate. In-house staff members are more apt to develop detailed knowledge of the library's particular needs, problems, and methods of operations than outside computing staff and are less likely to have work assignments not pertaining to the library.

Ideally, the in-house system staff members should have both library and computing background, and more library schools are addressing this. There is no equivalent to an ALA-accredited degree for data processing, so library managers must use their own judgment in assessing MIS staff credentials. To obtain in-house computing staff, libraries may have to offer salaries competitive with those of business and industry.

Following are some types of MIS staff. In many cases several functions can be performed by the same person. The number and size of staff will depend on the system size and complexity.

The *database administrator or systems manager* insures the continued availability and integrity of the database, and has experience with the logical and physical structure of data files and backup and recovery procedures.

The *systems analyst* has training in one or more systems analysis methodologies; this position is often an auxiliary duty of an applications programmer.

The *applications programmer* has experience and training in one or more programming languages. This position will be required if modifications or custom programming is desired.

The *systems programmer* maintains the operating systems software and has experience in a machine language, such as Assembler.

The *computer operator* performs the day-to-day operational tasks such as turning on the computer, loading paper on the printer, removing and bursting output from the printer, keeping the problem log and reporting problems to the vendor, and storing and retrieving system backup files.

The article by Weber (1971), although published long ago and directed to a systems design project, provides a good reference for the types of staff required, the attributes and organization of the project team, and other elements contributing to a successful automation project team.

Consultants

Areas of automation in which consultants or outside expertise can be effectively used are to define system requirements, to prepare or review the RFP, and to assist in system evaluation, system implementation, data conversion, and hardware sizing and configuration. Usually outside assistance is a wise investment, although consultants are expensive. Merilees (1983) suggests 2.5 to 5% of the total system acquisition cost is an appropriate amount to spend on assistance. If the library finds it is using outside assistance very frequently, it may be more economical to create a staff position to perform those tasks.

Malinconico (1983c,f) lists several professional associations which help in determining available consultation services and supply names of consultants. He also suggests that the most common source of difficulty is the library's failure to clarify its expectations and constraints and to provide sufficient feedback. The library should ascertain the consultant's competency, interview former clients, and draw an explicit contract before engaging the person.

Budgeting

Budgeting is one of the most difficult aspects of automation project management. It is hard to anticipate all expenditures and computer costs and capabilities are constantly changing. The section on additional cost factors gives a list of costs to consider in budgeting for automation.

Very little exists in the literature to assist in planning and identifying budget items. One particularly good article is by Lawrence et al. (1983). The article discusses initial and ongoing hardware, software, personnel, and other costs for several types of systems (i.e., turnkey, service bureaus, custom). Next the article provides a method for estimating project costs from the prices of its components. The authors give actual dollar amounts for various (though not all) system parts, such as the computer, storage, terminals, software, and maintenance. Then, using these component costs (which the authors call "building blocks"), the authors present cost estimates for differently sized libraries, and an "average" academic and an "average" research library. The figures are geared to the online catalog and they must be adjusted to reflect today's prices, especially since some of the costs come from a research study conducted in 1980. However, it is the most useful study available.

Some broader questions that should be addressed include: Does sufficient funding exist for the acquisition and continued maintenance of the needed hardware and/or software? Does the project manager have unambiguous guidelines and authority for budget spending and adequate administrative support for controlling the budget? Are the controls for spending the budget adequate, clearly specified, and consistent with the governing body's policies and procedures? Are there defined procedures for exception situations such as the need to contract for outside support or overspending for specific items? Are auditors and the financial control departments properly informed about the project?

TRENDS

Microcomputers

Microcomputers or micros are a good investment because they cost no more than many "dumb" terminals but can do so much more. They can be used as stand-alones to manipulate data (perhaps downloaded from another computer) with their own database management or word-processing software. They can also be networked (Levert, 1985) to allow sharing of data or use of expensive peripherals (i.e., a high-quality laser printer), and they can be used simply as terminals for larger computers.

Micros and the advent of the new generation of high-level, easy-to-use software are helping to bridge the communication gap between data-processing persons and librarians, as librarians become more computing-knowledgeable. However, micros are not the complete solution to a library's automation problems.

Currently, micros perform best with single and relatively small applications. For example, an IBM PC floppy disk will hold fewer than 500 bibliographic records; very few sci/tech libraries have less than 500 items. Large-volume applications could work on micros by splitting data across multiple hard disks or multiple micros, but this can be very inconvenient and is not a good idea. Complex applications, such as those based on inverted files with many indexes and long bibliographic records, still work better on larger computers. Woods and Pope (1983) state that many mistakes made in the early years of library automation are being repeated because of micros' limited power and storage capabilities and because many micro applications were designed for fixed-length records. He has valid concerns regarding a reversal of the trend toward full bibliographic records and variable-length fields and the potential problems for networking and resource-sharing caused by the development of micro data files with abbreviated records. He also warns that the use of micro systems, by their sheer numbers and variety, may present problems for standardization of records. Matthews (1983a) states that libraries should carefully consider before deviating from standards (i.e., full MARC records), because no system will be truly stand alone in the future.

Interface with Office Automation

Library systems will increasingly be expected to interface with word processing, electronic mail, and other features of office automation. Users will want to electronically send requests for documents, to download data from the library system to their own word-processing systems, and to format and otherwise manipulate output from the library's system as they do with data on their own database management systems. The library will increasingly want the library system to connect with outside vendors for sending purchase orders, to download records from online vendors and networks, and to send search results via electronic mail. Many innovations in the interfaces between systems can be expected in the immediate future.

Continuing Education

Automation is one of the most rapidly changing areas in the library world. Keeping up with changes involves several tactics. Attendance at professional meetings is valuable. Often the exhibits are the first announcements of new systems, services, and products. Also, professional contacts allow "how is it working for you?" inquiries about a new product. The American Society for Information Science (ASIS), Special Libraries Association (SLA), American Library Association (ALA), and ALA's Library and Information Technology Association (LITA) are all useful professional organizations for keeping up with trends in library automation, as are users groups for specific products.

Keeping an eye on professional literature is also helpful. The *Library Journal* features regular columns with brief (2–3 pages) articles on various aspects of automation. Trends in software, equipment, micros, etc. can be pursued through *Computerworld, Infoworld, PC Magazine, PC Tech, PC Week,* and *PC World.* More technical publications are the *Commu-*

nications of the ACM (Association of Computing Machinery), *Proceedings* of the Institution of Electrical Engineers (IEE), and publications of the IEE special-interest groups.

Library Technology Reports and *Library Software Review* provide evaluations of particular hardware and software for libraries. For evaluations of non-library-specific hardware and software, consult *Data Source,* any of the Datapro guides, *PC World Annual Software Review, Computerworld Buyer's Guide to Software, IBM Personal Computer & XT— The Software Guide,* the *Ratings Newsletter,* or the directories listed by Dewey (1984).

Chapter 12

Equipment for Storage, Duplication, and Access

Information is being stored in an increasing number of formats, including paper, microforms (film, fiche, and microprint, in reductions from $8\times$ to $78\times$), video and audio tapes, records, slides, and machine-readable form (including magnetic tape, magnetic disks, floppy disks, and optical and video disks). Figure 12.1 shows that although information will continue to be stored on paper, as time has gone by increasing amounts have been stored on microforms. And although microforms will also survive, an increasing amount of information will be stored in machine-readable form.

Each format requires special equipment to access, store, and duplicate the information, and libraries are spending increasing amounts of time and resources to select, acquire, and maintain the various kinds of equipment, from photocopiers to microform reader/printers to microcomputers.

Although managers of libraries might find this maintenance of equipment a chore, our users are quick to remind us how important it is. Even with paper resources, a working photocopier is a necessity to the use of the collection, and with the newer formats, the equipment is even more necessary.

For any kind of library equipment, certain guidelines for selection and maintenance hold. In selecting equipment (or software), do not rely on the manufacturer's literature. Get demonstrations set up as close to the way you will be using the equipment as possible. In fact, try to get the equipment on loan so you can test it in your own environment. Talk to others who have the same equipment. Consider the technical aspects (e.g., how good are the copies or the display), but also consider the user aspects. Are the directions for use understandable? Are there buttons that should not be pushed? How often does it run out of paper or need toner? What happens if it is not used for long periods of time?

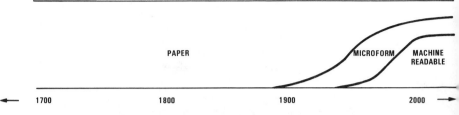

Fig. 12.1. The mix of storage formats as it has changed with time. [From Cruse (1985).]

If equipment is maintained by some central service of the university or corporation, get training for certain library staff so that as many problems as possible can be fixed without a repair call. Although most libraries get service contracts on their equipment, anything that can be fixed in-house will be fixed more quickly and thus distress fewer users.

PAPER

The primary equipment to store paper is shelving, file cabinets, and compact shelving, all of which are included in standard library and office-supply catalogs. The duplication of paper is primarily dependent on photocopiers. Photocopy machines were reviewed in *Library Technology Reports* (Buyers Laboratory, Inc., 1984). Some of the variables include the type of toner, the photoconductor used in the drum, the type of paper, and mechanical factors related to feeding the paper. Some machines make multiple copies quickly but single copies (the more common library use) more slowly. Reliability is perhaps the most important factor (after cost) for a library photocopier. The provision of enough photocopy machines, conveniently located and kept in good working order, is an essential library service. Sometimes, when the provision of photocopiers is controlled by a central organization rather than the sci/tech library itself, this agency must be convinced of the need for apparently large numbers of copiers. Estimates of expected volume can sometimes make the point; other times the cost of the photocopier may need to be compared to the value of the collection (and one of the functions convenient copiers perform is to protect the collection from vandalism).

Telecopiers are useful for transmitting paper over distance, and are covered in Chapter 8. Access to information in paper is direct (i.e., books, reports, journals are browsable) and also via catalogs and indexes.

MICROFORMS

Microforms contain images of textual documents or graphics reduced to the point where they are too small to be read with the unaided eye. Microforms allow almost 90% more information to be stored in the same space; they handle images as well as text; they can be updated and replaced conveniently; and they can be both duplicated and mailed more cheaply than paper. The most common formats are microfiche, 35-mm microfilm, and 16-mm microfilm, which is often available in cartridges. The 35-mm width offers the large image area necessary for the reproduction of newspapers, maps, charts, etc. at low to medium reduction. However, the 16-mm width is more popular for business uses: thus there is more variety and quality of equipment available to use it. The cartridges are more convenient than regular roll microfilm, but they are not interchangeable among machines. Microforms also come in various magnifications. The most common ones in sci/tech libraries are 24× and 48×, and all readers should be able to handle these two magnifications. In addition to interchangeable lenses, microform equipment for patron use should have good image quality, ease of operation, quality of copy, and ease of maintenance. Its use should also not scratch or otherwise harm the microform. Saffady (1985b, pp. 145–161) discusses methods of evaluating microform equipment. The two serials most likely to evaluate microform equipment are *Library Technology Reports* [which has frequently reviewed microform equipment (Howard White, 1982, 1983, 1984)], and *Micrographics Equipment Review,* an annual publication. Also, Spreitzer's guide to the selection of microform readers and reader/printers (1983) is useful, as is the guide prepared by the Texas A&M Library staff based on actual library experience with most of the reading and printing equipment on the market in 1983 (Michaels *et al.,* 1984).

In addition, three sources provide names and addresses of manufacturers of micrographics equipment: the *International File of Micrographics Equipment and Accessories,* 1983–84 (1983), which contains the complete catalogs of over 200 micrographics equipment vendors; the *International Microfilm Source Book* (1972–, annually); and the *AIIM Buying Guide: Registry of Equipment, Supplies, and Services* (annually). Libraries with microform equipment will want service contracts to maintain the equipment, and will also need a copy of *A Microform Reader Maintenance Manual* (Michaels *et al.,* 1984).

Equipment for storing microforms varies from boxes that can be interfiled with books in the stacks, to cabinets for film, film cartridges, or fiche, to lectrievers that can hold hundreds of thousands of microfiche.

CAR (computer aided retrieval) systems are available for large collections; these provide machine filing and access.

Microforms cannot be browsed, so they are dependent on indexing to provide access. And machines have to be used to read them or to reproduce hard copy from them. The machines vary from small lap-held microfiche readers for about $200 to microform reader–printers which may cost $10,000 and up.

More information on microforms is available in Volume 2, Chapter 7.

MACHINE-READABLE FORM

More and more information is being stored in machine-readable form. In performing online literature searches, we are accessing large numbers of databases for bibliographic citations and an increasing number of databases with full text online. Software is increasingly becoming available on floppy disks for microcomputers, and libraries are beginning to collect them (see Volume 2, Chapter 10). We are also beginning to use microcomputers in our own libraries in a variety of ways: as word processors, as aids to compiling budget information, and to help with library processes such as interlibrary loan, cataloging, circulation, acquisitions, etc. They are also appearing as terminals to access a variety of other computers (e.g., for online literature searching and electronic mail). Libraries with online catalogs are providing telecommunications connections so that their users can search the library catalog from their offices.

Although storage on microcomputers is limited in the mid-1980s, many experiments are being conducted with the various formats of optical and video disks for storage of text. Although the specific formats vary in their storage capacity and replication costs, as a general guideline a one-sided 12-inch optical digital disk can store between 10,000 and 20,000 pages of text, depending on the resolution required, and an analog digital disk (also called a videodisc) can store up to 54,000 separate black-and-white or color images. A CD-ROM (compact disc read-only memory) will store about 200,000 pages of text. Optical storage avoids wear, since only a beam of light touches the medium during playback, and information can be transferred to a new disk without any loss. However, none of these formats are yet erasable and reusable, as are magnetic tape and traditional magnetic disk drives. A detailed description of the technology and the various possible formats is available in *Videodisc and Optical Digital Disk Technologies and Their Applications in Libraries* (Information Systems

Consultants, Inc., 1985); Lunin and Paris (1983); Herther (1985); Slonim *et al.* (1986); and Roth (1986).

There are a number of experiments being conducted with the various types of optical storage media (Connolly, 1986a,b). For example, the Library of Congress is running an Optical Disk Pilot Program which has two aspects. Print materials are being stored on digital optical disks. These include high-use current periodicals in science, technology, and business. Nonprint or image-based materials are being stored on analog optical disks (videodisks) (Parker, 1985). The National Library of Canada successfully produced a videodisk in 1982 and has several other projects investigating this technology (Sonneman, 1983). Pergamon's Video PATSEARCH was also an example of the use of videodisc. Two government agencies are in the process of using optical disks to replace their paper files. These are the U.S. Patent and Trademark Office, which anticipates a totally paperless operation by 1990, and NTIS. There are also several experiments with LC MARC records on various optical disks. Carrollton Press's MARVYLS prototype is a digitally encoded videodisc containing 228,000 records from the bibliographic (Z) and education (L) classes and appropriate indexing (Gale, 1984). The Library Corporation is marketing Bibliofile, a cataloging system based on the CD-ROM. The MARC database is stored on less than two compact disks (R. Mason, 1985).

Brodart's LePac (local Public Access Catalog) uses CD-ROM technology to store up to a million MARC-formatted bibliographic records, and FAX-ON has made available a CD-ROM disk loaded with complete records for the entire MARC-S Serials file. Databases appearing in CD-ROM include subsets of COMPENDEX and NTIS marketed by DEC (Tenopir, 1986), a couple of Bowker databases (R. R. Bowker, 1986), and InfoTrac, a laserdisc periodical index from Information Access Company (Stephens, 1986).

Articles are appearing which compare costs for paper, microform, and optical or digital disks for various information systems and commonly find great advantages to the optical digital disks (e.g., Snyder, 1984). CD-ROM is particularly attractive because of its popularity for audio disks and its large storage capacity. Although no one format has yet emerged as a commercial success (Information Systems Consultants, Inc., 1985), this medium's potential for making the content of a complete sci/tech library available to an individual scientist or engineer on one shelf in his or her office is constantly before us. This technology is changing very quickly, and specific equipment needs will not be discussed here, since any discussion will be out of date immediately. For keeping up-to-date, *Videodisc and Optical Disk,* published by Meckler, and the journals of the AIIM are the primary sources. Meckler has also sponsored workshops on laser technology and optical publishing at the American Library Association conferences, and ALA's Library Resources and Technical Services Division (LRTS) has included video technology in its reviews (e.g., see Nadeski and Pontius, 1984).

It does seem reasonable to discuss the use of microcomputers in libraries and their equipment support needs, however, since Mitchem (1984) estimated that 34.5% of libraries of all types had one or more microcomputers in 1984 and the number is clearly growing. Pratt's review of microcomputers in ARIST shows that the increase in interest in microcomputers in libraries has paralleled the interest of the general public, and he predicts that before long microcomputers will become just another everyday tool like the typewriter or telephone. They will increasingly be available in technical processing departments, since three of the major bibliographic utilities have selected the IBM PC as their next terminal (Pratt, 1984).

Selection of the microcomputer, the software for the various applications, and the furniture to support it is a new challenge. Libraries which are beginning office automation have to plan not only their own automation, but effective interfaces with their users, who may be handling information in totally new ways (Spinrad, 1982).

The four common applications of microcomputers in the administrative environment are word processing, electronic spreadsheets, database management, and telecommunications. In addition, microcomputers have been applied to some library functions (including circulation, interlibrary loan, and online catalogs).

For the first two applications (word processing and spreadsheets), the general-purpose programs which are marketed to businesses and the general public are satisfactory for library use. R. Mason reviewed word-processing software (1983), and McIntyre (1983) reviewed Lotus 1-2-3, the most popular of the spreadsheet programs. Schuyler's review of Visicalc, the original spreadsheet program, is a good explanation for the layman of how spreadsheets work (Schuyler, 1982). Spreadsheets have immediate applications within libraries for graphing statistics and for budgeting, and Clark (1985) shows many of these applications.

In contrast, the popular database management systems (for example dBaseII) are not completely satisfactory for bibliographic data because they require fixed-length fields. Software specifically for personal libraries is being developed, however (see Chapter 7). Although none of these packages is completely satisfactory, improvements can be expected. Applications of microcomputers to library functions such as catalogs and circulation have been limited to small applications because of the limited storage capacity (a floppy disk will only hold about 500 records). These software packages have been reviewed by Nolan (1984). There are also a few unusual library applications of micros, such as that by Schoenly (1982) in which a library-orientation program was patterned on the popular adventure games. As storage capacity increases, use of microcomputers for library applications will also increase.

In the meantime, the development of local area networks (LANs) may provide some of the solution to limited storage.

Libraries need to be concerned with interfacing automated systems, since they will want their automated online catalogs to be able to be searched from their users' microcomputers. Boss (1984) gives a good overview of the problems of interfacing automated library systems, including local area networks (LANs). Matthews and Williams (1983) and Boss (1985) wrote basic guides to telecommunication for libraries, and Derfler and Stallings (1983) have provided a good introduction to the technicalities of local networks. Further information has been included in Chapter 11.

Two journals which are useful for keeping up-to-date with new microcomputer products for libraries are *Small Computers in Libraries* and *Access: Microcomputers in Libraries*. A list of microcomputing periodicals was compiled by Shirinian (1985), but of course the number is growing. Also, useful summaries are included in the Woods and Pope (1983) *Librarian's Guide to Microcomputer Technology and Applications,* in the *Directory of Microcomputer Applications in Libraries* (1984), and in *A Microcomputer Handbook for Small Libraries and Media Centers* (Costa and Costa, 1983). Since so much of library use is no different from other administrative uses, the reviews in *Byte, PC World,* etc. are also useful. Further information on automation of library functions can be found in the chapter on automation, and in the chapters on the various functions.

MICROCOMPUTER FURNITURE

There is evidence that the use of video display terminals (VDTs) causes health problems, particularly musculoskeletal problems and eyestrain, although the evidence is still inconclusive. Radiation does not appear to be a hazard (Murray, 1986). The choice of proper furniture and allowing for proper lighting can minimize these problems.

The design of the workstation can also affect productivity. Mason reported that furniture manufacturers have found that

> . . . ergonomically designed furniture can increase productivity by ten percent over conventional furniture, and studies by the U.S. government have demonstrated a difference in productivity of almost 25 percent between operators at "best designed" and operators at "worst designed" video display terminal (VDT) workstations. (Mason, 1984, p. 331)

Tables should be about 26 inches from the floor, about 4 inches shorter than a normal desk or table. The operator's elbows should be at 90 degrees and the wrists should not be bent (see Figs. 12.2 and 12.3). The ideal angle for viewing the terminal is 10–20 degrees below eye level (although people wearing bifocals may want it even lower). Also, a copy stand which gets the material the operator is looking at into the same plane as the screen eases back problems and eyestrain. Adjustable furniture is essential if more than one person is going to use the workstation.

Seating is also important. The chair should enable the operator to sit comfortably with feet flat on the floor. The chair should provide good support for the lower back, and the seat height, back height, and back-tilt tension each should be adjustable. Many operators find the new chairs which have knee rests and no back support comfortable for terminal work. However, chairs should allow an operator to work in a variety of positions, since he or she should change position at least every 30 minutes.

Improper lighting also contributes to eyestrain. Terminals should be positioned to reduce glare, and antiglare screens should be installed if necessary (Schliefer and Santer, 1986).

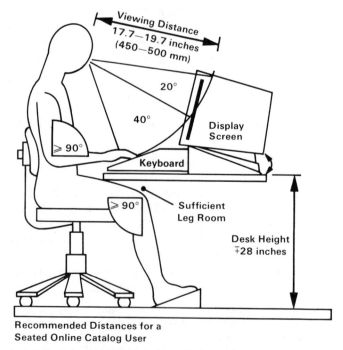

Recommended Distances for a
Seated Online Catalog User

Fig. 12.2. Proper arrangement for a seated terminal user.

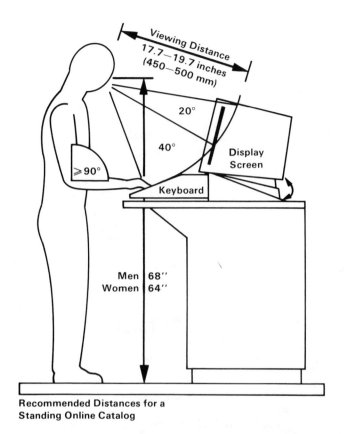

Recommended Distances for a
Standing Online Catalog

Fig. 12.3. Proper arrangement for a standing terminal user.

A good summary of workstation ergonomics is provided by Tijerina (1984). Further information about furniture can be found in the article by Farkas (1984) and about health problems in those by R. Miller (1983) and Arndt (1983).

Bibliography

Abernathy, William, and Hayes, Robert. (1980). Managing our way to decline. *Harv. Business Rev.* July/August, 67–77.

Adler, Anne G., ed. (1983). Automation in libraries: A LITA bibliography 1978–1982. Pierian, Ann Arbor, Mich.

Adler, Anne G., and Baber, Elizabeth A., eds. (1984). Retrospective conversion. Pierian, New York.

Ahrensfeld, Janet L., Christianson, E. B., and King, D. E. (1981). Special libraries: A guide for management. Special Libraries Association, New York.

"AIIM Buying Guide: Registry of Equipment, Supplies, and Services" (Annual). Association for Information and Image Management, Silver Spring, MD.

Akers, S. G. (1981). To what extent do the students of the liberal arts colleges use bibliographic items given on the catalogue card? *Library Q.* **1**, 394–408.

Allen, T. J. (1966). Performance of information channels in the transfer of technology. *Ind. Manage. Rev.* **8**(1), 87–98.

Allen, T. J. (1968). Organizational aspects of information flow in technology. *Aslib Proc.* **20**(11), 433–453.

Allen, T. J. (1977). "Managing the Flow of Technology: Technology Transfer and the Dissemination of Technological Information within the R&D Organization." MIT Press, Cambridge, Mass.

Allen, T. J., and Gertsberger, P. G. (1968). Criteria for selection of an information source. *J. Appl. Psych.* **52**(4), 272–279.

Alley, Brian, and Cargill, Jennifer. (1982). "Keeping Track of What You Spend: The Librarian's Guide to Simple Bookkeeping. Oryx, Phoenix, Ariz.

Almond, J. R., and Nelson, C. H. (1978). Improvements in cost effectiveness in on-line searching. I. Predictive model based on search cost analysis. *J. Chem. Inf. Comput. Sci.* **18**, 13–15.

Almond, J. R., and Nelson, C. H., (1979). Improvements in cost-effectiveness in on-line searching. II. File structure, searchable fields, and software contributions to cost-effectiveness in searching commercial data bases for U.S. patents. *J. Chem. Inf. Comput. Sci.* **19**, 222–227.

Aluminum Association. Committee on Technical Information. (1980). "Thesaurus of Aluminum Technology," 3d ed. The Aluminum Association, Washington, D.C.

Alvord, Katharine T., ed. (1985). "Document Retrieval: Sources and Services," 3d ed. Information Store. San Francisco, California.

American Library Association. (1985). Library education and personnel utilization, a statement of policy adopted by the Council of the American Library Association, June 30,

1970. American Library Association Office for Library Personnel Resources, Chicago, March.

American Library Association, ALA Editorial Committee, Subcommittee on the ALA Rules for Filing Catalog Cards. (1968). "ALA Rules for Filing Catalog Cards," 2d. ed. American Library Association, Chicago.

American Library Association. Bookdealer–Library Relations Committee. (1984). "Guidelines for Handling Library Orders for In-print Monographic Publications," 4th ed. American Library Association, Chicago, Ill.

American Library Association, Resources and Technical Services Division, Collection Development Committee. (1979). "Guidelines for Collection Development." American Library Association, Chicago.

American Library Association. (In preparation). "Selection of Library Materials," 3 volumes. Note: There will be chapters for each discipline, including annotated lists of major review journals in each field.

"American Library Directory." (Annual). Bowker, New York.

American National Standards Institute. (1968). American National Standard abbreviation of titles of periodicals. American National Standards Institute. ANSI Z39.4-1968(R1974).

American National Standards Institute. (1977). American National Standard for bibliographic references. American National Standards Institute, Inc., New York, ANSI Z39.29-1977.

American National Standards Institute (1980). American National Standard guidelines for thesaurus structure, construction, and use. American National Standards Institute, Inc., ANSI Z39.19-1980.

American National Standards Institute. (1982). American National Standard for order form for single titles of library materials in 3-inch by 5-inch format. American National Standards Institute. ANSI Z39.30-1982.

American National Standards Institute. (1983). American National Standard for library and information sciences and related publishing practices—library statistics. American National Standards Institute, ANSI Z39.7-1983.

Amsden, D. (1968). Information problems of anthropologists. *College & Research Libraries.* **29**(2), 117–131.

Andreasen, Alan R. (1980). Advanced library marketing. *J. Library Admin.* **1**(3), 17–32.

Andrews, Theodora. (1968). The role of departmental libraries in operations research studies in a university library. Part 1. Selection for storage problems. *Special Libraries* **59**, 519–524.

"Anglo-American Cataloguing Rules." (1978). 2d. ed. American Library Association, Chicago.

Archer, Mary Ann E. (1978). The "make or buy" decision: Five main points to consider. *ONLINE* **2**(3), 24–25.

ARL at Colorado Springs. (1985). *Library J.* April 1, 16.

Arndt, Robert. (1983). Working posture and musculoskeletal problems of video display terminal operators—Review and reappraisal. *Am. Ind. Hygiene Assoc. J.* **44**, 437–446.

Arnett, Thomas. (1981). Intelligent copiers—Printers can save you time and labor. *The Office,* November, 155–158.

Arnovick, George N., and Gee, Larry G. (1978). Design and evaluation of information systems. *Inf. Processing Manage.* **14**, 369–80.

Ash, Lee. (1963). "Yale's Selective Book Retirement Program," Linett Books, Hamden, Conn.

Association of Research Libraries. (1983). Office of Management Studies. System and Procedures Exchange Center. Fund raising in ARL Libraries. Association of Research Libraries, Washington, D.C. (SPEC Kit 94).

Association of Research Libraries. (1984). ARL Spec Kit Flyer. Electronic mail in ARL Libraries, July/August. (Spec Kit 106).

Atherton, Pauline. (1978). BOOKS are for use. Final Report to the Council on Library Resources. Syracuse University, School of Information Studies, Subject Access Project, Syracuse, N.Y., ED156131.

Atherton, Pauline, and Christian, Roger. (1977). "Librarians and Online Services." Knowledge Industry Publications, Inc., White Plains, N.Y.

Atkinson, Hugh C. (1984a). Strategies for change: Part II. Library J. March 15, 556–557.

Atkinson, Hugh C. (1984b). Two reactions to change. Library J. August, 1426–1427.

Attig, John. (1982). The US/MARC formats—Underlying principles. Inf. Technol. Libraries 1(2), 169–170.

Auer, Joseph, and Harris, Charles Edison. (1981). "Computer Contract Negotiations." Van Nostrand Reinhold, New York.

Auster, Ethel, and Lawton, Stephen B. (1984). Search interview techniques and information gain as antecedents of user satisfaction with online bibliographic retrieval. J. Am. Soc. Info. Sci. 35(2), 90–103.

Austin, Derek, and Digger, Jeremy A. (1977). PRECIS: The preserved context index system. Library Res. Tech. Services (LRTS) 21, 13–30.

Aveney, Brian H., and Butler, Brett, eds. (1984). Online catalogs, online reference: Converging trends: Proceedings of a Library and Information Technology Preconference Institute 1983, June 23–24. American Library Association, Chicago.

Aveney, Brian, and Ghikas, Mary F. (1979). Reactions measured: 600 Users meet the COM catalog. Am. Libraries 10(2), 82–83.

Avram, Henriette D. (1984). Authority control and its place. J. Academic Librarianship 9(6), 331–335.

Ayres, F. H., et al. (1968). Author versus title: A comparative survey of the information which the user brings to the library catalogue. J. Documentation. 24, 266–272.

Baatz, Wilmer H. (1978). Collection development in nineteen libraries of the Association of Research Libraries. Library Acquisitions: Practice and Theory 11(2), 85–121.

Backyard satellite dishes as checkout counters. (1985). LJ Hotline XIV-5, 4.

Bailey, Martha J. (1981). Supervisory and middle managers in libraries: Scarecrow, Metuchen, N.J.

Baker, J. S. K. (1981). The efficiency of interlibrary loan, a study of response time for ILL requests submitted by mail, TWX, and an automated system. Presented at the ACRL Second National Conference, Minneapolis, Minn.

Bambach, R. K. (1981). Keeping current in an interdisciplinary field: Paleontology. Geosci. Inf. Soc. Proc. 11, 11–22.

Banks, Paul N. (1972). Preservation of library materials. In "Encyclopedia of Library and Information Science." Vol. 7, pp. 69–120. Marcel Dekker, New York.

Banks, Paul N. (1974). Environmental standards for storage of books and manuscripts. Library J. 99, 339–343.

Bates, M. (1979a). Information search tactics. J. Am. Soc. Inf. Sci. 30, 205–214.

Bates, M. J. (1979b). Idea tactics. J. Am. Soc. Inf. Sci. 30, 280–289.

Bates, Marcia. (1977). Factors affecting subject cataloging search success. J. Am. Soc. Inf. Sci., 28, 161–169.

Baumol, William J. (1983). Electronics, the cost disease, and the operation of libraries. J. Am. Soc. Inf. Sci. 34, 181–191.

Beaumont, Jane, and Krueger, Donald. (1983). Microcomputers for libraries: How useful are they? Canadian Library Association, Ottawa.

Becker, Joseph, and Katzenstein, Gary. (1983). Word processors for libraries. *Library Technol. Rep.* March/April, 121–190.
Becket, Margaret, and Smith, Henry Bradford. (1986). Designing a reference station for the information age. *Library J.* **15**, 42–46.
Beckman, Margaret. (1982a). Library buildings and the network context. Presented at 48th International Federation of Library Associations (IFLA) General Conf., Montreal.
Beckman, Margaret M. (1982b). Online catalogs and library users. *Library J.* November 1, 2043–2047.
Bedsole, Dan T. (1963). A library survey of 117 corporations. *Special Libraries* **54**(10), 615–622.
Beiser, Karl. (1983). Microcomputer periodicals for libraries. *Am. Libraries* January, 43–48.
Bellardo, Trudi. (1985a). An investigation of online searcher traits and their relationship to search outcome. *J. Am. Soc. Info. Sci.* **36**(4), 241–250.
Bellardo, Trudi. (1985b). Telecommunications and networking. *Library J.* **110**(13), 52.
Belli, Frank G., Bement, J. H., Majcher, M. D., and Neal, A. A. (1985). Total library integration at Xerox Corporation Technical Information Center—Part I: system selection and specifications. *ONLINE* March, 23–28.
Berger, Mary C. (1977) Starting up an online user group—A case history. *ONLINE* **1**, 32–35.
Berger, Mary C. (1980). Online user groups. *In* "The Library and Information Manager's Guide to Online Services" (Ed. Ryan E. Hoover), pp. 233–249. Knowledge Industry Publications, Inc., White Plains, N.Y.
Bernier, C. L. (1976). Microthesauri. *In* "Encyclopedia of Library and Information Science," Vol. 18, pp. 114–119. Marcel Dekker, New York.
Berul, L. H., *et al.* (1965). DoD user needs study—Phase I. The Auerbach Corporation, Philadelphia, Pa., Final Technical Report 1151-TR-3; AD 615501.
Besant, Larry. (1982). Online catalog users want sophisticated subject access. *Am. Libraries* **14**(3), 160.
BISAC and SISAC: Computer-to-computer ordering. (1984). *Library Hi Tech* **1**(2), 9.
Bishop, David F. (1983). The CLR OPAC study: analysis of ARL user responses. *Inf. Technol. Libraries* **2**(3), 315–321.
Bivans, Margaret M. (1974). A comparison of manual and machine literature searches. *Special Libraries* **65**(5), 216–222.
Black, Don. (1985). Don't write it, say it. *Inf. Today* February, 21.
Blackburn, Robert. (1979). Two years with a closed catalog. *J. Academic Librarianship* **4**(6), 424–429.
Blackwell, Gerry (1985). Cracking down on telephone tag. *InfoAge* **4**(1), 27–30.
Blagden, J. F. (1968). Thesaurus compilation methods: A literature review. *Aslib proc.* **20**(8), 345–359.
Blick, A. R., Gaworska, S. J., and Magrill, D. S. (1982). A comparison of on-line databases with a large in-house information bulletin in the provision of current awareness. *J. Inf. Sci.* **4**(2/3), 79–82.
Blood, Richard W. (1983). Evaluation of online searches. *RQ* Spring, 266–277.
Bogardus, Janet. (1960). Basic aids in business libraries. Delivered at Annual Conf., Special Libraries Association, June 8.
Bond, Marvin A. (1984). Planning, budgeting and personnel management in a scientific library of the federal government: National Bureau of Standards. *Sci. Technol. Libraries* **4**(3/4), 45–60.

Bonk, Wallace John, and Magrill, Rose Mary. (1979). "Building Library Collections," 5th ed. Scarecrow, Metuchen, N.J.

Bonn, George S. (1974). Evaluation of the collection. *Library Trends* January, 265–304.

Borgman, Christine L. (1983). End user behavior on the Ohio State University Libraries' online catalog: A computer monitoring study. OCLC, Dublin, Ohio, OCLC/OPR/RR-83/7.

Borgman, Christine L. (1984). Psychological research in human–computer interaction. *Annu. Rev. Inf. Sci. Technol. (ARIST)* **19**, 33–64.

Borgman, Christine L., and Case, Donald. (1984). Database guides: Are they doing a good job? *Proc. 47th ASIS Annu. Meeting* **21**, 158–162.

Borko, Harold, and Bernier, Charles L. (1975). "Abstracting Concepts and Methods." Academic Press, New York.

Borko, Harold, and Bernier, Charles L. (1978). "Indexing concepts and methods." Academic Press, New York.

Borovansky, Vladimir T., and Machovec, George S. (1985). Microcomputer-based faculty profile. *Inf. Technol. Libraries* December, 300–305.

Boss, Richard. (1979). "The Library Manager's Guide to Automation." Knowledge Industry Publications, Inc., White Plains, N.Y.

Boss, Richard W. (1982). "Automating Library Acquisitions: Issues and Outlook." Knowledge Industry Publications, Inc., White Plains, N.Y.

Boss, Richard W. (1984). Interfacing automated library systems. *Library Technol. Rep.* September/October, 615–703.

Boss, Richard W. (1985). "Telecommunications for Library Management." Knowledge Industry Publications, Inc., White Plains, New York.

Boss, Richard, and Marcum, Deanna B. (1980). The library catalog: COM and on-line options. *Library Technol. Rep.* **16**(5), 443–556.

Boss, Richard W., and Marcum, Deanna B. (1981). On-line acquisitions systems for libraries. *Library Technol. Rep.* **17**(2), 115–202.

Boss, Richard W., and McQueen, Judy. (1982). Automated circulation control systems. *Library Technol. Rep.* **18**(2), 125–266.

Boss, Richard, and McQueen, Judy. (1983). High-speed telfacsimile in libraries. *Library Technol. Rep.* **19**(1), 7–111.

Boucher, V. (1976). Nonverbal communication and the library reference interview. *RQ* **16**, 27–32.

Bourne, C. P., and Robinson, J. (1980). Education and training for computer-based reference services: review of training efforts to date. *J. Am. Soc. Inf. Sci.* **31**(1), 25–35.

Bourne, Charles. (1965). Some user requirements stated quantitatively in terms of the 90 percent library. *In* "Electronic Information Handling" (Eds. Allen Kent and Orrin E. Taulbee), pp. 93–110. Spartan Books, Washington, D.C.

Bourne, Charles. (1980). Online systems: History, technology and economics. *J. Am. Soc. Inf. Sci.* May, 155–160.

Boyce, B. R., and Gillen, E. J. (1981). Is it more cost-effective to print on- or offline? *Ref. Q.* **21**, 117–120.

Brandchoff, Susan. (1985). A new director tests her management mettle. *Am. Libraries* January, 47–48.

Branin, Joseph J., *et al.* (1985). The national shelflist count project: its history, limitations and usefulness. *Library Res. Tech. Services* October/December, 333–342.

Breton, Ernest J. (1981). Why engineers don't use databases. *Bull. Am. Soc. Inf. Sci.* **7**(6), 20–23.

Brogan, Linda L., and Lipscomb, Carolyn E. (1982). Moving the collections of an academic health sciences library. *Bull. Med. Library Assoc.* **70**, 377.

Brookes, B. C. (1970). The growth, utility, and obsolescence of scientific periodical literature. *J. Documentation* **26**, 283–294.

Brough, Kenneth J. (1953). "Scholars workshop: Evolving conceptions of library service." University of Illinois, Urbana (Illinois Contributions to Librarianship, No. 5).

Bruman, Janet L. (1983). Communications software for microcomputers. CLASS, San Jose, Calif.

Bruno, Michael J. (1971). Decentralization in academic libraries. *Library Trends* **19**, 311–317.

Buckland, M. K., *et al.* (1970). "Systems analysis of a university library." University of Lancaster Library, Lancaster, England.

Buckland, Michael K. (1972). Are obsolescence and scattering related? *J. Documentation* **28**, 242–246.

Buckland, Michael K. (1975). "Book availability and the library user." Pergamon, New York.

Bunge, Charles A. (1982). Strategies for updating knowledge of reference resources and techniques. *RQ* Spring, 228–232.

Bunge, Charles A. (1984). Planning, goals and objectives for the reference department. *RQ* Spring, 306–315.

Bunge, Charles A. (1985). Factors related to reference question answering success: The development of a data-gathering form. *RQ* Summer, 482–486.

Buntrock, R. E. (1979). The effect of the searching environment on search performance. *Online* **3**(4), 10–13.

Buntrock, Robert E. (1984). Cost effectiveness of on-line searching of chemical information: An industrial viewpoint. *J. Chem. Inf. Comput. Sci.* **24**, 45–57.

Burns, Rower W., Jr. (1968). Evaluation of the holdings in science–technology in the University of Idaho Library. Idaho University, Moscow Library, Moscow, Idaho, June, ED 021-579 (publication no. 2).

Burton, Hilary D. (1985). The changing environment of personal information systems. *J. Am. Soc. Inf. Sci.* **36**(1), 48–52.

Burton, R. E., and Kebler, R. S. (1960). The half-life of some scientific and technical literature. *Am. Documentation* **11**, 18–22.

Butler, Brett. (1983). Online public access: The sleeping beast awakes. *ASIS Bull.* **14**(2), 6–10.

Butler, Brett, Aveney, Brian, and Scholz, William. (1978). The conversion of manual catalogs to collection data bases. *Library Technol. Rep.* March/April 109–206.

Buyers Laboratory, Inc. (1984). Photocopiers. *Library Technol. Rep.* **20**(3), 289–422.

Byrd, Gary D., Thomas, D. A., and Hughes, Katherine E. (1982). Development using interlibrary loan borrowing and acquisitions statistics. *Bull. Med. Library Assoc.* **70**(1), 1–9.

Campbell, M. B. M. (1975). A survey of the use of science periodicals in Wolverhampton Polytechnical Library. *Res. Librarianship* **5**(26), 39–71.

Carter, Constance comp. (1984). Scientific and technical libraries: Administration and management. Science Reference Section, Science and Technology Division, Library of Congress, Washington, D.C. October (LC Science Tracer Bull. TB 84–5).

Casbon, Susan. (1983). Online searching with a microcomputer—Getting started. *ONLINE* November, 42–46.

Case, Donald, Borgman, Christine L., and Meadow, Charles T. (1986). Information seeking by energy researchers. *ASIS Bull.* December/January, 12–13.

"Catalog of the United States Geological Survey Library." (1964–1976). G. K. Hall & Co., Boston. Note: Vols. 1–25, 1964; vols. 1–11, 1973 (1st suppl.); vols. 1–4, 1974 (2d suppl.); vols. 1–6, 1976 (3d suppl.).

Cates, Jo. (1985). Sexual harassment: What every woman and man should know. *Library J.* July, 23–29.

Chaloner, Kathryn, and de Klerk, Ann. (1980). A comparison of two current awareness methods. In "Communicating Information: Proceedings of the 43rd ASIS Annual Meeting; October 5–10, 1980" (Eds. A. R. Benenfeld and E. J. Kazlauskas). Knowledge Industry Publications, Inc., for ASIS, White Plains, N.Y.

Chan, Lois Mai. (1981). "Cataloging and Classification: An Introduction." McGraw-Hill, New York.

Chapman, Edward A., St. Pierrè, Paul L., and Lubans, John, Jr. (1970). "Library Systems Analysis Guidelines." Wiley-Interscience, New York.

Chappell, Barbara, and Goodwin, George. (1984). The U.S. Geological Survey Library and the user community: A past and future view. *Geosci. Inf. Soc. Proc.* **14,** 17–36.

Chemical Abstracts Service Source Index (CASSI). (Quarterly). Chemical Abstracts Service, Columbus, Ohio.

Chen, Ching-Chih. (1974). How do scientists meet their information needs? *Special Libraries* **16,** 272–280. (Note: This is a study of physicists.)

Chen, Ching-Chih. (1976). "Biomedical Scientific and Technical Book Reviewing." Scarecrow, Metuchen, N.J.

Chen, Ching-Chih. (1978). Reviews and reviewing of scientific and technical materials. In "Encyclopedia of Library and Information Science," vol. 25, pp. 350–372. Marcel Dekker, New York.

Chen, Ching-Chih. (1979). Scientific and technical libraries. In "Encyclopedia of Library and Information Science," vol. 27, pp. 1–86. Dekker, New York.

Chen, Ching-Chih. (1980). "Zero-Base Budgeting in Library Management: A Manual for Librarians." Oryx, Phoenix, Ariz.

Chen, Ching-Chih, and Bressler, Stacey E. (1982). "Microcomputers in Libraries." Neal-Schuman, New York.

Cheney, Frances Neel, and Williams, Wiley J. (1980). "Fundamental reference sources," 2d ed. American Library Association, Chicago.

Chiang, Katherine S., Curtis, H., and Stewart, L. G. (1985). Creating bibliographies for business use. *PC Magazine* **12,** 249–260.

Citron, Helen R., and Dodd, James B. (1984). Cost allocation and cost recovery considerations in a special academic library: Georgia Institute of Technology. *Sci. Tech. Libraries* **5**(2), 1–14.

Ciucki, Marcella. (1977). Recording for reference/information service activities: A study of forms currently used. *RQ* **16,** 273–283.

Clack, Mary E., and Williams, Sally F. (1983). Using locally and nationally produced periodical price indexes in budget preparation. *Library Res. Tech. Services* **27,** 345–356.

Clapp, Verner, and Jordan, R. T. (1965). Quantitative criteria for adequacy of academic library collections. *College Res. Libraries* **26,** 371–380. [Note: Also (1966) **27,** 72.]

Clark, Philip M. (1985). Microcomputer spreadsheet models for libraries: Preparing documents, budgets and statistical reports. American Library Association, Chicago.

Cleveland, Donald B., and Cleveland, Ana D. (1983). "Introduction to Indexing and Abstracting." Libraries Unlimited; Littleton, Colo.

Coch, L., and French, J. (1968). Overcoming resistance to change. In "Group dynamics," (Eds. D. Cartwright and A. Zander) Harper and Row, New York.

Cochrane, Pauline Atherton. (1981). Improving the quality of information retrieval—Online to a library catalog or other access service . . . or . . . Where do we go from here? *ONLINE* **5**(3), 30–42.

Cochrane, Pauline A. (1982). "Friendly" catalog forgives user errors. *Am. Libraries* **13**(5), 303–306.
Cochrane, Pauline Atherton. (1984a). Modern subject access in the online age: Part 3. *Am. Libraries* April, 250–255.
Cochrane, Pauline Atherton. (1984b). Modern subject access in the online age: Part 1. *Am. Libraries* February, 80–83.
Cochrane, Pauline Atherton. (1984c). Modern subject access in the online age: Part 5. *Am. Libraries* June, 438–443.
Cochrane, Pauline Atherton. (1984d). Modern subject access in the online age: Part 4. *Am. Libraries* May, 336–339.
Cochrane, Pauline Atherton. (1984e). Modern subject access in the online age: Part 2. *Am. Libraries* March, 145–150.
Cochrane, Pauline A., and Arret, Linda. (1985). Online library users—The library comes of age. *Online* September, 58–70.
Cochrane, Pauline A., and Markey, Karen. (1985). Preparing for the use of classification in online cataloging systems and in online catalogs. *Inf. Technol. Libraries* June, 91–111.
Coffee break and videodisks. (1984). *Am. Libraries* December, 768.
Cohen, Aaron, and Cohen, Elaine. (1979). "Designing and Space Planning for Libraries: A Behavioral Guide." Bowker, New York.
Cohen, Elaine, and Cohen, Aaron. (1981). "Automation, Space Management, and Productivity: A Guide for Libraries." Bowker, New York.
Cohen, Jackson B. (1975). Science acquisitions and book output statistics. *Library Res. Tech. Services* **19**(4), 370–379.
Colbert, Antionette. (1982). Document delivery. *ONLINE* March, 74.
Colbert, Antoinette. (1983). Document delivery. *ONLINE* July, 79.
Colbert, Antoinette. (1985). Document delivery. *ONLINE* July, 69.
Cole, Elliot. (1981). Examining design assumptions for an information retrieval service: SDI use for scientific and technical databases. *J. Am. Soc. Inf. Sci.* **32**(6), 444–450.
Cole, P. F. (1965). Journal usage versus age of journal. *J. Documentation* **19**, 1–11.
Coleman, Kathleen, and Dickinson, Pauline. (1977). Drafting a reference collection policy. *College Res. Libraries* May, 227–233.
Conger, Lucinda D. (1980). Multiple system searching: A searcher's guide to making use of the real differences between systems. *ONLINE* April, 10–21.
Conger, Lucinda. (1984). Types of databases—Some definitions. *DATABASE* February, 94–95.
Connolly, Bruce. (1986a). Laserdisk directory—Part I. *Database* June, 15–26.
Connolly, Bruce. (1986b). Laserdisk directory—Part II. *Online* July, 39–49.
Cooper, Marianne. (1968a). Criteria for weeding of collections. *LRTS* **12**(3), 339–351.
Cooper, Marianne. (1968b). Organizational patterns of academic science libraries. *College Res. Libraries* **29**, 357–363.
Cooper, Marianne. (1983). Collection development in science and technology: A focus on books. *Sci. Technol. Libraries* **3**(3), 15–20.
Cooper, W. S. (1970). The potential usefulness of catalog access points other than author, title and subject. *J. Am. Soc. Inf. Sci.* **21**, 112–127.
Coplen, Ron, and Regan, Muriel. (1981). Internship programs in special libraries. *Special Libraries* January, 31–38.
Copyright basics. (1983). Copyright Office, Library of Congress, Washington, D.C. (Circular R1).
Copyright basics. (1985). *Inf. Rep. Biblio.* **14**(1), 5–13.

Corbin, John B. (1981). "Developing Computer-Based Library Systems." Oryx Press, Phoenix, Ariz.

Corporate Author Authority List—1983. (1983). NTIS/CAAL-83;PB83-156034. National Technical Information Service, Springfield, Va.

Costa, Betty, and Costa, Marie. (1983). "A Micro Handbook for Small Libraries and Media Centers." Libraries Unlimited, Littleton, Colo.

Council on Library Resources. (1983). Document delivery in the United States, a preliminary report to the Council on Library Resources; July 28. Rev. October 19, 1983.

Crane, Diana. (1972). "Invisible Colleges." University of Chicago Press, Chicago.

Crane, Nancy B., and Pilachowski, David M. (1978). Introducing the online bibliographic service to its users: The online presentation. *ONLINE* 2(4), 20–29.

Crawford, S. Y. (1971). Informal communication among scientists in sleep research. *J. Am. Soc. Inf. Sci.* 22, 301–310.

Crawford, Susan. (1978). Information needs and uses. *Ann. Rev. Inf. Sci. Technol. (ARIST)* 13, 61–81.

Crawford, Walt. (1984). "MARC for Library Use." Knowledge Industry Publications, Inc., White Plains, N.Y.

Creth, Sheila, and Duda, Frederick. (1981). "Personnel Administration in Libraries." Neal-Schuman, New York.

Crissinger, John D. (1981). The use of journal citations in theses as a collection development methodology. *Geosci. Inf. Soc. Proc.* 11, 113–124.

Cronin, Mary J. (1985). Performance measurement for public services in academic and research libraries. Association of Research Libraries, Washington, D.C., ED 254–256.

Crowley, Terence. (1968). The effectiveness of information service in medium size public libraries. PhD dissertation, Rutgers University.

Crowley, Terence. (1985). Half-right reference: Is it true? *RQ* 25(1), 59–68.

Crowley, Terence, and Childers, Thomas. (1971). "Information Service in Public Libraries: Two Studies." Scarecrow, Metuchen, N.J.

Cuadra Associates, Inc. (1979–). "Directory of Online Databases" (semiannual). Cuadra Associates, Inc., Santa Monica, Calif.

Culnam, Mary J. (1985). The dimensions of perceived accessibility to information: implications for the delivery of information systems and services. *J. Am. Soc. Inf. Sci.* 36(5), 302–308.

Cunha, George M., and Cunha, Dorothy Grant. (1971). "Conservation of Library Materials: A Manual and Bibliography on the Care, Repair and Restoration of Library Materials," 2d ed. Scarecrow Press, Metuchen, N.J.

Curley, Arthur, and Broderick, Dorothy. (1985). "Building Library Collections," 6th ed. Scarecrow, Metuchen, N.J.

Curley, Arthur, and Varlejs, Jana. (1984). "Akers Simple Library Cataloging," 7th ed. Scarecrow, Metuchen, N.J.

Cutter, Charles Ammi. (1904). "Rules for a Dictionary Catalog," 4th ed. Government Printing Office, Washington, D.C. (Note: First published in 1876.)

Daehn, Ralph M. (1982). The measurement and projection of shelf space. *Collection Manage.* 4(4), 25–38.

Damchok, M. M. (1976). CRT displays in power plants. *Instrumentation Technol.* 32(10), 2936.

Daniels, Linda. (1978). A matter of form. *ONLINE* 2(4), 31–39.

Darling, Pamela. (1974). Developing a preservation microfilming program. *Library J.* 99, 2803–2809.

"Database Directory 1985–86." (1985). Knowledge Industry Publications, Inc. and ASIS, White Plains, N.Y.

Daval, Nicola. (1983). National inventory of collections launched. Association of Research Libraries, Washington, D.C. (Note: Press release, August 3.)

Davis, R. A. (1965). How engineers use literature. *Chem. Eng. Prog.* **61**(3), 30–34.

De Gennaro, Richard. (1981). Libraries & networks in transition: Problems and prospects for the 1980's. *Library J.* May 15, 1045–1049.

De Gennaro, Richard. (1983a). Library automation & networking perspective on three decades. *Library J.* April 1, 629–635.

De Gennaro, Richard. (1983b). Theory vs. practice in library management. *Library J.* July, 1318–1321.

De Gennaro, Richard. (1984). Shifting gears: Information technology and the academic library. *Library J.* June 15, 1204–1209.

De Gennaro, Richard. (1985). Integrated online library systems: Perspectives, perceptions & practicalities. *Library J.* February 1, 37–40.

DeProspo, E. R. Altman, Ellen, and Beasley, Kenneth E. (1973). Performance measures for public libraries. Public Library Association, Chicago.

Derfler, Frank J., and Stallings, William. (1983). "A Manager's Guide to Local Networks." Prentice Hall, Englewood Cliffs, N.J.

Derksen, Charlotte R. M. (1984). Citation overlap among Geoarchive, GeoRef, PASCAL and Chemical Abstracts. *Geosci. Inf. Soc. Proc.* **15**, 125–138.

DesChene, Dorice. (1985). Online searching by end users. *RQ* **25**(1), 89–95.

De Solla Price, D. J. (1963). "Little Science, Big Science." Columbia University Press, New York.

De Solla Price, Derek J. (1971). Some remarks on elitism in information and the invisible college phenomenon in science. *J. Am. Soc. Inf. Sci.* **22**, 74–75.

Dewey, Patrick R. (1984). Searching for software: A checklist of microcomputer software directories. *Library J.* March 15, 544–564.

Dickson, Jean. (1984). An analysis of user errors in searching an online catalog. *Cataloging and Classification Q.* **4**(3), 19–38.

"Directory of fee-based information services." (1985). Burwell Enterprises.

"Directory of Microcomputer Applications in Libraries." (1984). CLASS, San Jose, Calif.

"Directory of Special Libraries and Information Centers." (1985). 9th ed. Gale, Detroit.

Dirlam, Dona Mary. (1986). The use of microcomputers in selecting diamond terminology for indexing gemological literature. Geoscience Information Society Proceedings, American Geological Institute, Alexandria, Virginia. (in preparation).

The dishy side of interlending. (1985). *British Library Lending Div. Newslett.* March, 4.

Disc-O-Tech (1985). *British Library Lending Div. Newslett.* December, (4).

Dodson, Ann T., Philbin, Paul P., and Rastogi, Kunj B. (1982). Electronic interlibrary loan in the OCLC Library. *Special Libraries;* **73**, 12–20.

Dolan, Donna R. (1979a). Before you touch the terminal: Flowchart of the search formulation process. *DATABASE* **2**(4), 86–8.

Dolan, Donna R. (1979b). What databases cannot do. *DATABASE;* **2**(3), 85–87.

Dolan, Donna R. (1980). Hedges for online searching. *DATABASE* **3**(1), 79–82.

Dolan, Donna R. (1983). Conceptualizing online search requests. *DATABASE* **6**(1), 77–78.

Dolan, Donna, and Kremin, Michael C. (1979). The quality control of search analysts. *Online* **3**(2), 8–16.

Doszkocs, Tamas E. (1983). CITE NLM: Natural language searching in an online catalog. *Inf. Technol. Libraries* **2**(4), 364–380.

Dougherty, Richard M., and Heinritz, Fred J. (1982). "Scientific Management of Library Operations." Scarecrow Press, Metuchen, N.J.

Dowd, Sheila T. (1980). The formulation of a collection development policy statement. *In* "Collection Development in Libraries: A Treatise" (Eds. Robert D. Stueart and George B. Miller, pp. 67–87. JAI, Greenwich, Conn.

Dowlin, Kenneth E. (1980). On-line catalog user acceptance survey. *RQ* **20**(1), 44–47.
Dowlin, Kenneth. (1984). "The Electronic Library." Neal-Schuman, New York.
Downloading/Uploading Online Databases and Catalogs. (1985). Pierian Press, Ann Arbor, Mich. [Note: Proceedings of the St. John's University Congress for Librarians (February 18, 1985), with additions, policy statements, and an extensive bibliography.]
Doyle, James M., and Grimes, George H. (1972). "Reference Sources: A Systematic Approach." Scarecrow Press, Metuchen, N.J.
Doyle, Michael, and Straus, David. (1976). How to make meetings work. Wyden, Ridgefield, Conn.
Drabenstott, Jon (1986). Projecting library automation costs. *Library Hi Tech.* **11**, 111–119.
Drake, Miriam A. (1977). The management of libraries as professional organizations. *Special Libraries* **68**, 181–186.
Drake, Miriam A. (1978). Impact of on-line systems on library functions. *In* "The Online Revolution in Libraries" (Eds. Kent Allen and Thomas Galvin), pp. 95–117. Marcel Dekker, New York.
Drucker, Peter. (1974). "Management." Harper & Row, New York.
Dudley, Norman H. (1980). Organizational models for collection development. *In* "Collection Development in Libraries: A Treatise " (Eds. Robert B. Stueart and George B. Miller), pp. 19–33. JAI Press, Greenwich, Conn.
Dwyer, James R. (1979). The least well and supplement of the card catalog: Public response to an academic library microcatalogue. *J. Academic Librarianship* **5**, 132–141.
Dwyer, James R. (1980). Comments and complaints on COM: Users look at what works and what doesn't. *ASIS Bull.* **7**(1), 19–23.
East, H. (1980). Comparative costs of manual and online bibliographic searching: A review of the literature. *J. Inf. Sci.* **2**, 101–109.
Eddison, Betty. (1984). Database design. *Database* December, 107–110.
Elman, Stanley A. (1975). Cost comparison of manual and on-line computerized literature searching. *Special Libraries* **66**(1), 12–17.
Emmick, Nancy J., and Davis, Luella B. (1984). A survey of academic library reference service practices: preliminary results. *RQ* Fall, 67–81.
End-user searching may lead to growth in number of librarians, says Case. (1985). *Adv. Technol. Libraries* **14**(9), 1, 8–9.
Engle, Stephen E., and Granda, Richard E. (1975). Guidelines for man/display interfaces. IBM, Poughkeepsie, N.Y., December 19.
Epstein, Susan Baerg. (1983a). Converting records for automation at the copy level. *Library J.* April 1, 642–643.
Epstein, Susan Baerg. (1983b). Procurement without problems: Preparing the RFP. *Library J.* June 1, 1109–1110.
Epstein, Susan Baerg. (1984). Integrated systems: Dream vs. reality. *Library J.* July, 1302–1303.
Ertel, Monica M. (1984). A small revolution: Microcomputers in libraries. *Special Libraries* April, 95–101.
Evans, Edward G. (1979). "Developing Library Collections." Libraries Unlimited, Littleton, Colo.
Evans, G. Edward. (1983). "Management Techniques for Librarians," 2d ed. Academic Press, New York.
Evans, G. Edward, and Borko, Harold. (1970). Effectiveness criteria for medical libraries: Final report. Institute of Library Research, University of California, Los Angeles, April, EDO57 813.

Expedited Inter-library loan agreements in force (1986). University of Texas at Austin *Library Bull.* February 28, **15**(8).

Falk, Joyce Duncan. (1984). Costs and budgeting for online ready reference. Distributed at 1984 ALA Annual Conference, Dallas, Texas, June.

Farkas, David L. (1984). Computer furniture . . . An expert's guide on how to be comfortable at your micro. *Online* May, 43–48.

Farmer, Sharon Cline. (1982). RLIN as a reference tool. *Online* September, 14–22.

Farrell, David (1986). The NCIP option for coordinated collection management. *Library Resources Tech. Serv.* January/March, 47–56.

Fayen, Emily Gallup. (1983). "The Online Catalog: Improving Public Access to Library Materials." Knowledge Industry Publications, Inc., White Plains, N.Y.

Fayen, Emily Gallup. (1984). The online public access catalog in 1984: Evaluating needs and choices. *Library Technol. Rep.* January–February, 5–60.

Feller, Siegfried. (1980). Developing the serials collection. *In* "Collection Development in Libraries: A Treatise" (Eds. Robert B. Stueart and George B. Miller), pp. 497–523. JAI Press, Greenwich, Conn.

Fenichel, Carol H. (1980–1981). The process of searching online bibliographic databases: A review of research. *Library Res.* **2,** 107–127.

Ferguson, Cheryl L., and Ballard, Robert M. (1985). Downloading and copyright: A selected survey of special librarians and database suppliers. *Sci-Tech News* January," 17–19.

Ferguson, Elizabeth, and Mobley, Emily R. (1984). Special Libraries at Work." Library Professional Publications, Hamden, Conn.

Ferriero, David. (1981). Impact of OCLC on interlibrary loan at MIT. *Res. Libraries OCLC* July (3), 2.

Fetters, Linda K. (1985). A guide to seven indexing programs . . . plus a review of the Professional Bibliographic System. *Database* December, 31–38.

"Fish and Wildlife Reference Service Thesaurus." (1981). 2d ed. Denver Public Library Fish and Wildlife Reference Service. Denver, Colo.

Flynn, Karen L. (1985). The 3M experience: use of external databases in a large diversified company. *Special Libraries* Spring, 81–87.

"Food Science and Technology Abstracts Thesaurus." (1981). 2d ed. International Food Information Service, Reading, England.

Ford, Stephen. (1973). "Acquisition of Library Materials." American Library Association, Chicago.

Foskett, A. C. (1970). "Guide to Personal Indexes," 2d ed. Shoe String, Hamden, Conn.

Foskett, A. C. (1982). "The Subject Approach to Information," 4th ed. Linnet Books, Hamden, Conn.

Fraley, Ruth A., and Anderson, Carol Lee. (1985). "Library Space Planning." Neal-Schuman, New York.

Frarey, C. J. (1953). Studies of the use of the subject catalog: Summary and evaluation. *In* "Subject Analysis of Library Materials" (Ed. M. F. Tauber) pp. 147–166. School of Library Service, Columbia University, New York.

Freedman, Maurice J. (1984). Automation and the future of technical services. *Library J.* June 15, 1197–1203.

Fryser, Benjamin S., and Stirling, Keith H. (1984). The effect of spatial arrangement, upper–lower case letter combinations, and reverse video on patron response to CRT displayed catalog records. *J. Am. Soc. Inf. Sci.* **35**(6), 344–350.

Fussler, Herman H., and Simon, Julian L. (1969). "Patterns in the Use of Books in Large Research Libraries." University of Chicago (University of Chicago Studies in Library Science), Chicago.

Futas, Elizabeth, Editor. (1977). "Library Acquisition Policies and Procedures." ORYX, Phoenix, Ariz. (Note: A compendium of policies from 12 public and 14 academic libraries.)

Gale, John C. (1984). Use of optical disks for information storage and retrieval. *Inf. Technol. Libraries* 3(4), 379–381.

Galitz, Wilbert O. (1980). "Human Factors in Office Automation." Life Office Management Association, Atlanta.

Galitz, Wilbert O. (1981). "Handbook of Screen Format Design." Q.E.D. Information Sciences, Wellesley, Mass.

Ganning, Mary Kay Daniels. (1976). The catalog: Its nature and prospects. *J. Library Automation* 9(1), 48–66.

Gardner, Richard K. (1981). "Library Collections, Their Origins, Selections and Development." McGraw-Hill, New York.

Garman, Nancy. Downloading . . . still a live issue?: A survey of database producer policies for both online services and laserdisks. *Online* 10(4), 15–25.

Garoogian, Rhoda. (1982). Pre-written software: Identification, evaluation and selection. *Software Review* February, 11–34.

Garvey, W. D. (1979). "Communication: The Essence of Science." Pergamon, New York.

Gasaway, Laura. (1981). Equal pay for equal work. Special Libraries Association, New York.

Genaway, David C. (1984). "Integrated Online Library Systems: Principles, Planning and Implementation." Knowledge Industry Publications, White Plains, N.Y.

Georgi, Charlotte, and Bellanti, Robert, Guest editors (1985). Excellence in library management. *J. Library Admin.* 6(3).

Gers, Ralph, and Seward, Lillie J. (1985). Improving reference performance: Results of a statewide study. *Library J.* November 1, 32–35.

Geta, Malcolm, and Phelps, Doug. (1984). Labor costs in the technical operation of three research libraries. *J. Academic Librarianship* 10(4), 209–219.

Girard, A., and Moreau, M. (1981). An examination of the role of the intermediary in the online searching of chemical literature. *Online Rev.* 5(3), 217–225.

Godden, Irene P., Ed. (1984). "Library Technical Services: Operations and Management." Academic Press, New York.

Godden, Irene P., *et al.* (1982). "Collection Development and Acquisitions, 1970–1980: An Annotated, Critical bibliography." Scarecrow, Metuchen, N.J.

Goldberg, Alan L. (1985). Query. *Small Systems World* April, 50.

Goldstein, Margery Ziegler, and Sweeney, Carolyn Musselman. (1979). Aptitude requirements for library assistants in special libraries. *Special Libraries* 70, 373–376.

Gore, Daniel. (1976). Farewell to Alexandria: The theory of the no-growth, high-performance library. *In* "Farewell to Alexandria" (Ed. Daniel Gore), pp. 164–180. Greenwood, Westport, Conn.

Gorman, Michael. (1979). On doing away with technical services departments. *Am. Libraries* 13(7), 474–474.

Gorman, Michael. (1981). The most concise AACR2. *American Libraries* September, 499.

Gorman, Michael, and Hotsinpiller, Jami. (1979). ISBD: Aid or barrier to understanding. *College Res. Libraries* 40, 519–526.

Gouke, Mary Noel, and Pease, Sue. (1982). Title searches in an online catalog and card catalog: A comparative study of patron success in two libraries. *J. Academic Librarianship* 8(3), 137–43.

Gralewska-Vickery, A. (1978). Communication and information needs of earth science engineers. *Inf. Processing Manage.* 12, 251–282.

Gralewska-Vickery, A., and Roscoe, H. (1975). Earth science engineers: Communication and information needs, final report. Imperial College (research report no. 32), London.

Grathwol, M. (1971). Bibliographic elements in citations and catalog entries: A comparison. Master's dissertation, Graduate Library School, University of Chicago, Chicago.

Grieder, Ted. (1978). "Acquisitions: Where, What and How." Greenwood Press, Westport, Conn. (Contributions in Librarianship and Information Science, no. 22).

Griffiths, Jose-Marie. (1984a). Microcomputers and online activities. *ASIS Bull.* April, 11–14.

Griffiths, Jose-Marie. (1984b). Our competencies defined: A progress report and sampling. *Am. Libraries* January, 43–45.

Grogan, Denis J. (1976). "Science and Technology: An Introduction to the Literature," 3d ed. Clive Bingley, London.

Grosch, Audrey N. (1982). "Minicomputers in Libraries, 1981–82: The Era of Distributed Systems." Knowledge Industry Publications, Inc., White Plains, N.Y.

GTE Telemail. (1985). *Information Today* February, 2.

"A Guide to the Selection of Computer-based Science and Technology Reference Sources in the U.S.A." (1969). American Library Association, Chicago.

Hafter, Ruth. (1979). The performance of card catalogs: A review of research. *Library Res.* 1, 199–200.

Hagler, Ronald. (1982). "The Bibliographic Record and Information Technology." American Library Association, Chicago.

Hagstrom, Warren O. (1965). "The Scientific Community." Basic Books, New York.

Halperin, Michael, and Pagell, Ruth A. (1985). Free "do-it-yourself" online searching . . . What to expect. *ONLINE* March, 82–84.

Hamburg, Morris, *et al.* (1974). "Library Planning and Decision-making systems." MIT Press, Cambridge, Mass.

Hamlin, Jean Boyer. (1980). The selection process. *In* "Collection Development in Libraries: A Treatise" (Eds. Robert D. Stueart and George B. Miller), pp. 185–201. JAI Press, Greenwich, Conn.

Harter, S. P. (1983). The online information specialist: Behaviors, philosophies, and attitudes. Collected Papers of the 12th ASIS Mid-Year Meeting, Lexington, Ky., pp. 201–202.

Harvey, John H., and Dickinson, Elizabeth M., eds. (1983). "Librarians' Affirmative Action Handbook." Scarecrow, Metuchen, N.J.

Haselbauer, Kathleen. (1984). The making of a science librarian. *Sci. Technol. Libraries* 4(3/4), 111–116.

Hawkins, Donald T. (1976). Impact of on-line systems on a literature searching service. *Special Libraries* 67(12), 559–567.

Hawkins, D. T. (1980). Management of an online search service. *In* "The Library and Information Manager's Guide to Online Services" (Ed. R. E. Hoover). Knowledge Industry Publications, Inc., White Plains, N.Y.

Hawkins, Donald T. (1981). Six years of online searching in an industrial library network. *Sci. Technol. Libraries* 1(1), 57–67.

Hawkins, Donald T. (1982a). Online bibliographic search strategy development. *Online* May, 12–19.

Hawkins, Donald T. (1982b). To download or not to download online searches. Online 82 Conf. Proc. 3–5.

Hawkins, Donald T. (1985). A review of online physical sciences and mathematics databases. *DATABASE* February, 14–18.

Hawkins, Donald T. (1986). Front end software for online database services: Part 3: Product selection chart and bibliography. *Online* May, 49–58.

Hawkins, Donald T., and Brown, Carolyn P. (1980). What is an online search? *ONLINE* **4**(1), 12–18.

Haykin, David Judson. (1951). Subject headings: A practical guide. Government Printing Office, Washington, D.C.

Hazell, J. C., and Potter, J. S. (1968). Information practices of agricultural scientists. *Austr. Library J.* **17**(5), 147–159.

Heald, J. H. (1966). DoD manual for building a technical thesaurus. DDC AD633 279. Prepared by Project LEX. Defense Technical Information Center, Cameron Station, Va.

Henderson, Faye W., and Chase, Leslie R., eds. (1985). "Information Sources 1985: Annual Directory of the Information Industry Association. Information Industry Association, Washington, D.C.

Hendricks, Tom. (1985). Serials management and the new technology. *Serials Perspect. (University Microfilms International)* **2**(2), 2.

Hernon, Peter, and McClure, Charles R. (1986). Unobtrusive reference testing: The 55 percent rule. *Library J.* **15**, 37–41.

Herther, Nancy K. (1985). CD ROM technology: A new era for information storage and retrieval? *Online* November, 17–28.

Hewes, Jeremy Joan. (1985). Gateways to on-line services. *PC World* May, 150–156.

Hewitt, Joe A. (1984). Technical services in 1983. *Library Res. Tech. Services* July/September, 205–218.

High-speed communicating copier demonstrated. (1980). *The Office*, August, 101.

Highlights of the year. (1985). *British Library Lending Div. Newslett.* June (2), 1.

Hildreth, Charles R. (1982). "Online Public Access Catalogs: The User Interface." OCLC, Dublin, Ohio.

Hildreth, Charles R. (1985). Online public access catalogs. *Annu. Rev. Inf. Sci. Technol.* **20**, 233–285.

Hill, Linda L. (1985). Issues in network participation for corporate librarians. *Special Libraries* Winter, 2–10.

Hills, P. J. (1983). The scholarly communication process. *Annu. Rev. Inf. Sci. Technol. (ARIST)* **18**, 99–125.

Hinckley, W. A. (1968). On searching catalogs and indexes with inexact title information. Master's dissertation, Graduate Library School, University of Chicago, Chicago.

Hindle, A., and Buckland, M. K. (1978). In-library book usage in relation to circulation. *Collection Manage.* **2**, 265–78.

Hitchingham, Eileen, *et al.* (1984). A survey of database use at the reference desk. *ONLINE* March, 44.

Hock, R. E. (1983). Who should search? The attributes of a good searcher. *In* "Online Searching Technique and Management" (Ed. J. J. Maloney), pp. 83–88. American Library Association, Chicago.

Hodges, Pauline R. (1983). Keyword in title indexes: Effectiveness of retrieval in computer searches. *Special Libraries* **74**(1), 56–60.

Hoffman, Herbert H. (1976). "What Happens in Library Filing." Linnet Books, Hamden, Conn.

Holladay, Janice. (1981). Small libraries: Keeping the professional position professional. *Special Libraries* January, 63–66.

Holley, Robert P., and Killhefer, Robert E. (1982). Is there an answer to the subject access crisis? *Catalog. Classif. Q.* **1**(213), 125–133.

Holton, Felicia Antonelli. (1985). For scientists, business, and the public—A striking new science library. *The University of Chicago Magazine* **77**(2), 12–35.

Hoover, Ryan E., ed. (1980). "Library and Information Managers' Guide to Online Services." Knowledge Industry Publications, White Plains, N.Y.

Horowitz, Gary L., and Bleich, Howard L. (1981). PaperChase: : A computer program to search the medical literature. *N. Engl. J. Med.* 305, 924–930.

Horton, Carolyn. (1967). Cleaning and preserving bindings and related materials. American Library Association, Chicago (LTP publication no. 12).

Houghton, Bernard. (1972). "Technical Information Sources: A Guide to Patent Specifications, Standards and Technical Report Literature," 2d ed. Shoe String Press, Hamden, Conn.

How libraries obtain materials for their users. (1984). *Inf. Hotline* 16(4), 1.

Howitt, Doran. (1984). "Magazine's Databasics, your guide to online business information." Garland Publishing, Inc.,

Hubbard, Abigail (1985). Reprint file management software. *Online* November, 67–73.

Hunsaker, Phillip L. (1982). Strategies for organizational change: the role of the inside change agent. *Personnel* Sept./Oct., 18–28.

Hunter, Janne A. (1984). When your patrons want to search—The library as advisor to end user . . . A compendium of advice and tips. *ONLINE* May, 36–41.

Information on Demand, Inc. (1985). *KnowNews* 8(2).

Information Systems Consultants, Inc. (1985). Videodisc and optical disk technologies and their applications in libraries: A report to the Council on Library Resources. Council on Library Resources, Washington, D.C.

Information Today. (1985). 2(1), 1.

Inkellis, Barbara. (1982). Legal issues of downloading online search results. *ONLINE 82 Conf. Proc.* 91–92.

Intelsat lowers rates. (1985). *Albuquerque J.* March 31, G3.

International Conference on Cataloguing Principles, Paris, 1961. (1966). Statement of Principles, provisional ed. International Federation of Library Associations Secretariat, Sevenoaks, Kent, England.

International Federation of Library Associations. (1977). Working Group on the General International Bibliographic Description. ISBD(G): General international standard bibliographic description: annotated text. IFLA International Office for UBC, London.

"International File of Micrographics Equipment and Accessories, 1983–1984. (1983). Meckler, Westport, Conn.

"International Micrographics Source Book." (Annual). Microfilm Publishing, New Rochelle, N.Y.

International standard abbreviations of typical words in bibliographic references, ISO 832–1968. (1968). Available from American National Standards Institute.

International standard documentation—international standard serial numbering (ISSN), ISO 2108–1972(E). (1972). Available from American National Standards Institute, New York.

International standard documentation—international list of periodical title word abbreviations, ISO 833–1974. (1974). Available from American National Standards Institute, New York. Published supplements are available from Centre International d'Enregistrement des Publications en Serie, 20, rue Bachaumont, 75002 Paris, France.

Intner, Sheila S. (1984). "Access to Media: A Guide to Integrating and Computerizing Catalogs. Neal-Schuman, New York.

Irvine, Betty Jo. (1985). "Sex Segregation in Librarianship: Demographic and Career Patterns of Academic Library Administrators." Greenwood Press, Westport, Conn.

Ivantcho, Barbara, ed. (1983). Position descriptions in special libraries: A collection of examples: Special Libraries Association, New York.

Jackson, E. B., and Jackson, Ruth L. (1980). Characterizing the industrial special library universe. *J. Am. Soc. Inf. Sci.* May, 208–214.

Jackson, Eugene. (1978). "Industrial Information Services." Dowden, Hutchinson and Ross, Inc., Stroudsburg, Pa.

Jackson, Marilyn E. (1978). Planning a business information system for engineers and scientists. American Society for Information Science. Management of Information Systems. Proceedings; May 21–24, 1978; Rice University, Houston, Texas; Texas Chapter, ASIS.

Jackson, William J. (1982). How to train experienced searchers to use another system. *ONLINE* May, 27–35.

Jacob, Mel (1986). CD-ROM standardization activities. *ASIS Bull.* February/March, 19–20.

Jahoda, G. (1970). "Information Storage and Retrieval Systems for Individual Researchers." Wiley-Interscience, New York.

Jahoda, G., and Braunagel, J. S. (1980). "The Librarian and Reference Queries: A Systematic Approach." Academic Press, New York.

Jahoda, G., and Olson, P. E. (1972). Analyzing the reference process: Models of reference. *RQ* 12(Winter), 148–156.

Jamieson, Alexis, and Dolan, M. Elizabeth (1985). University library experience with remote access to online catalogs. Survey conducted by University of Western Ontario School of Library and Information Science.

Janke, Richard V. (1984). Online after six: End user searching comes of age. *ONLINE* November, 15–29.

Jenks, George M. (1976). Circulation and its relationship to the book collection and academic departments. *College Res. Libraries* 37, 145–152.

Jennerich, E. J. (1980). Before the answer: Evaluating the reference process. *RQ* 19(Summer), 360–366.

Jennerich, E. J., and Jennerich, E. Z. (1976). Teaching the reference interview. *J. Education for Librarianship* 17(Fall), 106–111.

Jensen, Rebecca, Asbury, Herbert O., and King, Radford G. (1980). Costs and benefits to industry of online literature searches. *Special Libraries* 71, 291–299.

Johnston, Donald F. (1978). "Copyright Handbook." Bowker, New York.

Jones, C. Lee. (1982). Status of bibliographic record system elements. *Inf. Technol. Libraries* June, 111–124.

Jones, Douglas. (1981). RLIN and OCLC as reference tools. *J. Library Automation* 14(3), 201.

Kaiser, John R. (1980). Resource sharing in collection development. *In* "Collection Development in Libraries: A Treatise" (Eds. Robert D. Stueart and George B. Miller), pp. 139–157. JAI Press, Greenwich, Conn.

Kaminecki, Ron. (1977). Comparison of selective dissemination of information systems. *On-Line Review* 1(3), 195–206.

Kantor, P. B. (1976a). Availability analysis. *J. Am. Soc. Inf. Sci.* 27, 311–319.

Kantor, P. B. (1976b). The library as an information utility in the university context: Evolution and measurement of service. *J. Am. Soc. Inf. Sci.* March–April, 102–112.

Kantor, Paul B. (1981a). Levels of output related to cost of operation of scientific and technical libraries: Part I: Techniques and cumulative statistics. *Library Res.* 3, 1–28.

Kantor, Paul B. (1981b). Levels of output related to cost of operation of scientific and technical libraries: Part II: A capacity model of the average cost formula. *Library Res.* 3, 141–154.

Kantor, Paul B. (1984a). Cost and usage of health science libraries: Economic aspects. *Med. Library Assoc. Bull.* 72, 274–286.

Kantor, Paul B. (1984b). "Objective Performance Measures for Academic and Research Libraries." Association of Research Libraries, Washington, D.C.

Kaplan, Denise P. (1984). Creating copy specific records for local databases. *Library Hi Tech* 7: 19–24.

Kaske, Neal K., and Sanders, Nancy P. (1983). A comprehensive study of online public access catalogs: An overview and application of findings. OCLC Online Computer Library Center, Inc., Columbus, Ohio.

Kaske, Neal K., and Sanders, Nancy P. (1980a). Evaluating the effectiveness of subject access: The view of the library patron. *Proc. Annu. ASIS Conf.*, 43rd, **17**, 323–325.

Kaske, Neal K., and Sanders, Nancy P. (1980b). On-line subject access: The human side. *RQ* **20**(1), 52–58.

Katayama, Jane H. (1983). The library committee: How important is it? *Special Libraries* **74**(1), 44–48.

Katz, William. (1982). "Introduction to Reference Work, Volume 2: Reference Services and Reference Processes," 4th ed. McGraw-Hill, New York.

Kazlauskas, Edward. (1976). An exploratory study: A kinesic analysis of academic public service points. *J. Academic Librarianship* July, 130–134.

Kennedy, Robert A. (1978). Bell Laboratories Library Network. *In* "Industrial Information Systems," (Eds. E. B. Jackson and Ruth Jackson), pp. 165–176. Dowden, Hutchinson & Ross, Stroudsburg, Penn.

Kennedy, Robert A. (1982). Computer-derived management information in a special library. *In* "Library Automation as a Source of Management Information" (Ed. F. Wilfred Lancaster). Graduate School of Library and Information Science, University of Illinois at Urbana-Champaign, Urbana, Ill.

Kent, Allen. (1978). Resource sharing in libraries. *In* "Encyclopedia of Library and Information Science," vol. 25, pp. 293–307. Marcel Dekker, New York.

Kent, Allen, and Galvin, Thomas. (1977). "Library Resource Sharing." Marcel Dekker, New York.

Kern-Simirenko, Cheryl. (1983). OPAC user logs: Implications for bibliographic instruction. *Library Hi Tech* **1**(3), 27–35.

Kilgour, Frederick G. (1984). The online catalog revolution. *Library J.* February 15, 319–321.

King, D. W. (1976). "Statistical Indicators of Scientific and Technical Communication 1960–1980." King Research, Washington, D.C.

King, D. W., and Bryant, E. C. (1971). "The Evaluation of Information Services and Products." Information Resources Press, Washington, D.C.

King, Donald W., and Griffiths, Jose-Marie. (1985). "The Use and Value of Special Libraries." Knowledge Industry Publications, Inc., for ASIS, White Plains, N.Y.

King, Donald W., Griffiths, Jose-Marie, Roderer, Nancy K., and Wiederkehr, Robert R. V. (1982). Value of the Energy Data Base. King Research, Inc., Washington, D.C., DOE/OR/11232-1;DE82014250.

King, Donald W., Griffiths, J. M., Sweet, E. A., Wiederkehr, R. R. V., and Roderer, N. K. (1984). A study of the value of information and the effect on value of intermediary organizations, timeliness of services & products, and comprehensiveness of the EDB. King Research, Inc., Washington, D.C., DOE/NBM-1078;DE85003670.

King, Geraldine B. (1972). Open & closed questions: The reference interview. *RQ* Winter, 157–160.

King, John Leslie, and Schrems, Edward L. (1978). Cost–benefit analysis in information systems development and operation. *Comput. Surv.* March, 19–34.

King Research, Inc. (1982). Libraries, publishing and photocopying: Final report of surveys conducted for the United States Copyright Office; King Research, Inc., Rockville, Md.

King Research, Inc. (1985). Interlibrary loan cooperation plan for the Commonwealth of Pennsylvania, final report. March 15.

King Research study compares ILL to document delivery costs. (1985). *Clearinghouse Q.* 1(2), 1.

Kirtland, Monika, and Cochrane, Pauline. (1982). Critical views of LCSH—Library of Congress Subject Headings, a bibliographic and bibliometric essay. *Catalog. Classif. Q.* 1(2/3), 71–91.

Knapp, Sara D., and Schmidt, C. James. (1979). Budgeting to provide computer based reference services: A case study. *J. Academic Librarianship* 5(1), 9–13.

Koenig, Michael E. D. (1980). "Budgeting Techniques for Libraries and Information Centers." Special Libraries Association, New York.

Koenig, Michael E. D. (1983). Education for special librarianship. *Special Libraries* 74(2), 182–196.

Koenig, Michael, E. D. (1984). Budgeting and financial planning for science and technology libraries. *Sci. Technol. Libraries* 4(3/4), 87–103.

Kohl, David F. (1985). "Acquisitions, Collection Development, and Collection Use: A Handbook for Library Management." ABC-CLIO, Santa Barbara, Calif.

Korfhage, Robert R. (1974). Informal communication of scientific information. *J. Am. Soc. Inf. Sci.* 25: 25–32.

Kosa, Giza. (1975). Book selection tools for subject specialists in a large research library: An analysis. *Library Resources and Technical Services* 19(1), 13–18.

Kotler, Philip. (1982). "Marketing for Nonprofit Organizations," 2d ed. Prentice Hall, Englewood Cliffs, N.J.

Kraft, Donald H. (1979). Journal selection models: Past and present. *Collection Manage.* 3(2/3), 163–185.

Kraft, Donald H., and Placsek, Richard A. (1978). A journal-worth measure for a journal-selection decision model. *Collection Manage.* 2(2), 129–139.

Kramer, Joseph. (1971). How to survive in industry: Cost justifying library services. *Special Libraries* 62, 487–89.

Kreutz, D. M. (1978). On-line searching—Specialist required. *J. Chem. Inf. Comput. Sci.* 18(1), 4.

Kriekelas, J. (1972). Catalog use studies and their implications. *In* "Advances in Librarianship" (Ed. M. J. Voigt), vol. 3, pp. 195–220. Seminar Press, New York.

Kriekelas, James. (1980–1981). Searching the library catalog—A study of users' access. *Library Res.* 2, 215–230.

Kriz, Harry M. (1978). Subscriptions vs. books in a constant dollar budget. *College Res. Libraries* 39(2), 105–109.

Krulee, G. K., and Nadler, E. B. (1960). Studies of education for science and engineering: Student values and curriculum choice. *IEEE Trans. Eng. Manage.* 7, 146–158.

Kruzas, A. T., and Schmittroth, J., Ed. (1981). "Encyclopedia of Information Systems and Services," 4th ed. Gale Research Co., Detroit.

Kwan, Julie, Pruett, N., and Yokote, G. (1980). Training manual for computer reference services. University of California, Los Angeles, unpublished.

Ladendorf, Janice M. (1970). Information flow in science, technology and commerce: A review of the concepts of the sixties. *Special Libraries* 61, 215–222.

Lamb, Connie. (1981). Searching in academia: Nearly 50 libraries tell what they're doing. *ONLINE* April, 78–81.

Lamberton, Donald M. (1984). The economics of information and organization. *Annu. Rev. Inf. Sci. Technol.* 19, 3–30.

Lancaster, F. W. (1968). "Information Retrieval Systems: Characteristics, Testing and Evaluation." Wiley, New York.

Lancaster, F. W. (1972). "Vocabulary Control for Information Retrieval." Information Resources Press, Washington, D.C.

Lancaster, F. W. (1977). "The Measurement and Evaluation of Library Services." Information Resources Press, Washington, D.C.

Lancaster, F. W. (1982a). Evaluating collections by their use. *Collection Manage.* **4**(1/2), 15–43.

Lancaster, F. Wilfrid. (1982b). Library automation as a source of management information. Graduate School of Library and Information Science, University of Illinois at Urbana–Champaign, Urbana, Ill.

Larason, Larry, and Robinson, Judith Schiek. (1984). The reference desk: service point or barrier? *RQ* **23**(3), 332–338.

Larson, Ray R. (1983). Users look at online catalogs: Part 2. Interacting with online catalogs. Final report to the Council on Library Resources. University of California, Berkeley, ERIC:ED 231401.

Larson, Signe E. (1983). Reference and information services in special libraries. *Library Trends* **31**, 475–493.

Lavendal, Giuliana A. (1981). SDI in scientific and technical libraries: An overview of the options. *Sci. Technol. Libraries* **2**(1), 3–16.

Lawrence, Gary S. (1982). Users look at online catalogs: Results of a national survey of users and non-users of online public access catalogs: Final report, Council on Library Resources, ED 231 395.

Lawrence, Gary S. (1985). System features for subject access in the online catalog. *Library Res. Tech. Services* **29**(1) January/March, 16–32.

Lawrence, Gary S., Matthews, J. R., and Miller, Charles E. (1983). Costs and features of online catalogs: the state of the art. *Inf. Technol. Libraries* **2**(4), 409–449.

Learmont, Carol L., and Van Houten, Stephen. (1984). Placements & salaries 1983: Catching up. *Library J.* October 1, 1805–1811.

Legg, Jean. (1965). Death of the departmental library. *Library Res. Tech. Services* **9**, 351–355.

Levert, Virginia M. (1985). Applications of local area networks of microcomputers in libraries. *Inf. Technol. Libraries* March, 9–18.

Library and Information Resources for the Northwest. (1985). Directory of telefacsimile sites in libraries in the United States, Preliminary ed., April.

Library of Congress. (1956). "Filing Rules for the Dictionary Catalogs of the Library of Congress." Processing Department, Washington, D.C.

Library of Congress. (1969). "Conversion of Retrospective Catalog Records to Machine Readable Form." Library of Congress, Washington, D.C.

Library of Congress Environmental Policy Division. (1975). "The U.S. Geological Survey." Government Printing Office, Washington, D.C. (Note: Prepared by the Environmental Policy Division, Congressional Research Service, Library of Congress, by Allen F. Agnew.)

Library of Congress, MARC Development Office. (1972). "Books: A MARC format," 5th ed. Library of Congress, Washington, D.C.

Library of Congress, MARC Development Office. (1974). "Information on the MARC System," 4th ed. Library of Congress, Washington, D.C.

"Library of Congress Subject Headings." (1980). 9th ed. Library of Congress, Washington, D.C.

"Library Statistics: a Handbook of Concepts, Definitions and Terminology." (1966). American Library Association, Chicago.

Library Systems Newsletter (1983), **3**, 6–7.

Line, M. B. (1970). The 'half-life' of periodical literature: apparent and real obsolescence. *J. Documentation* **26**, 46–54.

Line, Maurice B. (1973). The ability of a university library to provide books wanted by researchers. *J. Librarianship* **5**, 37–51.

Line, Maurice B. (1974). Does physics literature obsolesce? A study of variation of citation frequency with time for individual journal articles in physics. *BLL Rev.* **2**, 84–91.

Line, Maurice B. (1978). Rank lists based on citations and library uses as indicators of journal usage in individual libraries. *Collection Manage.* 2(4), 313–316.

Line, Maurice B. (1982). "Library Surveys: An Introduction to Their Use, Planning, Procedures and Presentation," 2d ed. Clive Bingley, London.

Line, Maurice B. (1986). Access to resources: The international dimension. *Library Resources Tech. Serv.* January/March, 4–12.

Lipetz, Ben-Ami. (1970). "User Requirements in Identifying Desired Works in a Large Library." Yale University Library, New Haven, Conn.

Lipetz, Ben-Ami. (1972). Catalog use in a large research library. *Library Q.* **42**, 129–139.

Lisowski, Andrew, and Sessions, Judith. (1984). Selecting a retrospective conversion vendor. *Library Hi Tech* Spring, 65–68.

Logging on. (1984). *Database Update* June, 5–6.

Lopez Munoz, J. (1977). The significance of nonverbal communication in the reference interview. *RQ* **16**(Spring), 220–224.

Lowell, Howard P. (1982). Sources of conservation information for the librarian. *Collection Manage.* 4(3), 1–18.

Lowry, Glenn R. (1981). Training of users of online services: A survey of the literature. *Sci. Technol. Libraries* **1**(3), 27–39.

Lubans, John, Jr., ed. (1974). "Educating the Library User." Bowker, New York.

Lufkin, J. M., and Miller, E. H. (1966). The reading habits of engineers—A preliminary survey. *IEEE Trans. Education* **E-9**(4), 179–82.

Lundeen, Gerald, and Tenopir, Carol. (1985). Microcomputer software for in-house databases . . . Four top packages under $2000. *Online* September, 30–38.

Lunin, Lois F., and Paris, Judith. (1983). Videodisc and optical disk: Technology, research and applications. *J. Am. Soc. Inf. Sci.* **34**, 405–440.

Lynch, Mary Jo. (1978). Reference interview in public libraries. *Library Quarterly;* **48**(April): 119–142.

Lynch, Mary Jo, and Eckard, Helen, eds. (1981). "Library Data Collection Handbook." American Library Association, Chicago.

Lynden, Frederick C. (1977). Sources of information on the costs of library materials. *Library Acquisitions: Practice and Theory* **1**, 101–116.

Lyon, Sally. (1984). End-user searching of online databases: A selective annotated bibliography. *Library Hi-Tech* **6**, 47–50.

MacRae, Duncan. (1969). Growth and decay curves in scientific citations. *Am. Sociolog. Rev.* **34**, 631–635.

Magson, M. S. (1973). Techniques for the measurement of cost-benefit in information centers. *ASLIB Proc.* **25**, 164–185.

Magson, M. S. (1980). Modelling on-line cost-effectiveness. *ASLIB Proc.* **32**, 35–41.

Maizell, Robert E. (1960). Information gathering patterns and creativity: A study of research chemists in an industrial research laboratory. *Am. Documentation* **11**, 9–17.

"Making the DIALOG Connection with a Personal Computer". (1983). DIALOG Information Services, Palo Alto, Calif.

Malinconico, S. Michael. (1977). The economics of computer output media. *In* Proc. 1976 Clinic on Library Applications of Data Processing (Ed. J. L. Divilbiss), pp. 145–162. University of Illinois, Urbana–Champaigna.

Malinconico, S. Michael. (1983a). Hearing the resistance. *Library J.* January 15, 111–113.

Malinconico, S. Michael. (1983b). Listening to the resistance. *Library J.* February 15, 353–355.

Malinconico, S. Michael. (1983c). Managing consultants. *Library J.* November 1, 2032–2034.

Malinconico, S. Michael. (1983d). People and machines: Changing relationships? *Library J.* December 1, 2222–2224.

Malinconico, S. Michael. (1983e). Planning for failure. *Library J.* April 15, 798–800.

Malinconico, S. Michael. (1983f). The use & misuse of consultants. *Library J.* March 15, 558–560.

Malinconico, S. Michael. (1984a). Catalogs and cataloging: Innocent pleasures and enduring controversies. *Library J.* June 15, 1210–1213.

Malinconico, S. Michael. (1984b). Decisions under uncertainty. *Library J.* November 15, 2129–2131.

Malinconico, S. Michael. (1984c). Managing organizational culture. *Library J.* April 15, 791–793.

Malinconico, S. Michael. (1984d). Planning for obsolescence. *Library J.* February 14, 333–335.

Malinconico, S. Michael, and Fasano, Paul J. (1979). "The Future of the Catalog: The Library's Choices." Knowledge Industry Publications, White Plains, N.Y.

Malinowsky, H. Robert, and Richardson, Jeanne M. (1980). "Science and Engineering Literature: A Guide to Reference Sources," 3d ed. Libraries Unlimited, Littleton, Colo.

Maloney, James J. (1983). "Online Searching Techniques and Management." American Library Association, Chicago.

Maltby, A. (1971). Measuring catalogue utility. *J. Librarianship* **3**, 180–189.

Maltby, A. (1973). U.K. catalogue use survey: A report. Library Association, London.

Maltby, A., and Duxbury, A. (1972). Description and annotation in catalogues: an attempt to discover readers' attitudes and requirements. *New Library World* **73**, 260–262, 273.

Maltby, A., and Sweeney, R. (1972). The U.K. catalogue use survey. *J. Librarianship* **4**, 188–204.

Mandel, Carol A. (1981). Subject access in the online catalog. Council on Library Resources, Washington, D.C., August.

Mandel, Carol A. (1984). Enriching the library catalog record for subject access. *Library Res. Tech. Services* January/March, 5–15.

Mandel, Carol A., and Herschman, Judith. (1983). Online subject access—Enhancing the library catalog. *J. Academic Librarianship* **9**(3), 148–55.

Manthey, Teresa, and Brown, Jeanne Owen. (1985). Evaluating a special library using public library output measures. *Special Libraries* **76**(4), 282–289.

Marchant, Maurice P. (1982). Participative management, job satisfaction, & service. *Library J.* April 15, 782–784.

Markey, Karen. (1980). Analytical review of catalog use studies. OCLC Inc., Columbus, Ohio, February, research report OCLC/OPR/RR-80/2; ED 186041.

Markey, Karen. (1983a). Online catalog use: results of surveys and focus group interviews in several libraries. OCLC Online Computer Library Center, Inc., Dublin, Ohio, research report OCLC/OPR/RR-83/3. (Final report to the Council on Library Resources: Volume 2.)

Markey, Karen. (1983b). The process of subject searching in the library catalog: Final report on the subject access research project. OCLC Online Computer Library Center, Dublin, Ohio, February 4, OCLC/OPR/RR-83/1.

Markey, Karen. (1983c). Thus spake the OPAC user. *Inf. Technol. Libraries;* 2(4), 381–387.

Markey, Karen. (1984a). Offline and online user assistance for online catalog searchers. *Online* May, 54–66.

Markey, Karen. (1984b). "Subject Searching in Library Catalogs." OCLC, Dublin, Ohio.

Markey, Karen. (1985). Subject-searching experiences and need of online catalog users: Implications for library classification. *Library Res. Tech. Services* January/March, 35–51.

Markuson, Barbara Evans, Wanger, J., Schatz, S., and Black, D. V. (1972). "Guidelines for Library Automation: A Handbook for Federal and Other Libraries." Systems Development Corporation, Santa Monica, Calif.

Marquis, D. G., and Allen, T. J. (1966). Communication patterns in applied technology. *Am. Psychologist* 21, 1052–1060.

Marron, Harvey. (1963). Science libraries: Consolidated or departmental? *Phys. Today* 16, 34–39.

Martin, James. (1973). "Design of Man–Computer Dialogues." Prentice Hall, Englewood Cliffs, N.J.

Martin, Lowell A. (1984). "Organization Structure of Libraries." Scarecrow Press, Metuchen, N.J.

Martin, Murray S. (1978). "Budgetary Control in Academic Libraries." JAI Press, Greenwich, Conn.

Martin, Murray S. (1980). The allocation of money within the book budget. *In* "Collection Development in Libraries: A Treatise" (Eds. Robert D. Stueart and George B. Miller), pp. 35–66. JAI Press, Greenwich, Conn.

Martin, Susan K. (1980). "The Professional Librarian's Reader in Library Automation." Knowledge Industry Publications, Inc., White Plains, N.Y.

Martin, Susan K. (1984). The new technologies and library networks. *Library J.* June 15, 1194–1196.

Mason, Ellsworth D. (1980). "Mason on Library Buildings." Scarecrow Press, Metuchen, N.J.

Mason, Robert M. (1983). Choosing software for text processing. *Library J.,* September 1: 1665–1666.

Mason, Robert M. (1984). Ergonomics: the human and the machine. *Library J.;* February 15: 331–332.

Mason, Robert M. (1985a). Laser disks for micros. *Library J.* February 15, 124–125.

Mason, Robert M. (1985b). Prospects for 1985. *Library J.* January, 60–61.

Mason, Robert M. (1985c). Should you consider a PC local area network? *Library J.* June 15, 42–43.

Massman, Virgil F., and Patterson, Kelly. (1970). A minimum budget for current acquisitions. *College Res. Libraries* 31, 83–88.

Matarazzo, James M. (1981). "Closing the Corporate Library: Case Studies on the Decision-Making Process." Special Libraries Association, New York.

Matheson, Nina W. (1984). The academic library nexus. *College Res. Libraries* May, 207–213.

Matheson, Nina W., and Cooper, J.A.D. (1982). Academic information in the academic health science center: roles for the library in information management. *J. Med. Education* October, 57(10, part 2).

Mathews, Anne J. (1983). Communicate!: A librarian's guide to interpersonal relations. American Library Association, Chicago.

Mathur, V. (1983). SDI (selective dissemination of information)—A need for special facilities in commercial systems. *Online Rev.* 7(4), 321–327.

Matthews, Joseph R. (1980). "Choosing an Automated Library System: Planning Guide." American Library Association, Chicago.

Matthews, Joseph R. (1981). "Comparative Information for Automated Circulation Systems: Turnkey & Other Systems," 2d ed. J. Matthews & Associates, Grass Valley, Calif.

Matthews, Joseph R. (1982a). On-line public catalogs: Assessing the potential. *Library J.* June 1, 1067–1071.

Matthews, Joseph R. (1982b). 20 Qs & As on automated integrated library systems. *Am. Libraries* June, 367–371.

Matthews, Joseph R. (1983a). The automated library system marketplace, 1982: Change and more change. *Library J.* March 15, 547–553.

Matthews, Joseph R. (1983b). "A Reader on Choosing an Automated Library System." American Library Association, Chicago.

Matthews, Joseph R. (1983c). Review of Markey. *Online* September, 118–119.

Matthews, Joseph R. (1984). Competition and change: The 1983 automated library marketplace. *Library J.* May 1, 853–860.

Matthews, Joseph R. (1985a). "Public Access to Online Catalogs," 2d ed. Neal-Schuman, New York.

Matthews, Joseph R. (1985b). Unrelenting change: The 1984 automated library system marketplace. *Library J.* April 1, 31–40.

Matthews, Joseph R. (1985c). "Directory of Automated Library Systems," Neal-Schuman, New York.

Matthews, Joseph R. (1986). Growth and consolidation: The 1985 automated library system marketplace. *Library J.* April 1, 25–37.

Matthews, Joseph R., and Hegarty, Kevin E., Eds. (1984). "Automated Circulation: An Examination of Choices." American Library Association, Chicago.

Matthews, Joseph R., and Williams, Joan Frye. (1982). The bibliographic utilities: Progress and problems. *Library Technol. Rep.* November–December, 609–653.

Matthews, Joseph R., and Williams, Joan Frye. (1983). Telecommunication technologies for libraries: A basic guide. *Library Technol. Rep.* July–August, 337–396.

Matthews, Joseph R., and Williams, Joan Frye. (1984). The user friendly index: A new tool. *Online* May, 31–34.

Matthews, Joseph R., Lawrence, Gary S., and Ferguson, Douglas K., Eds. (1983). "Using Online Catalogs: A Nationwide Survey." Neal-Schuman, New York.

Mauerhoff, G. R. (1974). Selective dissemination of information. *Adv. Librarianship* 5, 25–62.

Maxwell, R. (1983). Measured benefits of an SDI program in a corporate environment. *In* "Productivity in the Information Age," pp. 99–100. Knowledge Industry Publications for ASIS, Washington, D.C. (American Society for Information Science Proceedings; 20).

McClure, Charles L., and Reifsnyder, Betsy. (1984). Performance measures for corporate information centers. *Special Libraries* July, 193–204.

McGrath, William E. (1971). Correlating the subjects of books taken out and of books used within an open-stack library. *College Res. Libraries* 32, 280–285.

McGrath, William E. (1978). Relationships between hard/soft, pure/applied and life/nonlife disciplines and subject book use in a university library. *Inf. Processing Manage.* 14, 17–28.

McGregor, Douglas. (1967). "The Professional Manager." McGraw-Hill, New York.

McIntyre, Donald M. (1983). Spreadsheets, database management, and graphics: As easy as 1–2–3. *Online* November, 31–40.

McLellan, Vin. (1985). Easy does it. *Digital Rev.* June, 25–26.

McQueen, Judy, and Boss, Richard W. (1985). Sources of machine-readable cataloging and retrospective conversion. *Library Technology Reports* 21(6), 601–732.

McVicker, Jennifer M. (1979). A comparison of on-line SDI with an in-house current awareness bulletin. City University Centre for Information Science, London.

Meadows, A. J. (1974). "Communication in Science." Butterworths, London.

Medical Subject Headings. (1983). PB82-232752 National Library of Medicine, Bethesda, Md. (1983).

Melin, Nancy Jean. (1985). New York technology at ALA. *Inf. Today* 2(2), 26.

Mellon, Constance A. (1986). Library anxiety: A grounded theory and its development. *Coll. Res. Libraries* March, 160–165.

Merilees, Bobbie. (1983). RFPs and on-line library system selection. *Can. Library J.* February, 15–19.

"Metals Information Thesaurus of Metallurgical Terms." (1981). 5th ed. American Society for Metals, Metals Park, Ohio.

Metcalf, Keyes D. (1965). "Planning Academic and Research Library Buildings." McGraw-Hill, New York. (Note: to be updated by David C. Weber and Philip D. Leighton and to be published by ALA.)

Meyer, Alan. (1977). Some important findings in catalog use studies. *In* "The Measurement and Evaluation of Library Services" (Ed. F. W. Lancaster), pp. 69–72. Information Resources Press, Washington, D.C.

Michaels, George H., Kerber, Mindy S., and Hall, Hal W. (1984). "A Microform Reader Maintenance Manual." Meckler, Westport, Conn.

Mildren, K. W., and Meadows, N. G. (1976). "Use of Engineering Literature." Butterworths, London.

Miller, Jerome. (1979). "Applying the New Copyright Law; A Guide for Educators and Librarians." American Library Association, Chicago.

Miller, R. Bruce. (1983). Radiation, ergonomics, ion depletion and VDTs: Healthful use of visual display terminals. *Inf. Technol. Libraries* June, 151–158.

Miller, William. (1984). What's wrong with reference. *Am. Libraries* May, 303–321.

Milstead, Jessica L. (1984). "Subject Access Systems: Alternatives in Design." Academic Press, Orlando, Fla.

Mischo, William. (1982). Library of Congress subject headings: A review of the problems and prospects for improved subject access. *Catalog. Classif. Q.* 1(2/3), 105–124.

Mitchem, Terri. (1985). The Bowker national library microcomputer usage study, 1984. *Bowker Annu.* 29, 426–433.

Montague, E. A. (1967). Card catalog use studies 1949–1965. Master's dissertation. Graduate Library School, University of Chicago, Chicago.

Morris, Leslie R., and Brautigam, Patsy Fowler. (1984). "Interlibrary Loan Policies Directory," 2d ed. American Library Association, Chicago.

Morris, Leslie R., Castle, R., and Brautigam, P. F. (1985). Interlibrary loan policies directory: Data from 832 libraries. *RQ* 25(2), 229–233.

Morse, Philip M., and Elston, Caroline A. (1969). A probabilistic model for obsolescence. *Operations Res.* 17, 36–47.

Mortenson, Erik. (1984). Downloading: potentials and restrictions in online searching. *Proc. 47th ASIS Annu. Meeting* 21, 166–169.

Mosher, Paul H. (1979). Collection evaluation in research libraries: The search for quality, consistency, and system in collection development. *Library Res. Tech. Services* **23**(1), 16–32.

Mosher, Paul H. (1980a). Managing library collections: The process of review and pruning. *In* "Collection Development in Libraries: A Treatise" (Eds. Robert D. Stueart and George B. Miller), pp. 159–181. JAI Press, Greenwich, Conn.

Mosher, Paul H. (1980b). Collection evaluation or analysis: Matching library acquisitions to library needs. *In* "Collection Development in Libraries: A Treatise" (Eds. Robert D. Stueart and George B. Miller), pp. 527–545. JAI Press, Greenwich, Conn.

Mosher, Paul H. (1982). Collection development to collection management. *Collection Manage.* **4**(4), 41–48.

Mostecky, V., ed. (1958). "Catalog Use Study." American Library Association, Chicago.

Mote, L. J. B. (1971). Personal patterns of information use in an industrial research and development laboratory. *J. Documentation* **27**, 200–204.

Mount, Ellis. (1966). Communication barriers and the reference question. *Special Libraries* October, 575–578.

Mount, Ellis. (1976). "Guide to Basic Information Sources in Engineering." Jeffrey Norton, New York.

Mount, Ellis, ed. (1972). "Planning the Special Library." Special Libraries Association, New York (SLA monograph no. 4).

Mount, Ellis (1982). "Ahead of its Time: The Engineering Societies Library, 1913–1980." Linnet, Hamden, Conn.

Mount, Ellis. (1985). "University Science and Engineering Libraries," 2d ed. Greenwood Press, Westport, Conn.

Murdock, John W., and Brophy, Charles A., Jr. (1966). A comparison of the functions of libraries and information centers. *Library Trends* **14**(3), 347–353.

Murray, William E. (1986). Video display terminals: Radiation issues. *Library Hi Tech* **12**, 43–47.

Myers, Marcia J., and Jirjees, Jassim M. (1983). "The Accuracy of Telephone Reference/ Information Services in Academic Libraries: Two Studies." Scarecrow Press, Metuchen, N.J.

Myers, Margaret. (1984). "Guide to library placement sources. *Bowler Annu.* **29**, 290–306.

Naber, G. (1985). Online versus manual literature retrieval: A test case shows interesting results in retrieval effectiveness and search strategy. *DATABASE* February, 20–24.

Nadeski, Karen, and Pontius, Jack. (1984). Developments in micrographics, video technology, and "fair use," 1983. *Library Res. Tech. Services* July/September, 219–238.

"NASA Thesaurus." (1982). National Aeronautics and Space Administration, Scientific and Technical Information Branch, NASA SP-7051.

National Federation of Science Abstracting and Indexing Services. (1971). Data element definitions for secondary services. Philadelphia, June (report no. 3).

National Commission on New Technological Uses of Copyrighted Works (CONTU). (1976). Guidelines for the proviso of subsection 108(G) (2).

Neal, James G. (1984). And the walls came tumblin' down: Distributed cataloging and the public/technical services relationship—The public services perspective. *Proc. 47th ASIS Annu. Meeting* **21**, 114–116.

Nelson, Carnot E., and Pollock, Donald K., Eds. (1970). "Communication among Scientists and Engineers." D. C. Heath, Lexington, Mass.

Network advisory committee holds April program on delivery systems. (1984). *LC Inf. Bull.* June 18, 214–215.

Neufeld, M. Lynne, and Cornog, Martha. (1983). Secondary information systems and services. *Annu. Rev. Inf. Sci. Technol. (ARIST)* **18**, 151–183.

Neufeld, M. Lynne and Cornog, Martha. (1986). Database history: from dinosaurs to compact discs. *J. Am. Soc. Inf. Sci.* **37**(4), 183–190.

New international standard for determining document subjects and choosing indexing terms. (1985). *Inf. Hotline* April, 4.

Newman, Wilda B. (1982). Acquiring technical reports in the special library: Another package for information transfer. *Sci. Technol. Libraries* **2**(4), 45–67.

Nichol, Kathleen M. (1983). Database proliferation: Implications for librarians. *Special Libraries* April, 110–118.

Nolan, J. M., ed. (1984). "Micro Software Evaluations." Nolan Information Management Services, Torrance, Calif.

Nolan, Jeanne M., Ed. (1983a). "Micro Software Report," Vol. II. Nolan Information Management Services, Torrance, Calif.

Nolan, Jeanne M. (1983b). Microcomputer software for libraries: A survey. *Electronic Library* October, 275–278.

North Atlantic Treaty Organization. Advisory Group for Aerospace Research and Development. (1974). How to obtain information in different fields of science and technology: A user's guide. Neuilly-sur-Seine, France, AGARD; AGARD LS-69; AD 780061.

NTIS moves ahead on disk project. (1983). *Micrographics Newslett.* **15**, 5.

NTIS/SR-77/04. (1977). NTIS subject classification (past and present). NTIS, Springfield, Va, PB270 575; NTIS/SR-77/04.

Number of Libraries in the U.S. and Canada. (1984). *Bowker Annu.* 436.

Nyren, Karl, ed. (1976). Library space planning. Bowker, New York (*LJ* special report 1).

OCLC and RLG. (1980). On-line public access to library bibliographic data bases: Developments issues and priorities. Final report to the Council on Library Resources. Council on Library Resources, Washington, D.C., September, ED 195275.

Ojala, Marydee. (1986). Views on end-user searching. *J. Am. Soc. Inf. Sci.* **37**(4), 197–203.

Oklahoma Department of Libraries. (1982). "Performance Measures for Oklahoma Public Libraries." American Library Association, Chicago.

Olsgaard, John N. (1982). Characteristics of managerial resistance to library management information systems. *In* "Library Automation as a Source of Management Information" (Ed. F. Wilfrid Lancaster), pp. 92–110. University of Illinois Graduate School of Library and Information Science, Champaign/Urbana, Ill.

Online training sessions: Suggested guidelines. (1981). *RQ* **20**(4), 353–357.

Opello, Olivia, and Murdock, Lindsay. (1976). Acquisitions overkill in science collections—And an alternative. *College Res. Libraries* September, 452–456.

Orr, R. H., Olson, Edwin E., Pings, V. M., and Pizer, I. H. (1968). Development of methodologic tools for planning and managing library services. *Bull. Med. Library Assoc.* **56**, 235–267.

Ortopan, LeRoy D. (1985). National shelflist count: A historical introduction. *Library Res. Tech. Services* October/December, 328–332.

Owens, Frederick H. (1983). From library to information service. *ChemTech* August, 464–469.

Palmer, Richard P. (1972). "Computerizing the Card Catalog in the University Library: A Survey of User Requirements." Libraries Unlimited, Littleton, Colo.

Palmer, Roger C. (1984). "dBASE II and dBASE III: An Introduction for Information Services," 2d ed. Pacific Information Inc., Studio City, Calif.

Palmer, Roger C. (1982). "Online Reference and Information Retrieval." Libraries Unlimited, Littleton, Colo.

Palmour, Vernon E., Bellassai, Marcia C., and DeWath, Nancy V. (1980). "A Planning Process for Public Libraries." American Library Association, Chicago.

Paper and its preservation: Environmental controls. (1983). Library of Congress, Washington, D.C., October (preservation leaflet no. 2).

Parker, Elisabeth Betz (1985). The Library of Congress Nonprint Optical Disk Pilot Program. *Info. Tech. Libraries* December, 289–299.

Parkhurst, Carol. (In press). "Retrospective Conversion." Neal-Schuman, New York.

Patent Office plans switch to optical disk memories. (1983). *Micrographics Newslett.* June 1, 3.

Peck, T. (1975). Counseling skills applied to reference services. *RQ* **14**, 233–235.

Pemberton, Jeff. (1983). Faults and failures—25 Ways that online searching can let you down. *ONLINE* September, 6–7.

Pemberton, Jeff. (1984). An essay on nomenclature . . . What should we call an online search service? *DATABASE* February, 6–7.

Penniman, W. D., and Dominick, W. D. (1980). Monitoring and evaluation of online system usage. *Inf. Processing Manage.* **16**, 17–35.

Pensyl, Mary E. (1982). The online policy manual. *ONLINE,* May, 46–49. [Note: Also *In* "Reference and Online Service Handbook." (1982). Eds. B. Katz and A. Clifford, pp. 25–32. Neal-Schuman, New York.]

Pensyl, Mary E., and Woodford, Susan E. (1980). Planning and implementation guidelines for an academic online experience: The M.I.T. experience. *Sci. Technol. Libraries* **1**(1), 17–45.

Perrine, R. H. (1967). Catalog use study. *RQ* **6**, 115–119.

Perrine, R. H. (1968). Catalog use difficulties. *RQ* **7**, 169–174.

Person, Ruth J., Ed. (1983). "The Management Process: A Selection of Readings for Librarians." American Library Association, Chicago.

Peters, Stephen H., and Butler, Douglas J. (1984). A cost model for retrospective conversion alternatives. *Library Res. Tech. Services* **28**(2), 149–162.

Peters, Thomas J., and Waterman, Robert H., Jr. (1982). "In Search of Excellence: Lessons from America's Best Run Companies." Harper and Row, New York.

"Petroleum Exploration and Production Thesaurus." (1985). 7th ed. Tulsa Petroleum Abstracts Advisory Committee, Tulsa, Okla.

Phinney, Hartley K. (1983). Map accessibility. *GIS Newslett.* **85**, 4–5.

Pierce, Thomas J. (1978). An empirical approach to the allocation of the university library book budget. *Collection Manage.* **39**, 39–56.

Pierce, William S. (1980). "Furnishing the Library Interior." Dekker, New York.

Pollet, Miriam. (1982). Criteria for science book selection in academic libraries. *Collection Building* **4**(3), 42–47.

Popovich, Marjorie, and Miller, Betty. (1981). Online ordering with DIALORDER. *ONLINE* April, 65.

Poulton, E. C. (1968). Rate of comprehension of an existing teleprinter output and of possible alternatives. *J. Appl. Psychol.* **52**, 16–21.

Powell, James R., Jr., and Slach, June E. (1985). How to evaluate library automation systems. *Online* March, 30–36.

Powell, Ronald R. (1984). Reference effectiveness: A review of research. *Library Inf. Sci. Res.* **6**, 3–10.

Pratt, Allan D. (1984). Microcomputers in libraries. *Annu. Rev. Inf. Sci. Technol. (ARIST)* **19**, 247–269.

Price, Bennett J. Printing and the online catalog. (1984). *Inf. Technol. Libraries* March, 15–20.

Primack, Alice Lefler. (1984). "Finding Answers in Science and Technology." Van Nostrand Reinhold, New York.

Pruett, N. J. (1982a). Online databasics (or . . . the Holocene online scene). *Geosci. Inf. Soc. Newslett.* **75**, 3–4.

Pruett, Nancy J. (1982b). Data on the rocks: A cross-section of user needs. Proceedings, Second International Conference on Geological Information, pp. 29–43 Oklahoma Geological Survey; Norman, Oklahoma. (special publication 82–4.)

Public Law 94-553, Copyright Revision Act. (1976). October 19.

Pyhrr, Peter A. (1973). "Zero-Base Budgeting: A Practical Tool for Evaluating Expenses." Wiley, New York.

Quinn, K. (1985). Using DIALOG as a book selection tool. *Library Acquisitions* 9(1), 79–82.

R. R. Bowker databases on CD-ROM (1986). *Library Hi Tech News* February 24, 1.

Rabbitt, Mary C. (1979). "A Brief History of the U.S. Geological Survey." Government Printing Office, Washington, D.C.

Radke, Barbara, Klemperer, K. E., and Berger, M. G. (1982). The user-friendly catalog: Patron access to MELVYL. *Inf. Technol. Libraries* December, 358–371.

Raffel, Jeffrey A., and Shishko, Robert. (1969). "Systematic Analysis of University Libraries. An Application of Cost–Benefit Analysis to the MIT Libraries." M.I.T. Press, Cambridge, Mass.

Randall, Gordon E. (1975). Randall's rationalized ratios. *Special Libraries* January, 6–11.

Rao, I. K. Ravichandra. (1973). Obsolescence and utility factors of periodical publications: A case study. *Library Sci. with a Slant to Documentation* 10, 297–307.

RASD/MARS Costs and Financing Committee. (1983). Online reference services: costs and budgets. December.

RASD/MARS Education and Training of Search Analysts Committee. (1981). Online training sessions: Suggested guidelines. *RQ* 20(4), 353–357.

RASD MARS Direct Patron Access Committee. (1984). Examples of ready reference use of bibliographic utilities. Distributed at the 1984 American Library Association Annual Conference, Dallas, June.

Rathbun, Loyd. (1974). The small library's large problem: "I'm ready and eager, but where are the clients?" *Special Libraries* 65, 223–226.

Rather, John C. (1972). Filing arrangement in the Library of Congress catalogs. *Library Res. Tech. Services (LRTS)* 16, 249–256.

Rawles, Beverly A., and Wessells, Michael B. (1984). "Working with Library Consultants." Shoe String, Hamden, Conn.

RBMS Ad Hoc Committee for Developing Transfer Guidelines. (1985). Guidelines on the selection of general collection materials for transfer to special collections. *C&RL News* July/August, 349–352.

"Reference Manual for Machine-Readable Bibliographic Descriptions." (1981). 2d ed. General Information Programme and UNISIST, Unesco, Paris.

Relyea, Harold C. (1986). National security controls and scientific information. *Inf. Rep. Bibliographies* 14(6), 2–13.

Reneker, Maxine H. (1983). Funding levels and changes in the process of scholarly communication: Critical issues for management of academic science libraries. *Sci. Technol. Libraries* 4(3/4), 19–32.

Reneker, Maxine H., and Fedunok, Suzanne. (1983). The acquisition of monographs in large academic scientific research libraries: Current issues and problems. *Sci. Technol. Libraries* 3(3), 31–51.

Reproduction of copyrighted works by educators and librarians. (1978). Copyright Office, Library of Congress, Washington, D.C., September (circular R21).

Reynolds, Dennis. (1984). Telefacsimile as a mechanism for document delivery in interlibrary loan. *Action for Libraries;* 10(5), 1.

Rice, Barbara A. (1978). Weeding in academic and research libraries: An annotated bibliography. *Collection Manage.* 2(1), 65–71.

Richards, Berry G., *et al.* (1981). Manual SDI services in an academic library: Case studies of sci-tech topics. *Sci. Technol. Libraries* **2**(1), 31–42.

Riggs, Donald E. (1984). "Strategic Planning for Library Managers." Oryx, Phoenix, Ariz.

Ritti, R. R. (1971). "The Engineer in the Industrial Corporation." Columbia University Pres. New York.

Rizzo, John R. (1980). "Management for Librarians: Fundamentals and Issues." Greenwood, Westport, Connecticut.

Roberts, Justine. (1984). Stack capacity in medical and science libraries. *College Res. Libraries* **45**(4), 306–314.

Rochell, Carlton. (1985). The knowledge business: Economic issues of access to bibliographic information. *College Res. Libraries* January, 5–12.

Rogers, Rutherford D., and Weber, David C. (1971). "University Library Administration." Wilson, New York.

Rollins, Stephen. (1981a). Annual Report of the Circulation Department University of New Mexico General Library, Albuquerque, NM. June 30.

Rollins, Stephen. (1981b). You get what you ask for: The University of New Mexico's document delivery program. Presented at the American Society for Information Science 10th Mid-Year Meeting, Durango, Colo., May.

Rollins, Stephen. (1983a). Annual Report of the Circulation Department University of New Mexico General Library, Albuquerque, NM. June 30.

Rollins, Stephen. (1983b). The new Tower of Babel. *Am. Libraries* **14**(4), 233.

Rollins, Stephen. (1984a). Annual Report of the Circulation Department, University of New Mexico General Library, June 30.

Rollins, Stephen. (1984b). Chemical Abstracts' document delivery service. *Online Rev.* **8**(2), 183–191.

Rollins, Stephen. (1985). Annual Report of the Circulation Department University of New Mexico; June 30.

Roose, Tina. (1985). Online or print: Comparing costs. *Library J.* September 15, 54–55.

Rorvig, Mark E. (1981). "Microcomputers and Libraries: A Guide to Technology and Application." Knowledge Industry Publications, Inc., White Plains, N.Y.

Rosenberg, V. (1967). Factors affecting the preferences of industrial personnel for information gathering methods. *Inf. Storage Retrieval* **3**(3), 119–127.

Rosenbloom, Richard S., and Wolek, Francis W. (1967). Technology, information and organization: A report to the National Science Foundation. Harvard University Graduate School of Business Administration, Boston.

Roth, Dana Lincoln. (1985). The role of subject expertise in searching the chemical literature . . . And pitfalls that await the inexperienced searcher. *DATABASE* February, 43–46.

Roth, Judith Paris, Ed. (1986). Essential guide to CD-ROM. Meckler Publishing, Westport, Conn.

Rothstein, Samuel. (1985). Why people really hate library schools. *Library J.* April 1, 41–48.

Roughton, Karen G. (1985). Browsing with sound: Sound-based codes and automated authority control. *Inf. Technol. Libraries* June, 130–136.

Rovelstad, Howard. (1983). Guidelines for planning facilities for sci-tech libraries. *Sci. Technol. Libraries* **3**(4), 3–19.

Rowe, Gladys E. (1984). Microform informing: Use of DIALOG SDI to produce a microfiche announcement bulletin. *ONLINE* **8**(2), 70–75.

Rush, James E., ed. (1984). Acquisitions. "Library systems evaluation guide," Vol. 4. James E. Rush Associates, Powell, Ohio.

Rutledge, Mary Ellen. (1982). In search of online user group information: A retrospective and current view. *ONLINE* November, 29–33.

Sadow, Arnold. (1970). Book reviewing media for technical libraries. *Special Libraries* **61**,

194–197. [Note: Commentary in *Special Libraries* by Maurice H. Smith, (1970), November, 515–516.]

Saffady, William. (1983). "Introduction to Automation for Libraries." American Library Association, Chicago.

Saffady, William. (1985a). Availability and cost of online search services. *Library Technol. Rep.* 21(1), 1–111.

Saffady, William. (1985b). "Micrographics," 2d ed. Libraries Unlimited, Littleton, Colo.

Saffady, William. (1985c). Communications software packages for the IBM Personal Computer and compatibles. *Library Technol. Rep.* 21(4), 355–456.

Saldinger, Jeffrey. (1985). On the relationship between information brokers and academic/ public libraries. Presented at the American Library Association Annual Conference, July 9.

Salmon, Steven R. (1983). Characteristics of online public catalogs. *Library Res. Tech. Services* 27, 53.

A sampler of forms for special libraries. (1982). Special Libraries Association, Washington D.C. Chapter/Social Science Group.

Sampson, Gary S. (1978). Allocating the book budget: Measuring for inflation. *College Res. Libraries* 39, 381–383.

Sanders, Nancy P., (1983). A review of selected sources in budgeting for collection managers. *Collection Manage.* 5(3/4), 151–159.

Schad, Jasper G. (1978). Allocating materials budgets in institutions of higher education. *J. Academic Librarianship* 3, 328–332.

Schiller, Anita R. (1965). Reference service: Instruction or information. *Library Q.* 35, 52–60.

Schleifer, Lawrence M., and Sauter, Steven L. (1986). Controlling glare problems in the VDT work environment. *Library Hi Tech* 12, 21–25.

Schoenly, Steven B. (1982). A library tour and orientation program for small microcomputers. *Software Rev.* 1(1), 44–57.

Schrader, Alvin M. (1980/1981). Performance measures for public libraries: Refinements in methodology and reporting. *Library Res.* 2, 129–155.

Schuman, Patricia Glass. (1984). Women, power and libraries. *Library J.* January, 42–47.

Schuyler, Michael. (1982). A review of VISICALC. *Software Rev.* 1(1), 68–77.

Schwarz, Philip. (1983). Selecting and implementing a computer based library system: An outline of the process and annotated bibliography. Wisconsin Department of Public Instruction, Madison, Wisc. ED231298.

Seal, Alan W. (1984). The development of online catalogs. *In* "Introducing the online catalogue" (Ed. Alan W. Seal), pp. 1–15. Centre for Catalogue Research, University of Bath, Bath, England.

Search aids for use with DIALOG databases, SA-1. (1985). DIALOG Information Service, Palo Alto, Calif., January.

The search for online data (1986). Part one—Database directories online and off. *Database End-User* February, 20–27.

Seba, Douglas B., and Forrest, Beth. (1978). Using SDI's to get primary journals—A new online way. *ONLINE* January, 10–15.

Seiden, Peggy, and Kibbey, Mark. (1985). Information retrieval systems for microcomputers. *Library Hi Tech* 9, 41–54.

Selenk, Margaret. (1976). An academic science/engineering library's experience with a N.Y. loan network. *Special Libraries* May/June, 239.

Selmer, Marsha L. (1982). Draft standards for university map libraries. *Special Libraries Assoc. Geogr. Map Div. Bull.* 129, 2–4.

Settel, Barbara, editor. (1977). Subject description of books: A manual of procedures for augmenting subject descriptions in library catalogs. Research Study 3. Syracuse University, School of Information Studies, Syracuse, N.Y.

Settel, Barbara, and Cochrane, Pauline A. (1982). Augmenting subject descriptions for books in online catalogs. *Database* December, 29–37.

Seymour, C. A., and Schofield, J. L. (1973). Measuring reader failure at the catalogue. *Library Res. Tech. Services (LRTS)* **17**, 6–24.

Shaw, Donald R. (1981). "Your Small Business Computer: Evaluating, Selecting, Financing, Installing and Operating the Hardware and Software That Fits." Van Nostrand Reinholt Company, New York.

Shaw, W. M., Jr. (1978). A practical journal usage technique. *College Res. Libraries* **39**, 479–484.

Sheehy, Eugene P. (1976). "Guide to Reference Books," 9th ed. American Library Association, Chicago.

Shera, J. H. (1961). How much is a physicist's inertia worth? *Phys. Today* **14**, 42–43.

Sherwin, C. W., and Inemson, R. S. (1966). First interim report on Project Hindsight (summary). Office of the Director of Defense Research and Engineering, Washington, D.C., October 13.

Shirinian, G. N. (1985). Microcomputer publications: An overview. *Serials Librarian* **9**(3), 19–24.

Schneiderman, Ben. (1980). "Software Psychology: Human Factors in Computer and Information Science." Winthrop, Cambridge, Mass.

Shroder, Emelie J. (1982). Online reference service—How to begin: A selected bibliography. *RQ* Fall, 70–75.

Simmons, P. (1970). Improving collections through computer analysis of circulation records in a University Library. *Proc. Am. Soc. Inf. Sci.* **7**, 59–63.

Simonds, Michael J. (1984). Database limitations and online catalogs. *Library J.* February 15, 329–330.

Skolnik, Herman. (1979). A classification system for polymer literature in an industrial environment. *J. Chem. Inf. Comput. Sci.* **19**, 76–79.

Skolnik, Herman. (1982). "Literature Matrix of Chemistry." Wiley, New York.

Slater, M., and Fisher, P. (1969). Use made of technical data. ASLIB Occasional Publication no. 2.

Slonim, Jacob, Mole, Dennis, and Bauer, Michael. (1986). Write-once laser disc technology. *Library Hi Tech* **12**, 27–42.

Slote, Stanley J. (1982). "Weeding Library Collections—II," 2d ed. Libraries Unlimited, Littleton, Colo.

Smith, Dennis. (1984). Forecasting price increase needs for library materials: The University of California Experience. *Library Res. Tech. Services* April/June, 136–148.

Smith, L. C. (1976). Artificial intelligence in information retrieval systems. *Inf. Processing Manage.* **12**, 189–222.

Smith, Linda C. (1980). Data base directories: A comparative review. *Ref. Services Rev.* Oct./Dec., 15–21.

Smith, R. Jeffrey (1986). Internal audit scores contracting by DoD Labs. *Science* **31**, 449.

Somerville, A. (1977). The place of the reference interview in computer searching: the academic setting. *ONLINE* **1**(October), 14–23.

Somerville, A. N. (1982). The pre-search reference interview—A step by step guide. *DATABASE* February, 32–38.

Sonneman, Sabine S. (1983). The videodisc as a library tool. *Special Libraries* **74**(1), 7–13.

Snyder, Barton E. (1984). A cost-benefit analysis method for various record storage media. *J. Inf. Image Man.*, **17**(5), 41–47.

Spinrad, R. J. (1982). Office automation. *Science* **214**, 808–813.

Sprague, Robert J., and Freudenreich, L. Ben. (1978). Building better SDI profiles for users of large, multidisciplinary data bases. *J. Am. Soc. Inf. Sci.* November, 278–282.

Spreitzer, Frances F. (1983). Selecting microform readers and reader-printers. Association for Information and Image Management, Silver Spring, Md.

Spyers-Duran, Peter, and Gore, Daniel, editors. (1972). "Economics of Approval Plans." Greenwood Press, Westport, Conn.

Stankus, Tony, and Schlessinger, Bernard S. (1979). Scientific-technical libraries in academic institutions in the United States. *Sci-Tech News* October, 85–86.

Stankus, Tony, and Schlessinger, Bernard S. (1980). Scientific-technical libraries in academic institutions in the United States. *Sci-Tech News* January, 5–6.

Stayner, Richard A. (1983). Economic characteristics of the library storage problem. *Library Q.* **53**, 313–327.

Steers, Richard M., and Porter, Lyman W. (1983). "Motivation and Work Behavior," 3d ed. McGraw-Hill, New York.

Steinke, Cynthia A. (1985). Seventy five years of sci-tech reference: What's next? *Sci. Technol. Libraries* **5**(4), 83–87.

Stephens, Kent (1986). Laserdisc technology enters mainstream. *Am. Libraries* April, 252.

Stevens, Norman D. (1980). The catalogs of the future: A speculative essay. *J. Library Automation* **13**(2), 88–95.

Stevens, R. E. (1950). A summary of the literature on the use made by the research worker of the university library catalog. Occasional paper no. 13. University of Illinois Library School, Urbana, Ill.

Stewart, A. K. (1978). The 1200 baud experience. *ONLINE* **2**: 13–18.

Stewart, T. F. M. (1976). Displays and the software interface. *Appl. Ergonomics* **7**(3), 137–146.

Stibic, V. (1980). "Personal Documentation for Professionals: Means and Methods." North-Holland, New York.

Stoan, Stephen K. (1984). Research and library skills: An analysis and interpretation. *College Res. Libraries* **45**(2), 99–109.

Strain, Paula M. (1982). Evaluation by the numbers. *Special Libraries* July, 165–172.

Strauss, Lucille J., Shreve, Irene M., and Brown, Alberta L. (1972). "Scientific and Technical Libraries: Their Organization and Administration," 2d ed. Wiley-Becker-Hayes, New York.

Striedleck, Suzanne. (1984). And the walls came tumblin' down: Distributed cataloging and the public/technical services relationship—The technical services perspective. *Proc. 47th ASIS Annu. Meeting* **21**, 117–120.

Stueart, Robert D. (1980). Mass buying programs in the development process. *In* "Collection Development in Libraries: A Treatise" (Eds. Robert D. Stueart and George B. Miller), pp. 203–217. JAI Press, Greenwich, Conn.

Stueart, Robert D. (1985). Weeding of library materials—Politics and policies. *Collection Manage.* **7**(2), 47–58.

Stueart, Robert D., and Eastlick, John Taylor. (1981). "Library Management." Libraries Unlimited, Littleton, Colo.

Stueart, Robert D., and Miller, George B. (1980). "Collection development in Libraries: A Treatise." JAI Press, Greenwich, Conn.

Stursa, Mary Lou (1985). Some thoughts on the invisible network. *Sci-Tech News* October, 100–101. (Note: A practical justification for SLA membership and networking.)

Subramanyam, K. (1976). Core journals in computer science. *IEEE Trans. Professional Commun.* **PC-19,** 22–25.

Subramanyam, K. (1979). Scientific literature. *In* "Encyclopedia of Library and Information Science," vol. 26, pp. 376–548. Marcel Dekker, New York.

Subramanyam, K. (1980a). Technical literature. *In* "Encyclopedia of Library and Information Science," vol. 30, pp. 144–209. Marcel Dekker, New York.

Subramanyam, Kris. (1980b). Citation studies in science and technology. *In* "Collection Development in Libraries: A Treatise" (Eds. Robert D. Stueart and George B. Miller), pp. 345–372. JAI Press, Greenwich, Conn.

Subramanyam, K. (1981). Scientific and technical information resources. Marcel Dekker, New York.

Sutherland, Stuart. (1980). "PRESTEL and the User: A Survey of Psychological and Ergonomic Research." Central Office of Information, London.

Swanson, D. R. (1972). Requirements study for future catalogs. *Library Q.* **42,** 302–315.

Swartzburg, Susan G. (1980). Preserving library materials: A manual. Scarecrow Press, Metuchen, N.J.

Sweetman, Peter, and Wiedemann, Paul. (1980). Developing a library book-fund allocation formula. *J. Academic Librarianship* **6,** 268–276.

Tagliacozzo, Renata. (1970). Types of catalog searches and their relationship to some characteristics of the users. *In* Integrative mechanisms in literature growth: Final report (Ed. Manfred Kochen), IV, pp. 27–39. University of Michigan, Mental Health Research Institute, Ann Arbor, Mich.

Tagliocozzo, R., and Kochen, M. (1970). Information-seeking behavior of catalog users. *Inf. Storage Retrieval* **6,** 363–381.

Tagliacozzo, R., Kochen, M., and Rosenberg, L. (1970). Access and recognition: From users' data to catalogue entries. *J. Documentation* **26,** 230–249.

Tague, Jean M. (1981). User-responsive subject control in bibliographic retrieval systems. *Inf. Processing Manage.* **17**(3), 149–159.

Tallman, Johanna E. (1980). The impact of the OCLC Interlibrary Loan Subsystem on a science oriented academic library. *Sci. Technol. Libraries* **1**(2), 27–34.

Tannehill, Robert S. (1982). Factors in document delivery: An analysis based on experience at Chemical Abstracts Service. *Sci. and Technol. Libraries* **2**(4), 3–25.

Tauber, Maurice F. (1972). "Library Binding Manual: A Handbook of Useful Procedures for the Maintenance of Library Volumes." Library Binding Institute, Boston.

Taylor, Arlene G. (1984). Authority files in online catalogs. *Catalog. and Classif. Q.* **4**(3), 1–17.

Taylor, R. S. (1968). Question negotiation and information seeking in libraries. *College Res. Libraries* **29,** 178–194.

Tedd, Lucy A. (1983). Software for microcomputers in libraries and information units. *Electronic Library* **1**(1), 31–48.

Tenopir, Carol. (1980). Evaluation of library retrieval software. *Proc. Am. Soc. Inf. Sci.* **17,** 64–67.

Tenopir, Carol. (1982a). Evaluation of database coverage: A comparison of two methodologies. *Online Rev.* **6**(5), 423–441.

Tenopir, Carol. (1982b). An in-house training program for online searchers. *ONLINE* **6,** 20–26.

Tenopir, Carol. (1983a). In-house databases I: Software sources. *Library J.* April 1, 639–641.

Tenopir, Carol. (1983b). In-house databases II: Evaluating and choosing software. *Library J.* May 1, 885–887.

Tenopir, Carol. (1983c). Databases: Catching up & keeping up. *Library J.* February 1, 180–182.

Tenopir, Carol. (1983d). Dialog's Knowledge Index and BRS/After Dark: Database searching on personal computers. *Library J.* March 1, 471–473.

Tenopir, Carol. (1983e). More publications about databases. *Library J.* November 15, 2140–2141.

Tenopir, Carol. (1984a). Database access software. *Library J.* October 1, 1828–1829.

Tenopir, Carol. (1984b). The database industry today: Some vendors' perspectives. *Library J.* February 1, 156–7.

Tenopir, Carol. (1984c). Identification and evaluation of software for microcomputer-based in-house databases. *Inf. Technol. Libraries* March, 21–34.

Tenopir, Carol. (1984d). In-house training and staff development. *Library J.* May 1, 870–871.

Tenopir, Carol. (1985a). Database subsets. *Library J.* May 15, 42–43.

Tenopir, Carol. (1985b). Database directories: the rest. *Library J.* September 15, 56–57.

Tenopir, Carol. (1985c). Conferences for online professionals. *Library J.* January, 62–63.

Tenopir, Carol (1986). Databases on CD-ROM. *Library J.* March 1, 68–69.

"Thesaurus of Engineering and Scientific Terms (TEST)." (1967). U.S. Department of Defense, Washington, D.C., AD 672000.

Thomas, Diana, Hinckley, Ann T., and Eisenback, Elizabeth. (1981). "The Effective Reference Librarian." Academic Press, New York.

Thompson, Godfrey. (1977). "Planning and Design of Library Buildings," 2d ed. Architectural Press, London.

Tijerina, Louis. (1984). Video display terminal workstation ergonomics. OCLC, Columbus, Ohio.

Tinker, M. A. (1963). "Legibility of Print." Iowa State University Press, Ames, Iowa.

Titles classified by the Library of Congress Classification: National Shelflist Count. (1977). General Library, University of California, Berkeley.

Tolle, John E. (1983a). Current utilization of online catalogs: Transaction log analysis. OCLC, Inc., Dublin, Ohio.

Tolle, John E., (1983b). Determining the required number of online catalog terminals: A research study. *Inf. Technol. Libraries* 2(3), 261–265.

Tolle, John E. (1984). Public access terminals: Determining quantity requirements. OCLC, Inc., Dublin, Ohio.

Training users of online catalogs. (1983). Council on Library Resources, Washington, D.C.

Trueswell, , R. W. (1966). Determining the optimum number of volumes for a library core collection. *LIBRI* 16, 49–60.

Trueswell, Richard W. (1965). A quantitative measure of user circulation requirements and its possible effect on stack thinning and multiple copy determination. *Am. Documentation* 16(1), 20–25.

Trueswell, R. W. (1969a). Some behavioral patterns of library users: The 80/20 rule. *Wilson Library Bull.* 43, 458–561.

Trueswell, R. W. (1969b). User circulation satisfaction vs. size of holdings of three academic libraries. *College Res. Libraries* 30, 204–213.

Trueswell, R. W. (1976). Growing libraries: Who needs them?: A statistical basis for the no-growth collection. *In* "Farewell to Alexandria" (Ed. Daniel Gore), pp. 72–104. Greenwood Press, Westport, Connecticut.

Truett, Carol. (1984). Evaluating software reviews. *Library Software Rev.* 3(3), 371–378.

Trumpeter, Margo C., and Rounds, Richard S. (1985). "Basic Budgeting Practices for Librarians." American Library Association, Chicago.

Tucci, Valerie. (1982). Online ordering of sci-tech materials. *Sci. Technol. Libraries* 2(4), 27–43.

Tullis, Thomas S. (1981). An evaluation of alphanumeric, graphic and color information displays. *Human Factors* 23(5), 541–50.

UNESCO. (1973). Guidelines for the establishment and development of monolingual science and technology thesauri for information retrieval. UNESCO, Paris, July (SC/MD/20).

UNESCO. (1974). UNISIST reference manual for machine-readable bibliographic descriptions. UNESCO, Paris (SC.74/WS120).

University of Arizona Science–Engineering Library. (1983). Final report 1982–83 planning study. Tucson, Ariz.

University of Chicago, Graduate Library School. (1968). Requirements study for future catalogs. Progress report no. 2. University of Chicago, Graduate Library School, Chicago.

Urquhart, J. A., and Schofield, J. L. (1971). Measuring readers' failure at the shelf. *J. Documentation* 27, 273–276.

Urquhart, J. A., and Schofield, J. L. (1972). Measuring readers' failure at the shelf in three university libraries. *J. Documentation* 28, 233–241.

U.S. Department of Energy. (1984). "Energy Data Base Subject Thesaurus." Technical Information Center, Office of Scientific and Technical Information, November, DOE/TIC-7000-R6; DE84010568. Oak Ridge, Tennessee.

U.S. Federal Council for Science and Technology Committee on Scientific and Technical Information. (1965). COSATI subject category list. CFSTI, Springfield, Va., October, AD612200.

U.S. Geological Survey Library. (1972). Classification scheme and index. USGS Library, Reston, Virginia.

U.S. Geological Survey. (1984a). "The U.S. Geological Survey's Library." Government Printing Office, Washington, D.C.

U.S. Geological Survey. (1984b). "United States Geological Survey Yearbook, Fiscal Year 1984." Government Printing Office, Washington, D.C.

Utterback, J. M. (1969). The process of technical innovation in industrial firms. Alfred P. Sloan School of Management, MIT, Cambridge, Mass.

Van Camp, A. (1979). Effective search analysts. *Online* 3(2), 18–20.

Vaughan, D. (1972). Titles as sources of subject headings. *In* Requirements study for future catalogs. Final Report. Appendix I, pp. 1–10. Graduate Library School, University of Chicago, Chicago.

Veanor, Allen B. (1970). Major decision points in library automation. *College Res. Libraries* September, 299–312.

Vitz, P. C. (1966). Performance for different amount of visual complexity. *Behav. Sci.* II, 105–14.

Voigt, Melvin J. (1961). Scientists approaches to information. American Library Association, Chicago (ACRL monograph no. 24).

Waldhart, Thomas J., and Zweifel, Leroy G. (1973). Organizational patterns of scientific and technical libraries: An examination of three issues. *College Res. Libraries* 34(6), 426–435.

Wall, Edward C. (1986). Computer-to-computer communications: A review of library-related activities. *Library HiTech News* April (26), 1,6.

Wallace, Danny P. (1984). The user friendliness of the library catalog. University of Illinois, Graduate School of Library and Information Science, Champaign, Ill., occasional paper 163, February.

Walsh, Robert R. (1969). Branch library planning in universities. *Library Trends* 18, 210–222.

Walton, Robert A. (1983). "Microcomputers: A Planning and Implementation Guide for Librarians and Information Professionals." Oryx Press, Phoenix, Ariz.

Wanat, Camille. (1985). Management strategies for personal files: The Berkeley seminar. *Special Libraries* 76(4), 253–260.

Wanger, Judith *et al.* (1976). Impact of on-line retrieval services: A survey of users, 1974–75. System Development Corporation, Santa Monica, Calif.

Warden, Carolyn L. (1978). An industrial current awareness service: a user evaluation study. *Special Libraries* 69, 459–467.

Warden, Carolyn L. (1981). User evaluation of a corporate library online search service. *Special Libraries* 72, 113–117.

Warrick, Thomas S. (1984). Large databases, small computers and fast modems . . . An attorney looks at the legal ramifications of downloading. *ONLINE* July, 58–70.

Weber, David C. (1971). Personnel aspects of library automation. *J. Library Automation* March, 27–37.

Weil, Ben H. (1980). Benefits from research use of the published literature at the Exxon Research Center. *In* "Special Librarianship: A New Reader" (Ed. Eugene B. Jackson), pp. 586–594. Scarecrow Press, Metuchen, N.J.

Weinberg, A. M., *et al.* (1963). Science, government and information; The responsibilities of the technical community and the government in the transfer of information. Report of the President's Science Advisory Committee, January 10, Washington, D.C.

Weinberg, Bella Hass. (1982). Multiple sets of human indexing for civil engineering documents: Comparison of structure and occurrence rates in full text. *Sci. Technol. Libraries* 2(3), 13–33.

Wellisch, Hans H. (1984). Indexing and abstracting, 1977–1981: An international bibliography: ABC-Clio, Santa Barbara, Calif.

Wells, D. A. (1961). Individual departmental libraries vs. consolidated science libraries. *Phys. Today* 14, 40–41.

Wenger, Charles B., and Childress, Judith. (1977). Journal evaluation in a large research library. *J. Am. Soc. Inf. Sci.* 28, 293–99.

Werner, Gloria. (1979). Use of on-line bibliographic retrieval services in health sciences libraries in the United States and Canada. *Bull. Med. Library Assoc.* 67(1), 1–14.

Wessel, Carl J. (1972). Deterioration of library materials. *In* "Encyclopedia of Library and Information Science," vol. 7, pp. 69–120. Marcel Dekker, New York.

West, S. S. (1960). The ideology of academic scientists. *IRE Trans. Eng. Manage.* EM-7(2), 54–62.

White, Herbert S. (1979). Cost-effectiveness and cost-benefit determination in special libraries. *Special Libraries* 70, 163–169.

White, Herbert S. (1981). Perceptions by educators and administrators of the ranking of library school programs. *College Res. Libraries* 42(3), 111.

White, Herbert S. (1984a) "Managing the Special Library: Strategies for Success within the Larger Organization." Knowledge Industry Publications, Inc., White Plains, N.Y.

White, Herbert S. (1984b). Special libraries and the corporate political process. *Special Libraries* April, 81–86.

White, Herbert S. (1985a). Cost benefit analysis & other fun & games. *Library J.* February 15, 118–121.

White, Herbert S. (1985b). Participative management is the answer, but what was the question? *Library J.* August, 62–63.

White, Herbert S. (1985c). The use and misuse of library user studies. *Library J.* December, 70–71.

White, Howard S. (1982). Computer output microform catalog readers. *Library Technol. Rep.* 18, 579–607.

White, Howard S. (1983). Microfiche readers for libraries. *Library Technol. Rep.* May/June, 223–325.

White, Howard S. (1984). Microform reader/printers for libraries. *Library Technol. Rep.* **20**(6), 707–862.

White, Howard S., Ed. (1985). Annual survey of library automated system vendors. *Library Systems Newslett.* April, 25–32.

White, Martha, and Kirk, Thomas G. (1984). Teaching how to teach science reference materials—a workshop for librarians who serve the undergraduate (CE 206). Association for College and Research Libraries, Chicago. *ACRL 6756–6*

White, Marilyn Domas. (1985). Evaluation of the reference interview. *RQ* **25**(1), 76–83.

Wilkinson, J. P., and Miller, W. (1978). The step approach to reference service. *RQ* **17**(Summer), 293–200.

Willet, Linda, and Finnigan, Georgia. (Undated). Interlibrary loan/document retrieval, what does it really cost?: American Cyanamid Company Report,

Williams, Martha E., Ed. (1985a). "Computer-Readable Databases: A Directory and Data Sourcebook." American Library Association, Chicago.

Williams, Martha E. (1985b). Electronic databases. *Science* **36**, 445–455.

Williams, Martha E. (1986). Transparent information systems through gateways, front ends, intermediaries, and interfaces. *J. Am. Soc. Inf. Sci.* **37**(4), 204–214.

Williams, Sally F. (1984). Budget justification: Closing the gap between request and result. *Library Res. Tech. Services* **28**(2), 129–135.

Wilson, Alexander. (1979). "The Planning Approach to Library Management." The Library Association, London.

Wood, D. N. (1971). User studies: A review of the literature from 1966 to 1970. *ASLIB Proc.* **23**(1), 11–23.

Wood, D. N., and Hamilton, D. R. L. (1967). "The Information Requirements of Mechanical Engineers—Report on a Recent Survey." Library Association, London.

Wood, James. (1982a). Document delivery: The current status and the near-term future. Network Planning Paper, Library of Congress Network Development Office (Network Development Office no. 7).

Wood, James L. (1982b). Factors influencing the use of technical standards in a nationwide library and information service network. *Library Trends;* Fall, 343–358.

Woods, Lawrence A., and Pope, Nolan F. (1983). "The Librarian's Guide to Microcomputer Technology and Applications." Knowledge Industry Publications, Inc. for ASIS, White Plains, New York.

"World guide to special libraries." (1983). 6th ed. K. G. Saur, Munich.

Worthen, Dennis B. (1979). Ideal qualifications for professional personnel in the technical information center. *In* "Innovative Management of the Technical Information Function," pp. 123–129. General Electric Co., Schenectady, N.Y.

Wright, William F. (1982). Information management: A bibliography. *Special Libraries* **73**, 298–310.

Wulfekoetter, Gertrude. (1961). "Acquisition Work: Processes Involved in Building Library Collections." University of Washington Press, Seattle.

Yalonis, Chris, and Padgett, Anthony. (1985). Tapping into on-line data bases. *PC World* May, 120–126.

Young, William F. (1985). Methods for evaluating reference desk performance. *RQ* **25**(1), 69–75.

Ziman, John. (1968). "Public Knowledge: An Essay Concerning the Social Dimension of Science." University Press, Cambridge, England.

Zweizig, Douglas, and Rodger, Eleanor Jo. (1982). "Output Measures for Public Libraries: A Manual of Standardised Procedures." American Library Association, Chicago.

Index